Lecture Notes in Mathematics

Edited by A. Dold and B. Eckmann

1210

Probability Measures on Groups VIII

Proceedings of a Conference held in Oberwolfach, November 10–16, 1985

Edited by H. Heyer

Springer-Verlag
Berlin Heidelberg New York London Paris Tokyo

Editor

Herbert Heyer
Mathematisches Institut, Universität Tübingen
Auf der Morgenstelle 10
7400 Tübingen, Federal Republic of Germany

Mathematics Subject Classification (1980): 60B15, 60J15, 60H25, 43A05, 60A10, 60B11, 43A10, 43A33, 46L50, 47D05

ISBN 3-540-16806-0 Springer-Verlag Berlin Heidelberg New York
ISBN 0-387-16806-0 Springer-Verlag New York Berlin Heidelberg

Printed in Germany

Printing and binding: Druckhaus Beltz, Hemsbach/Bergstr.
2146/3140-543210

PREFACE

The methods developed in structural probability theory during the last three decades proved to be of increasing importance in various fields of mathematics: from harmonic analysis to quantum theory. This remarkable feature serves as a convincing motivation for the continuation of the meetings on Probability Measures on Groups initiated in 1970. The eighth conference held as usual in the inspiring atmosphere of the Lorenzenhof attracted a fine group of mathematicians from Europe and abroad who were eager to contribute to this traditional exchange of ideas on open problems and new directions of research.

Among the *open problems* which recently enjoyed significant progress we mention

(i) a deepening of the embedding problem for measures on a Lie group,

(ii) stable measures on groups and contractive automorphisms,

(iii) the transience problem for convolution semigroups on a hypergroup,

(iv) compactness of measures on generalized translation algebras,

as well as

(v) Martin boundaries and isoperimetric inequalities for random walks on graphs.

The six expository talks presented at the conference reflect further developments, in particular those advances which are related to the central limit theorem for products of random matrices, the method of entropy in the theory of random walks on groups, and the theory of random fields in connection with C*-algebras and their duality.

Topics which led to *new direction of research* concern

(i) convolution semigroups on the diffeomorphism group over a manifold,

(ii) Fürstenberg's boundary theory in connection with the notion of the Lyapunov exponent,

(iii) Dirichlet space methods in the theory of stochastic differential equations,

(iv) semi-Markovian chains and their reneval theory

as well as

(v) quantum stochastic increment processes.

The editor extends his thanks to the contributors, in particular to those who agreed to present an expository lecture and to offer it in reworked and enlarged form to the Proceedings. The expository talks were highlights of the meeting which at the same time gave shape to the organization of the program. The entire group of participants deserves special appreciation for its enthusiasm in commenting on their colleagues' work and asking profound questions which occasionally led to evening sessions.

Needless to note, the week in Oberwolfach turned out to be far too short to settle the most urgent problems. It is good to know that our joint efforts can be resumed in early 1988 on the occasion of "Probability Measures on Groups IX".

Herbert Heyer
Tübingen, April 1986

C O N T E N T S

Survey articles

The authors J.-L. Dunau and H. Senateur kindly provided their manuscript although they did not participate at the conference.

PAPERS GIVEN AT THE CONFERENCE

BUT NOT INCLUDED IN THIS VOLUME

M. Babillot	Semi-markovian chains and renewal theory for random walks on groups
P. Baxendale	Convolution semigroups of measures on the diffeomorphism group
C. Berg	Hoeffding inequalities and Schur increasing functions
T. Byczkowski	Random series and seminorms
S.G. Dani	Dynamics of boundary actions and stochastic harmonic functions
T. Drisch	Stable laws on the motion and diamond groups
F. Hirsch	Density of Wiener functionals and applications to stochastic differential equations
F. Kinzl	Gleichverteilung der Folge der Faltungspotenzen eines Wahrscheinlichkeitsmaßes auf lokalkompakten Halbgruppen
G. Letac	The various Cauchy types of distributions in $\mathbb{R}^d$, and the functions which preserve them

LIST OF PARTICIPANTS

M. Babillot	Paris, France
P. Baxendale	Aberdeen, England
C. Berg	København, Denmark
M.S. Bingham	Hull, England
W.R. Bloom	Perth, Australia
Ph. Bougerol	Paris, France
T. Byczkowski	Wrocław, Poland
H. Carnal	Bern, Switzerland
S.G. Dani	Bombay, India
Y. Derriennic	Brest, France
T. Drisch	Dortmund, Germany
L. Elie	Paris, France
B.-J. Falkowski	Neubiberg, Germany
L. Gallardo	Nancy, France
O. Gebuhrer	Strasbourg, France
P. Gerl	Salzburg, Austria
W. Hazod	Dortmund, Germany
H. Heyer	Tübingen, Germany
F. Hirsch	Cachan, France
A. Janssen	Siegen, Germany
E. Kaniuth	Paderborn, Germany
F. Kinzl	Salzburg, Austria
J. Kisyński	Lublin, Poland
G. Letac	Toulouse, France
M. McCrudden	Manchester, England
M. Reményi	Erlangen, Germany
H. Rindler	Wien, Austria
L. Schmetterer	Wien, Austria
R. Schott	Nancy, France
M. Schürmann	Heidelberg, Germany
E. Siebert	Tübingen, Germany

G.J. Székely	Budapest, Hungary
K. Urbanik	Wrocław, Poland
W. von Waldenfels	Heidelberg, Germany
M.E. Walter	Boulder, USA
W. Woess	Leoben, Austria
K. Ylinen	Turku, Finland
Hm. Zeuner	Tübingen, Germany

On the assumptions of a central limit theorem for approximate martingale arrays on a group

MICHAEL S. BINGHAM
Department of Statistics
University of Hull
Hull, England

In Bingham (1986) a central limit theorem was proved for approximate martingale arrays of random variables with values in a locally compact second countable abelian group. This result is stated as Theorem 1 below. The purpose of the present paper is to elucidate some connections between the assumptions of that theorem and certain alternative conditions that one might wish to assume instead. The discussion is necessarily rather technical.

Throughout, G denotes a locally compact (Hausdorff) second countable abelian group and $\hat{G}$ denotes the dual group of G. The value of the character $y \in \hat{G}$ at the point $x \in G$ is denoted by $\langle x,y \rangle$ and g denotes a local inner product function on $G \times \hat{G}$ as in Bingham (1986). All the random variables in this paper are assumed to be Borel measurable and defined on the same probability space $(\Omega,\mathcal{F},P)$; equalities and inequalities between random variables will generally be assumed true a.e.P. For the theory of locally compact abelian groups and their duals the reader is referred to Hewitt and Ross (1963, 1970) or Rudin (1962) and references for the probability theory and existence of g include Heyer (1977) and Parthasarathy (1967).

Let (k_n) be a sequence of positive integers increasing to infinity and, for each $n \geq 1$, let $\{\mathcal{F}_{nj}: 1 \leq j \leq k_n\}$ be a filtration in $\mathcal{F}$, so that each $\mathcal{F}_{nj}$ is a sub-σ-field of $\mathcal{F}$ and $\mathcal{F}_{nj} \subseteq \mathcal{F}_{n,j+1}$ for $1 \leq j < k_n$. For each n, let $\{S_{nj} : 1 \leq j \leq k_n\}$ be a sequence of G-valued random variables adapted to $\{\mathcal{F}_{nj} : 1 \leq j \leq k_n\}$; thus S_{nj} is $\mathcal{F}_{nj}$-measurable for every n,j. Define $X_{nj} := S_{nj} - S_{n,j-1}$ where $S_{n0} = e$, the identity of G, and, for each n, let $\mathcal{F}_{n0}$ be a sub-σ-field of $\mathcal{F}_{n1}$. Then we shall call $\{S_{nj}, \mathcal{F}_{nj} : 0 \leq j \leq k_n, n \geq 1\}$ an adapted triangular array of G-valued random variables with differences $\{X_{nj}\}$. If the adapted triangular array satisfies

$$\sum_{j=1}^{k_n} \left| E[g(X_{nj},y) \mid \mathcal{F}_{n,j-1}] \right| \xrightarrow{P} 0 \text{ as } n \to \infty \tag{1}$$

for every $y \in \hat{G}$, where $\xrightarrow{P}$ denotes convergence in probability, we call the array an approximate martingale array.

By a random continuous nonnegative quadratic form on $\hat{G}$ we mean a real-valued function Φ on $\hat{G} \times \Omega$ such that, for each $y \in \hat{G}$, $\omega \to \Phi(y,\omega)$ is a nonnegative random variable on Ω and, for each ω(a.e.P), $y \to \Phi(y,\omega)$ is a continuous function on $\hat{G}$ satisfying

$$\Phi(y_1 + y_2,\omega) + \Phi(y_1 - y_2,\omega) = 2\Phi(y_1,\omega) + 2\Phi(y_2,\omega)$$

for all y_1, $y_2 \in \hat{G}$. Whenever it appears, Φ will denote such a function. The ω will be suppressed in the notation and $\Phi(y,\omega)$ written as $\Phi(y)$.

The central limit theorem proved in Bingham (1986) can now be stated.

Theorem 1

Let $\{S_{nj}, \mathcal{F}_{nj} : 0 \le j \le k_n, n \ge 1\}$ be an approximate martingale array of G-valued random variables on a probability space $(\Omega,\mathcal{F},P)$ with differences $\{X_{nj}\}$. Suppose that for every neighbourhood N of the identity in G

$$P(X_{nj} \notin N \text{ for some } j = 1, 2, \ldots, k_n) \to 0 \text{ as } n \to \infty$$

and that for every $y \in \hat{G}$

$$\sum_{j=1}^{k_n} g(X_{nj},y)^2 \xrightarrow{P} \Phi(y) \quad \text{as } n \to \infty$$

where Φ is a random continuous nonnegative quadratic form on $\hat{G}$. Suppose also that the filtrations $\{\mathcal{F}_{nj}\}$ are nested; that is

$$\mathcal{F}_{nj} \subseteq \mathcal{F}_{n+1,j} \quad \text{for } 1 \le j \le k_n, \ n \ge 1.$$

Then, for every $F \in \mathcal{F}$ with $P(F) > 0$, the conditional distribution of S_{nk_n} given F converges weakly to the mixture of Gaussian distributions on G with the characteristic function

$$y \to E[\exp(-\tfrac{1}{2}\Phi(y)) \mid F], \quad y \in \hat{G}.$$

From now on assume that $\{S_{nj}, \mathcal{F}_{nj} : 0 \le j \le k_n, n \ge 1\}$ is an adapted triangular array of G-valued random variables with differences $\{X_{nj}\}$. Introduce the following notations for each $y \in \hat{G}$.

$$U_n(y) := \sum_j g(X_{nj},y)^2$$

$$V_n(y) := \sum_j E[g(X_{nj},y)^2 \mid \mathcal{F}_{n,j-1}]$$

$$f_n(y) := \prod_j E[\langle X_{nj},y\rangle \mid \mathcal{F}_{n,j-1}]$$

$$h_n(y) := \prod_j E[\exp(ig(X_{nj},y)) \mid \mathcal{F}_{n,j-1}].$$

Here and everywhere else, when the range of a sum or product etc. is not specifically indicated, the index j will be assumed to go from 1 to k_n. The function f_n corresponds to what is sometimes referred to in the literature for real martingale arrays as the "conditional characteristic function" and $V_n(y)$ is an analogue of the "conditional variance".

Proposition 1

Let $\{S_{nj}, \mathcal{F}_{nj} : 0 \leq j \leq k_n, n \geq 1\}$ be an approximate martingale array with differences $\{X_{nj}\}$. Assume that for every neighbourhood N of the identity in G

$$\max_j P(X_{nj} \notin N \mid \mathcal{F}_{n,j-1}) \overset{P}{\to} 0 \quad \text{as } n \to \infty \tag{2}$$

and that

$$\text{for every } y \in \hat{G}, \quad V_n(y) \to \Phi(y) \text{ in } L^1 \text{ as } n \to \infty \tag{3}.$$

Then for every $y \in \hat{G}$ conditions (4) and (5) below are equivalent and each of them is implied by (6):

for every $\varepsilon > 0$

$$\sum_j E[g(X_{nj},y)^2 1(|g(X_{nj},y)| > \varepsilon)] \to 0 \quad \text{as } n \to \infty \tag{4}$$

where 1(F) denotes the indicator of the event F;

$$h_n(y) \to \exp[-\tfrac{1}{2}\Phi(y)] \text{ in } L^1 \quad \text{as } n \to \infty \tag{5};$$

$$U_n(y) \to \Phi(y) \text{ in } L^1 \text{ as } n \to \infty \tag{6}.$$

Remark

Condition (2) is a conditional uniform infinitesimality assumption and (4) is a Lindeberg type condition. It is implicit that $\Phi(y)$ has finite expectation. Proposition 1 is based on Theorem 3.5 in Hall and Heyde (1980).

Proposition 1 is a consequence of Lemmas 1, 2 and 3 below.

Lemma 1

If an approximate martingale array satisfies (2) and (3) then (4) implies (5).

Proof

Fix $y \in \hat{G}$. Denote $Y_{nj} := g(X_{nj},y)$ and $E_{nj}(.) := E(. \mid \mathcal{F}_{n,j-1})$. Given $\varepsilon > 0$, choose a neighbourhood N of the identity in G such that $|g(x,y)| < \varepsilon$ for all x in N. Then

$$E_{nj}(Y_{nj}^2) \leq \varepsilon^2 + c.P(X_{nj} \notin N \mid \mathcal{F}_{n,j-1})$$

where $c = \sup_{x \in G} g(x,y)^2 < \infty$. Using (2) and then letting $\varepsilon \searrow 0$ we obtain

$$\max_j E_{nj}(Y_{nj}^2) \overset{p}{\to} 0 \quad \text{as } n \to \infty \qquad (7).$$

Define the functions A and B of the real variable x by

$$e^{ix} = 1 + ix - \tfrac{1}{2}x^2 + \tfrac{1}{2}x^2A(x) \quad (x \neq 0), \quad A(0) = 0$$

$$B(x) = \min\left(\frac{x}{3}, 2\right).$$

Then $|A(x)| \leq B(|x|)$. Moreover

$$\left|h_n(y) - \exp(-\tfrac{1}{2}\sum_j E_{nj}(Y_{nj}^2))\right|$$

$$\leq \sum_j \left|E_{nj}[\exp(iY_{nj})] - \exp[-\tfrac{1}{2}E_{nj}(Y_{nj}^2)]\right|$$

$$= \sum_j \left|\left(1 + iE_{nj}(Y_{nj}) - \tfrac{1}{2}E_{nj}(Y_{nj}^2) + \tfrac{1}{2}E_{nj}[Y_{nj}^2A(Y_{nj})]\right)\right.$$

$$\left. -\{1 - \tfrac{1}{2}E_{nj}(Y_{nj}^2) + R_{nj}\}\right|$$

$$\leq \sum_j \left|E_{nj}(Y_{nj})\right| + \tfrac{1}{2}\sum_j \left|E_{nj}[Y_{nj}^2A(Y_{nj})]\right| + \sum_j |R_{nj}| \qquad (8).$$

Because $e^{-x} - (1-x) \leq x^2$ for $x \geq 0$, the remainder terms R_{nj} satisfy

$$|R_{nj}| \leq \tfrac{1}{4}[E_{nj}(Y_{nj}^2)]^2$$

whence

$$\sum_j |R_{nj}| \leq \tfrac{1}{4}[\max E_{nj}(Y_{nj}^2)] \sum_j E_{nj}(Y_{nj}^2) \overset{p}{\to} 0$$

by (3) and (7).

For any $\varepsilon > 0$

$$\sum_j \left|E_{nj}[Y_{nj}^2A(Y_{nj})]\right|$$

$$\leq \sum_j E_{nj}[Y_{nj}^2B(|Y_{nj}|)(1[|Y_{nj}| \leq \varepsilon] + 1[|Y_{nj}| > \varepsilon])]$$

$$\leq \frac{\varepsilon}{3}\sum_j E_{nj}(Y_{nj}^2) + 2\sum_j E_{nj}[Y_{nj}^2\, 1(|Y_{nj}| > \varepsilon)]$$

$$\overset{p}{\to} \frac{\varepsilon}{3}\Phi(y) \quad \text{by (3) and (4).}$$

As $\varepsilon > 0$ is arbitrary, it follows that

$$\sum_j \left|E[Y_{nj}^2A(Y_{nj})]\right| \overset{p}{\to} 0 \text{ as } n \to \infty .$$

Recalling the approximate martingale condition (1), we see that the right hand side of (8) converges to 0 in probability and therefore so does the left hand side. Apply the dominated convergence theorem to deduce (5).//

Lemma 2

If an approximate martingale array satisfies (2) and (3) then (5) implies (4).

Proof

Define

$$Z_{nj} := iE_{nj}(Y_{nj}) - \tfrac{1}{2} E_{nj}(Y_{nj}^2) + \tfrac{1}{2} E_{nj}[Y_{nj}^2 A(Y_{nj})]$$

so that

$$E_{nj}[\exp(iY_{nj})] = 1 + Z_{nj}.$$

Because of (1), (7) and the inequality $|A(.)| \le B(|.|) \le 2$,

$$\max_j |Z_{nj}| \overset{p}{\to} 0 \quad \text{as } n \to \infty \qquad (9).$$

Therefore $P(A_n) \to 1$ as $n \to \infty$ where $A_n := \{|Z_{nj}| < 1 \text{ for every } j\}$.

For complex z with $|z| < 1$ take the complex logarithm defined by

$$\log(1 + z) = \sum_{n=1}^{\infty} (-1)^{n+1} \frac{z^n}{n}$$

so that $$\left| \log(1+z) - z \right| \le \sum_{2}^{\infty} \frac{|z|^n}{n} \le \frac{1}{2} \frac{|z|^2}{1-|z|}.$$

Then on A_n

$$\mathcal{R} \log E_{nj}[\exp(iY_{nj})] = \mathcal{R} \log(1 + Z_{nj}) = \mathcal{R} Z_{nj} + R'_{nj}$$

where the remainder satisfies

$$|R'_{nj}| \le \frac{1}{2} \cdot \frac{|Z_{nj}|^2}{1 - |Z_{nj}|}.$$

Here $\mathcal{R}$ denotes the "real part". Therefore, on A_n, any version of the complex logarithm on the left satisfies

$$\mathcal{R} \log h_n(y) = \sum_j \mathcal{R} \log E_{nj}[\exp(iY_{nj})] = \sum_j \mathcal{R} Z_{nj} + \sum_j R'_{nj} \qquad (10)$$

where

$$\left|\sum_j R'_{nj}\right| \le \frac{1}{2} \sum_j \frac{|Z_{nj}|^2}{1 - |Z_{nj}|}$$
$$\le \frac{1}{2} (1 - \max_j |Z_{nj}|)^{-1} (\max_j |Z_{nj}|)(\sum_j |Z_{nj}|) \qquad (11).$$

But, using the definition of Z_{nj} and $|A(.)| \le 2$,

$$\sum_j |Z_{nj}| \le \sum_j |E_{nj}(Y_{nj})| + \frac{3}{2} \sum_j E_{nj}(Y_{nj}^2)$$

$\overset{p}{\to} \frac{3}{2}\Phi(y)$ by (1) and (3).

Thus (9) and (11) imply that

$$1(A_n) \sum_j R'_{nj} \overset{p}{\to} 0 \quad \text{as } n \to \infty.$$

Returning to (10), we have, on A_n,

$$\mathcal{R} \log h_n(y) = -\tfrac{1}{2} \sum_j E_{nj}(Y_{nj}^2) + \tfrac{1}{2}\mathcal{R} \sum_j E_{nj}[Y_{nj}^2 A(Y_{nj})] + R_n$$

where $1(A_n)R_n \overset{p}{\to} 0$ as $n \to \infty$. Therefore (3) and (5) imply that

$$\mathcal{R} \sum_j E_{nj}[Y_{nj}^2 A(Y_{nj})] \overset{p}{\to} 0 \quad \text{as } n \to \infty \qquad (12).$$

Because

$$\mathcal{R} \sum_j E_{nj}[Y_{nj}^2 A(Y_{nj})] \le 2 \sum_j E_{nj}(Y_{nj}^2) \to 2\Phi(y) \text{ in } L^1,$$

the convergence in (12) can be strengthened to L^1 convergence.

Finally, $\mathcal{R} A(x) = 1 - \frac{2}{x^2}(1 - \cos x)$ is an increasing function of $x > 0$ and is even, so

$$\mathcal{R} A(x) \ge \mathcal{R} A(\varepsilon) > 0 \quad \text{for } |x| > \varepsilon > 0,$$

whence

$$\sum_j E_{nj} [Y_{nj}^2 1(|Y_{nj}| > \varepsilon)] \le \frac{1}{\mathcal{R} A(\varepsilon)} \mathcal{R} \sum_j E_{nj}[Y_{nj}^2 A(Y_{nj})] \to 0 \text{ in } L^1.$$

Therefore (4) holds.//

Lemma 3

If an approximate martingale array satisfies (2) and (3) then (6) implies (4).

Proof

Consider the real martingale array $\{T_{nj}, \mathcal{F}_{nj} : 0 \le j \le k_n, n \ge 1\}$ with $T_{n0} := 0$ and

$$T_{nj} := \sum_{k=1}^{j} [Y_{nk}^2 - E_{nk}(Y_{nk}^2)], \quad 1 \le j \le k_n.$$

Using Doob's martingale inequality (Corollary 2.1 in Hall and Heyde (1980)), we have, for every $\varepsilon > 0$,

$$P(\max_j |T_{nj}| > \varepsilon) \le \frac{1}{\varepsilon} E\,|U_n(y) - V_n(y)| \to 0 \text{ by (3) and (6).}$$

Therefore

$$\max_j |Y_{nj}^2 - E_{nj}(Y_{nj}^2)| \le 2 \max_j |T_{nj}| \overset{p}{\to} 0.$$

But, as in the proof of Lemma 1, (2) implies (7). Therefore we conclude that $\max_j Y_{nj}^2 \overset{p}{\to} 0$ or, equivalently,

$$\sum_j Y_{nj}^2 \, 1(|Y_{nj}| > \varepsilon) \overset{p}{\to} 0 \text{ as } n \to \infty \tag{13}$$

for every $\varepsilon > 0$. Because the left hand side of (13) is dominated by $U_n(y)$, which converges in L^1 to $\Phi(y)$, the convergence in (13) can be strengthened to L^1 convergence. Condition (4) is an immediate consequence.//

The limiting behaviour of h_n and of the conditional characteristic function f_n are the same if we impose the following strong conditional uniform infinitesimality assumption:

for every neighbourhood N of the identity in G

$$\sum_j P(X_{nj} \notin N | \mathcal{F}_{n,j-1}) \overset{p}{\to} 0 \text{ as } n \to \infty \tag{14}.$$

Specifically, we have

Lemma 4

If an adapted triangular array satisfies (14) then, for every $y \in \hat{G}$, $f_n(y) - h_n(y) \to 0$ in L^1 as $n \to \infty$.

Proof

Given $y \in \hat{G}$ choose a corresponding neighbourhood N of the identity in G such that $\langle x,y \rangle = \exp[ig(x,y)]$ for all x in N. Such a neighbourhood exists by the defining properties of g; see Bingham (1986) or Parthasarathy (1967). Then

$$\begin{aligned} |f_n(y) - h_n(y)| &= \left| \prod_j E_{nj}[\langle X_{nj},y \rangle] - \prod_j E_{nj}[\exp(ig(X_{nj},y))] \right| \\ &\le \sum_j \left| E_{nj}[\langle X_{nj},y \rangle] - E_{nj}[\exp(ig(X_{nj},y))] \right| \\ &\le 2 \sum_j P(X_{nj} \notin N | \mathcal{F}_{n,j-1}) \overset{p}{\to} 0 \text{ as } n \to \infty. \end{aligned}$$

Thus $f_n(y) - h_n(y) \overset{p}{\to} 0$ as $n \to \infty$. Use the dominated convergence theorem to obtain L^1 convergence.//

If we impose (14) and a uniform integrability condition, we can complement Proposition 1 as follows.

Proposition 2

Consider an approximate martingale array which satisfies the strong conditional uniform infinitesimality assumption (14). If in addition for every $y \in \hat{G}$, $\{V_n(y) : n \geq 1\}$ is uniformly integrable (w.r.t.P) (15)

then conditions (3) and (5) are equivalent to each other.

Proof

As in the proof of Lemma 1, (7) and (8) hold and

$$\sum_j |R_{nj}| \leq \frac{1}{4} (\max_j E_{nj}(Y_{nj}^2)) V_n(y).$$

By (7) and (15), the right hand expression goes to 0 in probability as $n \to \infty$; note that (15) implies that $\sup_n V_n(y) < \infty$ a.s.P.

Given $\varepsilon > 0$ and $y \in \hat{G}$, choose a neighbourhood N of the identity in G such that $|g(x,y)| < \varepsilon$ for all x in N. Using the inequality $|A(x)| \leq B(|x|) \leq |x|/3$,

$$3 \sum_j |E_{nj}[Y_{nj}^2 A(Y_{nj})]| \leq \sum_j E_{nj}[|Y_{nj}|^3 (1[X_{nj} \in N] + 1[X_{nj} \notin N])]$$

$$\leq \varepsilon . V_n(y) + \sup_{x \in G} |g(x,y)|^3 \sum_j P(X_{nj} \notin N | \mathcal{F}_{n,j-1}) \qquad (16).$$

By using (14) and (15), and letting $\varepsilon \searrow 0$, we see that the left member of the inequalities (16) goes to 0 in probability as $n \to \infty$. From (8) we conclude that

$$h_n(y) - \exp[-\tfrac{1}{2} V_n(y)] \overset{p}{\to} 0 \text{ as } n \to \infty.$$

Thus $h_n(y) \overset{p}{\to} \exp[-\frac{1}{2}\Phi(y)]$ if and only if $V_n(y) \overset{p}{\to} \Phi(y)$ as $n \to \infty$. As both $(h_n(y))$ and $(V_n(y))$ are uniformly integrable, their convergence in probability is equivalent to their convergence in L^1. //

We shall call an adapted triangular array a strong approximate martingale array if it satisfies the following condition:

$$\text{for every } y \in \hat{G}, \ \sum_j | E[g(X_{nj},y) | \mathcal{F}_{n.j-1}] | \to 0 \text{ in } L^1 \text{ as } n \to \infty \qquad (17).$$

Theorem 2

Consider a strong approximate martingale array which satisfies the strong conditional uniform infinitesimality assumption (14) and the uniform integrability condition (15). Then the conditions (3) and (6) are equivalent to each other and to the following condition:

$$\text{for every } y \in \hat{G}, \ f_n(y) \to \exp[-\tfrac{1}{2}\Phi(y)] \text{ in } L_1 \text{ as } n \to \infty \qquad (18).$$

Proof

The equivalence of (3) and (18) follows from Proposition 2 and Lemma 4. The proof will be completed by showing that (3) and (6) are equivalent if (14), (15) and (17) hold.

Define $\overline{Y}_{nj} := E_{nj}(Y_{nj})$ and consider the zero mean real martingale array $\{T_{nj}, \mathcal{F}_{nj} : 0 \le j \le k_n, n \ge 1\}$ given by

$$T_{nj} := \sum_{k=1}^{j} (Y_{nk} - \overline{Y}_{nk}),\ 1 \le j \le k_n;\ T_{n0} = 0.$$

We shall apply Theorem 2.23 in Hall and Heyde (1980) to this array to deduce that

$$\sum_j (Y_{nj} - \overline{Y}_{nj})^2 - \sum_j E_{nj}[(Y_{nj} - \overline{Y}_{nj})^2] \to 0 \text{ in } L^1 \text{ as } n \to \infty \quad (19).$$

In order to do so we need to prove that, for each $y \in \hat{G}$,

$$\{\sum_j E_{nj}\ [(Y_{nj} - \overline{Y}_{nj})^2] : n \ge 1\} \quad \text{is uniformly integrable} \quad (20)$$

and that the following conditional Lindeberg condition holds:

$$\text{for every } \varepsilon > 0,\ \sum_j E_{nj}[(Y_{nj} - \overline{Y}_{nj})^2 1(|Y_{nj} - \overline{Y}_{nj}| > \varepsilon)] \xrightarrow{P} 0 \quad (21).$$

First observe that, using (17),

$$\left| \sum_j E_{nj}[(Y_{nj} - \overline{Y}_{nj})^2] - \sum_j E_{nj}(Y_{nj}{}^2) \right|$$

$$\le \sum_j (\overline{Y}_{nj})^2 \le (\sum_j |\overline{Y}_{nj}|) \sup_{x \in G} |g(x,y)| \xrightarrow{L^1} 0 \text{ as } n \to \infty \quad (22).$$

Therefore (15) implies (20).

For any $\varepsilon > 0$

$$\sum_j E_{nj}\ [(Y_{nj} - \overline{Y}_{nj})^2 1(|Y_{nj} - \overline{Y}_{nj}| > \varepsilon)]$$

$$\le (2 \sup_{x \in G} |g(x,y)|)^2 \sum_j P(|Y_{nj} - \overline{Y}_{nj}| > \varepsilon | \mathcal{F}_{n,j-1})$$

so, in order to establish (21), we need only prove that

$$\sum_j P(|Y_{nj} - \overline{Y}_{nj}| > \varepsilon | \mathcal{F}_{n,j-1}) \xrightarrow{P} 0 \text{ as } n \to \infty \quad (23).$$

By (17), $P(A_n) \to 1$ as $n \to \infty$, where $A_n := [\max_j |\overline{Y}_{nj}| < \varepsilon/2]$. Using the $\mathcal{F}_{n,j-1}$ measurability of $\overline{Y}_{nj}$ we also have

$$P(|Y_{nj} - \overline{Y}_{nj}| > \varepsilon | \mathcal{F}_{n,j-1}) \le P(|Y_{nj}| > \varepsilon/2 | \mathcal{F}_{n,j-1}) \text{ on } A_n$$

and therefore (23) follows from (14).

Thus, (20) and (21) are proved and, by the theorem cited above, (19) holds. Finally,

$$\left|\sum_j (Y_{nj} - \bar{Y}_{nj})^2 - \sum_j Y_{nj}^2\right| \leq \sum_j \left|\bar{Y}_{nj}^2 - 2Y_{nj}\bar{Y}_{nj}\right|$$

$$\leq (\sum_j |\bar{Y}_{nj}|)(\sup_j |\bar{Y}_{nj}| + 2 \sup_j |Y_{nj}|) \tag{24}$$

$$\leq 3 \sup_{x\in G} |g(x,y)| \sum_j |\bar{Y}_{nj}| \to 0 \text{ in } L^1 \text{ by (17).}$$

The results (19), (22) and (24) show that (3) and (6) are equivalent.//

We now turn to the possibility of replacing conditions involving g by conditions involving $\langle .,. \rangle$.

Lemma 5

Consider an adapted triangular array with the strong conditional uniform infinitesimality property (14). If either

$$\sup_n V_n(y) < \infty \tag{25}$$

or

$$\sup_n \sum_j E[1 - \mathcal{R}\langle X_{nj}, y\rangle \mid \mathcal{F}_{n,j-1}] < \infty \tag{26}$$

then

$$W_n(y) := \sum_j E[\,|\langle X_{nj}, y\rangle - 1 - ig(X_{nj},y) + \tfrac{1}{2}g(X_{nj},y)^2| \,|\mathcal{F}_{n,j-1}] \overset{p}{\to} 0$$

as $n \to \infty$.

Proof

We use the inequality

$$\left| e^{i\alpha} - 1 - i\alpha + \tfrac{1}{2}\alpha^2 \right| \leq \frac{|\alpha|^3}{6}$$

which holds for all real α. Fix $y \in \hat{G}$ and, as in Lemma 4, choose a neighbourhood N of the identity in G such that $\langle x,y\rangle = \exp[ig(x,y)]$ for all x in N. Then, putting $\alpha = g(X_{nj},y)$ for $X_{nj} \in N$,

$$\begin{aligned} W_n(y) &\leq \tfrac{1}{6} \sum_j E_{nj}[\,|g(X_{nj},y)|^3 1(X_{nj} \in N)] + R_n(y) \\ &\leq \tfrac{1}{6} \sup_{x\in N} |g(x,y)| \sum_j E_{nj}[Y_{nj}^2 1(X_{nj} \in N)] + R_n(y) \end{aligned} \tag{27}$$

where $|R_n(y)| \leq c \sum_j P(X_{nj} \notin N \,|\mathcal{F}_{n,j-1}) \overset{p}{\to} 0$ as $n \to \infty$ and

$$c = \sup_{x\in G} \left|\langle x,y\rangle - 1 - ig(x,y) + \tfrac{1}{2}g(x,y)^2\right| < \infty.$$

Since $1 - \cos\alpha \geq \alpha^2/4$ for real α sufficiently near 0, we see that, provided N is sufficiently small,

$$\sum_j E_{nj}[Y_{nj}^2 1(X_{nj} \in N)] \leq A_n(y)$$

where $\quad A_n(y) := \min(V_n(y),\ 4 \sum_j E_{nj}\ [1 - \mathcal{R}\langle X_{nj}, y\rangle])$.

By (25) and (26), $\sup_n A_n(y) < \infty$. Because $\sup_{x\in N} |g(x,y)| \to 0$ as N shrinks to the identity, the required result now follows from (27).//

Theorem 3

Consider an adapted triangular array with the strong conditional uniform infinitesimality property (14). Then, for any $y \in \hat{G}$,

(i) if (25) or (26) holds, the following conditions (28) and (29) are equivalent to each other:

$$M_n(y) := \sum_j \left| E[g(X_{nj},y) | \mathcal{F}_{n,j-1}] \right| \overset{p}{\to} 0 \text{ as } n \to \infty \tag{28}$$

$$M'_n(y) := \sum_j \left| E[\mathcal{J}\langle X_{nj},y\rangle | \mathcal{F}_{n,j-1}] \right| \overset{p}{\to} 0 \text{ as } n \to \infty \tag{29}$$

(ii) the following conditions (30) and (31) are equivalent to each other:

$$V_n(y) \overset{p}{\to} \Phi(y) \text{ as } n \to \infty \tag{30}$$

$$V'_n(y) := \sum_j E[1 - \mathcal{R}\langle X_{nj},y\rangle | \mathcal{F}_{n,j-1}] \overset{p}{\to} \tfrac{1}{2}\Phi(y) \text{ as } n \to \infty \tag{31}$$

Here $\mathcal{R}$ and $\mathcal{J}$ denote the taking of real and imaginary parts of complex numbers.

Proof

Noting that $\mathcal{R}\langle x,y\rangle - 1 + \frac{1}{2}g(x,y)^2$ and $\mathcal{J}\langle x,y\rangle - g(x,y)$ are just the real and imaginary parts of $\langle x,y\rangle - 1 - ig(x,y) + \frac{1}{2}g(x,y)^2$, and that

$$\left| M_n(y) - M'_n(y) \right| \le \sum_j E_{nj} \left[\left| \mathcal{J}\langle X_{nj},y\rangle - g(X_{nj},y) \right| \right]$$

we have

$$\max\left(\left|M_n(y) - M'_n(y)\right|, \left|\frac{V_n(y)}{2} - V'_n(y)\right| \right) \le W_n(y).$$

Because (30) and (31) imply (25) and (26) respectively, Lemma 5 implies the theorem.//

Theorem 4

Suppose that convergence in probability in each of (14), (28), (29), (30) and (31) is replaced by convergence in L^1 and that (25) and (26) are replaced by

$$\sup_n E\, V_n(y) < \infty \quad \text{and} \quad \sup_n E\, V'_n(y) < \infty$$

respectively. Then the statements in Theorem 3 remain true.

Proof

It is straightforward to modify the proof of Lemma 5 to deduce that, under the assumptions of the present theorem, $W_n(y) \to 0$ in L^1 as $n \to \infty$. The rest follows as in Theorem 3.//

Theorem 5

Consider an adapted triangular array which satisfies

$$\sum_j P(X_{nj} \notin N) \to 0 \quad \text{as } n \to \infty$$

for every neighbourhood N of the identity in G. Then

$$U_n(y) = \sum_j g(X_{nj},y)^2 \overset{P}{\to} \Phi(y) \quad \text{as } n \to \infty$$

if and only if

$$\sum_j [1 - \mathcal{R}\langle X_{nj},y\rangle] \overset{P}{\to} \tfrac{1}{2}\Phi(y) \quad \text{as } n \to \infty.$$

The same is true if convergence in probability is replaced by L^1 convergence.

Proof

If N is as in the proof of Lemma 5,

$$\left|\sum_j (1 - \mathcal{R}\langle X_{nj},y\rangle) - \tfrac{1}{2}\sum_j g(X_{nj},y)^2\right|$$

$$\leq \frac{1}{6} \sup_{x\in N} |g(x,y)| \sum_j g(X_{nj},y)^2 1(X_{nj} \in N) + c \sum_j 1(X_{nj} \notin N).$$

and the theorem follows easily from this inequality.//

References

Bingham M. S. (1986). 'A central limit theorem for approximate martingale arrays with values in a locally compact abelian group', to appear in Mathematische Zeitschrift.

Hall P. and Heyde C.C. (1980). 'Martingale Limit Theory and Its Application', Academic Press, New York.

Heyer H. (1977). 'Probability Measures on Locally Compact Groups', Springer Verlag, Heidelberg.

Hewitt E. and Ross K. A. (1963, 1970). 'Abstract Harmonic Analysis' Volumes I and II, Springer Verlag, Heidelberg.

Parthasarathy K. R. (1967). 'Probability Measures on Metric Spaces', Academic Press, New York.

Rudin W. (1962). 'Fourier Analysis on Groups', Wiley (Interscience), New York.

IDEMPOTENT MEASURES ON COMMUTATIVE HYPERGROUPS

Walter R. Bloom
Murdoch University
Perth, Western Australia, 6150
Australia

1. INTRODUCTION

The analysis in this paper will be carried out on a (locally compact) commutative hypergroup. We adhere to the following notation. Let X be a locally compact Hausdorff space.

$M(X)$	Space of bounded Radon measures on X.
$M^+(X), M^1(X)$	Subset of $M(X)$ consisting of those measures that are nonnegative, and those that are nonnegative with total variation one respectively.
$C_b(X), C_{00}(X)$	Space of bounded continuous functions on X, and those with compact support respectively.
supp μ	Support of $\mu \in M(X)$.
ε_x	Point measure at $x \in X$.
N, N′	Space of nonnegative integers, and positive integers respectively.
1_A	The characteristic function of the nonempty set $A \subset X$.

A nonvoid locally compact Hausdorff space K will be called a hypergroup if the following conditions are satisfied:

(1) $M(K)$ admits a binary operation $*$ under which it is a complex algebra.

(2) The bilinear mapping $* : M(K) \times M(K) \to M(K)$ given by $(\mu,\nu) \to \mu * \nu$ is nonnegative ($\mu * \nu \geq 0$ whenever $\mu,\nu \geq 0$) and its restriction to $M^+(K) \times M^+(K)$ is continuous when $M^+(K)$ is given the weak topology.

(3) Given $x,y \in K$, $\varepsilon_x * \varepsilon_y \in M^1(K)$ and supp $\varepsilon_x * \varepsilon_y$ is compact.

(4) The mapping $(x,y) \to \operatorname{supp} \varepsilon_x * \varepsilon_y$ of $K \times K$ into the space of nonvoid compact subsets of K is continuous, the latter space with the topology as given in Jewett [8], Section 2.5.

(5) There exists a (necessarily unique) element e of K such that $\varepsilon_x * \varepsilon_e = \varepsilon_e * \varepsilon_x = \varepsilon_x$ for all $x \in K$.

(6) There exists a unique involution (a homeomorphism $x \to x^-$ of K onto itself with the property $x^{--} = x$ for all $x \in K$) such that for $x,y \in K$, $e \in \operatorname{supp} \varepsilon_x * \varepsilon_y$ if and only if $x = y^-$, and $(\mu * \nu)^- = \nu^- * \mu^-$ for all $\mu,\nu \in M(K)$, where $\mu^- \in M(K)$ is defined by $\mu^-(A) = \mu(A^-)$ for Borel subsets A of K and $A^- = \{x^- : x \in A\}$.

We shall assume throughout that K is commutative, that is,

(7) $\varepsilon_x * \varepsilon_y = \varepsilon_y * \varepsilon_x$ for all $x,y \in K$.

In this case the dual $K^\wedge$ is given by

$$K^\wedge = \{\chi \in C_b(K) : \chi \neq 0,\ \chi(x^-) = \overline{\chi(x)}$$
$$\text{and} \int \chi \, d\varepsilon_x * \varepsilon_y = \chi(x)\chi(y) \quad \text{for all } x,y \in K\}.$$

For $\mu \in M(K)$ its Fourier transform $\hat{\mu}$ is the function on $K^\wedge$ defined by

$$\hat{\mu}(\chi) = \int_K \bar{\chi} \, d\mu .$$

It is clear that $\hat{\mu}$ is continuous, the mapping $\mu \to \hat{\mu}$ is linear and one-to-one, and $(\mu * \nu)^\wedge = \hat{\mu}\,\hat{\nu}$ for all $\mu,\nu \in M(K)$.

The harmonic analysis of commutative hypergroups and their duals follows the classical case for locally compact abelian groups : every commutative hypergroup admits a Haar measure, that is, a nonnegative (not necessarily bounded) measure ω satisfying $\varepsilon_x * \omega = \omega$ for all $x \in K$; there is a Plancherel theorem; and, when $K^\wedge$ is a hypergroup under pointwise operations, there is a (weak) duality. Reference to these results, and many others, may be found in the overview of Heyer [7].

A bounded measure μ is called idempotent if $\mu * \mu = \mu$. The set of idempotent measures will be denoted by $I(K)$; the symbols $+$ and 1 will be used to restrict attention to those idempotent measures that are nonnegative and of norm 1 respectively. The problem of characterising $I(K)$ is interesting and deep. For locally compact abelian groups G it was solved by Cohen [4], who showed that the idempotent measures are precisely those bounded measures μ for which $\hat{\mu}$ is the characteristic function 1_A, where A belongs to the ring generated by the open cosets in $G^\wedge$. However the problem of extending Cohen's result to commutative hypergroups remains open. The results in this direction are as follows:

$I_1^+(K)$ Dunkl [5], Theorem 1.13; Jewett[8], Theorem 10.2E (commutativity not assumed)

$I_1(K)$ Theorem 2 below

$I(K)$ for K "close" to a locally compact abelian group: Dunkl [6], Theorems 3.8, 3.9.

In this paper we give an alternative proof for the characterisation of $I_1^+(K)$, which turns out to be particularly simple as K is assumed to be commutative. We then consider the problem for $I_1(K)$, with a strengthening of the results when K has a hypergroup dual, and finally we indicate Dunkl's results for $I(K)$ when K is suitably restricted.

2. CHARACTERISATION OF $I_1(K)$

We commence with an alternative proof of the characterisation of idempotent probability measures obtained by Dunkl and Jewett. Recall that a nonvoid subset H of K is called a subhypergroup if $H^- = H$ and $H * H \subset H$, where

$$A * B = \cup\{\text{supp } \varepsilon_x * \varepsilon_y : x \in A, y \in B\},$$

$A, B \subset K$. It will be convenient to write

$$S(\mu) = \{\chi \in K^\wedge : \hat{\mu}(\chi) \neq 0\}$$

and

$$A(K^\wedge, H) = \{\chi \in K^\wedge : \chi(x) = 1 \text{ for all } x \in H\} .$$

The set $A(K^\wedge, H)$ is usually referred to as the annihilator of H in $\hat{K}$; for properties of this and related notions, see Bloom and Heyer [2], Proposition 3.1.

Theorem 1 For $\mu \in I_1^+(K)$ there is a compact subhypergroup H of K for which $\mu = \omega_H$ (the normalised Haar measure of H).

Proof From $\mu * \mu = \mu$ we have $\hat{\mu} = 0$ or 1. Since $\hat{\mu}$ is continuous it is the case that $S(\mu)$ is open, and hence $H = A(K, S(\mu))$

is compact by Bloom and Heyer [2], Proposition 3.1(h). Now for $\chi \in S(\mu)$ we have $\chi = 1$ on $\operatorname{supp}\mu$, using the property that $|\chi| \le 1$ and $\mu \in M^1(K)$. Thus $\operatorname{supp}\mu \subset H$. We show that $S(\mu) = A(K^\wedge, H)$.

Clearly $S(\mu) \subset A\big(K^\wedge, A(K, S(\mu))\big) = A(K^\wedge, H)$. In the other direction, consider $\chi \in A(K^\wedge, H)$, so that $\chi(x) = 1$ for all $x \in H$. Then $\chi(x) = 1$ for all $x \in \operatorname{supp}\mu$ and $\hat{\mu}(\chi) = 1$, which entails that $\chi \in S(\mu)$.

Finally, observe using Bloom and Heyer [1], Theorem 3.2.2, that

$$\hat{\omega}_H = 1_{A(K^\wedge, H)} = 1_{S(\mu)} = \hat{\mu}$$

and, by the uniqueness of the Fourier transform, $\mu = \omega_H$. //

In the proof of Theorem 1 the nonnegativity of $\mu \in M(K)$ (with $\|\mu\| = 1$) enters in an essential way, namely in the deduction that $\chi \in A(K^\wedge, \operatorname{supp}\mu)$ from $\hat{\mu}(\chi) = 1$. With the dropping of nonnegativity the situation becomes more complicated; in general, $\hat{\mu}(\chi) = 1$ is equivalent to $|\chi| = 1$ on $\operatorname{supp}\mu$, where $\|\mu\| = 1$. We first observe that if $\mu \in I(K)$ then $\|\mu\| \ge 1$. Thus the members of $I_1(K)$ are extreme in $I(K)$ (and will turn out to be the so-called elementary idempotents; see Theorem 2 below). We have the following lemma.

Lemma Suppose that $\|\mu\| = 1$ and $\hat{\mu}(\gamma) = 1$ for some $\gamma \in K^\wedge$. Then $\gamma\mu * \gamma\mu = \gamma(\mu * \mu)$.

Proof From the assumptions on μ we have, since $|\gamma| \le 1$, that $|\gamma| = 1$ on $\operatorname{supp}\mu$. Now, by Dunkl [5], Proposition 2.2(3), $\{x \in K : |\gamma(x)| = 1\}$ is a subhypergroup of K and, by the preceding sentence, this set contains $\operatorname{supp}\mu$. Thus $|\gamma| = 1$ on $H = \langle \operatorname{supp}\mu \rangle$, the subhypergroup of K generated by $\operatorname{supp}\mu$. Furthermore γ is constant with value $\gamma(x)\gamma(y)$ on $\operatorname{supp}\varepsilon_x * \varepsilon_y$ for all $x, y \in \langle \operatorname{supp}\mu \rangle$, where we have appealed to Dunkl [5], Proposition 2.2(3) once again.

Now, for $f \in C_{00}(K)$,

$$\begin{aligned}
\gamma\mu * \gamma\mu(f) &= \int_H \int_H \int_K f \, d\varepsilon_x * \varepsilon_y \, \gamma(x)\,\gamma(y)\, d\mu(x)\, d\mu(y) \\
&= \int_H \int_H \int_K \gamma f \, d\varepsilon_x * \varepsilon_y \, d\mu(x)\, d\mu(y) \\
&= \mu * \mu(\gamma f) \\
&= \gamma(\mu * \mu)(f),
\end{aligned}$$

and $\gamma\mu * \gamma\mu = \gamma(\mu * \mu)$. //

Corollary If $\mu \in I_1(K)$ then $\gamma\mu \in I_1(K)$ for all $\gamma \in S(\mu)$.

While both the lemma and corollary hold trivially for all bounded measures and for all characters in the case when the underlying space is a locally compact abelian group, they no longer hold in general if the requirement that $\hat{\mu}(\gamma) = 1$ is dropped. We illustrate this via an example introduced by Jewett [8] (Example 9.1D).

Example It was shown by Jewett that the conjugacy class hypergroup K of the alternating group on four letters is given by $K = \{e,a,b,c\}$, where $a^- = a$, $b^- = c$ and

$$\varepsilon_a * \varepsilon_a = \tfrac{1}{3}\varepsilon_e + \tfrac{2}{3}\varepsilon_a \qquad \varepsilon_a * \varepsilon_b = \varepsilon_c * \varepsilon_c = \varepsilon_b$$

$$\varepsilon_b * \varepsilon_b = \varepsilon_a * \varepsilon_c = \varepsilon_c \qquad \varepsilon_b * \varepsilon_c = \tfrac{1}{4}\varepsilon_e + \tfrac{3}{4}\varepsilon_a$$

The normalised Haar measure m on K is given by

$$m = \tfrac{1}{12}\varepsilon_e + \tfrac{1}{4}\varepsilon_a + \tfrac{1}{3}\varepsilon_b + \tfrac{1}{3}\varepsilon_c .$$

It is not difficult to identify $K^\wedge = \{1,\chi,\psi,\zeta\}$; the values of these characters are given in the following table.

	e	a	b	c
1	1	1	1	1
χ	1	1	z	$\bar{z}$
ψ	1	1	$\bar{z}$	z
ζ	1	$-1/3$	0	0

$z = e^{2\pi i/3}$

The Plancherel measure is just

$$\pi = \varepsilon_1 + \varepsilon_\chi + \varepsilon_\psi + 9\varepsilon_\zeta .$$

Computing products of characters, one sees that $K^\wedge$ is a hypergroup under pointwise operations (for example, $(\varepsilon_\chi * \varepsilon_\zeta)^\vee = \chi\zeta = \zeta$) so that, using Jewett [8], Theorem 12.4A, π is in fact a Haar measure for $K^\wedge$.

Consider the subhypergroup $H = \{e,a\}$ of K . Then $A(K^\wedge,H) = \{1,\chi,\psi\}$ is a subhypergroup of $K^\wedge$, the normalised Haar measure on H is easily computed as

$$\omega_H = \tfrac{1}{4}\varepsilon_e + \tfrac{3}{4}\varepsilon_a$$

and, of course, $\omega_H * \omega_H = \omega_H$. Thus $\omega_H \in I_1^+(K)$. However $\zeta\omega_H \notin I(K)$; indeed

$$(\zeta\omega_H)^\wedge = \tfrac{1}{3}1_{\{\zeta\}} \ .$$

Thus

$$(\zeta\omega_H)^\wedge(\zeta\omega_H)^\wedge = \tfrac{1}{9}1_{\{\zeta\}} = \tfrac{1}{3}(\zeta\omega_H)^\wedge \ ,$$

which shows that

$$(\zeta\omega_H) * (\zeta\omega_H) = \tfrac{1}{3}\zeta\omega_H = \tfrac{1}{3}\zeta(\omega_H * \omega_H) \ ,$$

and the factor of $\frac{1}{3}$ destroys any chance of the lemma (or corollary) holding in this case.

We now prove that the members of $I_1(K)$ are in fact given by $\gamma\omega_H$, where $\gamma \in K^\wedge$ and H is a compact subhypergroup of K; these are known as elementary idempotents.

Theorem 2 Let $\mu \in I_1(K)$. Then for all $\gamma \in S(\mu)$,

$$\mu = \gamma\omega_H$$

for some compact subhypergroup H of K.

Proof Let $\gamma \in S(\mu)$, so that $\hat{\mu}(\gamma) = 1$, and write $\sigma = \bar{\gamma}\mu$. Then, appealing to the corollary to the lemma (the proof holds with $\bar{\gamma}$ replacing γ), $\sigma \in I_1(K)$. We also have $\|\sigma\| = 1$ and

$$1 = \hat{\mu}(\gamma) = \hat{\sigma}(1) = \sigma(K) \le \|\sigma\| = 1 \ ,$$

so that $\|\sigma\| = \sigma(K)$. It follows that $\sigma \in M^1(K)$ and, by Theorem 1, $\sigma = \omega_H$ for some compact subhypergroup H of K. Thus

$$\mu = \gamma\bar{\gamma}\mu = \gamma\omega_H$$

since $|\gamma| = 1$ on $\operatorname{supp} \mu$, which gives the result. //

The conditions in Theorem 2 and the corollary can be related as follows.

Theorem 3 Let H be a compact subhypergroup of K. The following are equivalent:

(i) $|\gamma| = 1$ on H;

(ii) $(\gamma\omega_H)^\wedge = 1_\Lambda$, where $\Lambda = \{\chi \in K^\wedge : \chi|_H = \gamma|_H\}$;

(iii) $\gamma\omega_H \in I_1(K)$.

Proof (i) => (ii). Suppose that $|\gamma| = 1$ on H. Now for $\chi \in K^\wedge$

$$(\gamma\omega_H)^\wedge(\chi) = \int_H \bar{\chi}\gamma\, d\omega_H \ .$$

If $\chi|_H = \gamma|_H$ then $\bar{\chi}\gamma|_H = |\gamma|^2|_H = 1$ on H, so that $(\gamma\omega_H)^\wedge(\chi) = 1$.

If $\chi|_H \neq \gamma|_H$ then we use the property that $\bar{\chi}\gamma|_H \in H^\wedge$ (just follow the proof of Dunkl [5], Proposition 2.2(4)) and, as $\bar{\chi}\gamma|_H \neq 1$, it must be the case that

$$(\gamma\omega_H)^\wedge(\chi) = \int_H \bar{\chi}\gamma d\omega_H = 0 ,$$

using the orthogonality of the characters.

(ii) => (iii). This is clear using the uniqueness of the Fourier transform.

(iii) => (i). From $\gamma\omega_H * \gamma\omega_H = \gamma\omega_H$ we have $\|\gamma\omega_H\| \geq 1$. But $|\gamma| \leq 1$ always holds, so that $\|\gamma\omega_H\| = 1$ and $|\gamma| = 1$ on H.//

More can be said when $K^\wedge$ is a pointwise hypergroup. From Theorem 1 we see that, in this case, $S(\mu) = A(K^\wedge,H)$ is a subhypergroup of $K^\wedge$ (see Bloom and Heyer [2], Proposition 3.1(i)), where $\mu \in I_1^+(K)$ and $H = A(K,S(\mu))$. For elementary idempotents we have:

Theorem 4 Let $K^\wedge$ be a pointwise hypergroup, H a compact subhypergroup of K and write $\mu = \gamma\omega_H$ where $|\gamma| = 1$ on H. Then

$$\hat{\mu} = 1_{\{\gamma\}*A(K^\wedge,H)} .$$

Proof For each $\chi \in K^\wedge$,

$$\hat{\mu}(\chi) = \int_K \bar{\chi}\gamma d\omega_H = \int_{K^\wedge} \hat{\omega}_H d\varepsilon_{\chi^-} * \varepsilon_\gamma = \varepsilon_{\chi^-} * \varepsilon_\gamma(A(K^\wedge,H)) .$$

Now, noting that $A(K^\wedge,H)^- = A(K^\wedge,H)$,

$$\{\chi^-\} * \{\gamma\} \cap A(K^\wedge,H) \neq \phi \text{ if and only if } \chi \in \{\gamma\} * A(K^\wedge,H) .$$

Thus for $\chi \notin \{\gamma\} * A(K^\wedge,H)$, $\hat{\mu}(\chi) = 0$, so that

$$S(\mu) \subset \{\gamma\} * A(K^\wedge,H) .$$

For the reverse inclusion, use the property $\bar{\gamma}\mu = \omega_H$ to obtain

$$1_{A(K^\wedge,H)}(\eta) = \hat{\omega}_H(\eta) = (\bar{\gamma}\mu)^\wedge(\eta) = \int_K \bar{\gamma}\bar{\eta} d\mu = \int_{K^\wedge} \hat{\mu}^- d\varepsilon_{\gamma^-} * \varepsilon_{\eta^-}$$

and, for $\eta \in A(K^\wedge,H)$,

$$\int_{K^\wedge} \hat{\mu} d\varepsilon_\gamma * \varepsilon_\eta = 1 .$$

It follows that $\hat{\mu} = 1$ on $\{\gamma\}*\{\eta\}$ and hence $\hat{\mu} = 1$ on $\{\gamma\} * A(K^\wedge,H)$. This, coupled with the details above, gives

$$S(\mu) = \{\gamma\} * A(K^\wedge,H) ,$$

from which the result follows. //

Taking Theorems 3 and 4 together, we have:

Theorem 5 Let $K^\wedge$ be a hypergroup and $\gamma \in K^\wedge$ with $|\gamma| = 1$ on H, where H is a compact subhypergroup of K. Then $\chi \in \{\gamma\} * A(K^\wedge,H)$ if and only if $\chi|_H = \gamma|_H$.

For a result similar to Theorem 5, see Chilana and Ross [3], Theorem 3.6.

If, in addition, H is assumed to be open then we have a version of Theorem 5 without the given restriction on γ.

Theorem 6 Let $K^\wedge$ be a hypergroup, H an open compact subhypergroup of K, and choose $\gamma \in K^\wedge$. Then $\chi \in \{\gamma\} * A(K^\wedge,H)$ if and only if $\chi|_H = \gamma|_H$.

Proof Choose any $a \in H$ and consider the function on $K^\wedge$ defined by $\mathbf{a}(\chi) = \chi(a)$. Then $\mathrm{Re}(\mathbf{a})$ is continuous and achieves its maximum on the compact set $\{\gamma\} * A(K^\wedge,H)$, at η say. We also have that $\{\gamma\} * A(K^\wedge,H) = \{\eta\} * A(K^\wedge,H)$ by Jewett [8], Lemma 10.3A and, for $\psi \in A(K^\wedge,H)$,

$$\int_{K^\wedge} \chi(a)\, d\,\varepsilon_\eta * \varepsilon_\psi(\chi) = \eta(a)\psi(a) = \eta(a) .$$

Thus

$$\int_{K^\wedge} \mathrm{Re}\big(\chi(a)\big)\, d\,\varepsilon_\eta * \varepsilon_\psi(\chi) = \mathrm{Re}\big(\eta(a)\big)$$

and, as $\mathrm{supp}\ \varepsilon_\eta * \varepsilon_\psi = \{\eta\} * \{\psi\} \subset \{\eta\}*A(K^\wedge,H)$, we have by the choice of η that $\mathrm{Re}\big(\chi(a)\big) = \mathrm{Re}\big(\eta(a)\big)$ for all $\chi \in \{\eta\} * \{\psi\}$. There is no dependence on $\psi \in A(K^\wedge,H)$, so that

$$\mathrm{Re}(\mathbf{a}) = \mathrm{Re}\big(\eta(a)\big) \quad \text{on} \quad \{\eta\} * A(K^\wedge,H) .$$

Similarly, $\mathrm{Im}(\mathbf{a})$ is constant on $\{\eta\} * A(K^\wedge,H)$, and hence so is $\mathbf{a}$. This gives the "only if" part of the theorem.

In the other direction, suppose that $\chi|_H = \gamma|_H$. Then, using the compactness of H,

$$\int_H \chi|_H \bar{\gamma}|_H \, d\omega_H > 0 .$$

However, for all $x \in K$,

$$\chi(x)\bar{\gamma}(x) = \int_{K^\wedge} x \, d\varepsilon_\chi * \varepsilon_{\gamma^-}$$

and

$$\int_H \chi\bar{\gamma}\, d\omega_H = \int_H \int_{K^\wedge} x(\psi)\, d\varepsilon_\chi * \varepsilon_{\gamma^-}(\psi)\, d\omega_H(x)$$

$$= \int_{K^\wedge} \left[\int_H \psi(x)\, d\omega_H(x) \right] d\varepsilon_\chi * \varepsilon_{\gamma^-}(\psi)$$

$$= \int_{A(K^\wedge,H)} d\varepsilon_\chi * \varepsilon_{\gamma^-} .$$

For this integral to be nonzero we must have $\{\chi\} * \{\gamma^-\} \cap A(K^\wedge,H) \neq \phi$ so that $\chi \in \{\gamma\} * A(K^\wedge,H)$. //

It should be observed in the proof of Theorem 6 that, in the second part, there is no restriction that H be open and, in the first part, there is no requirement that H be compact. It should also be observed that Theorem 6 can be phrased dually:

Theorem 7 Let H be a compact subhypergroup of K, $x \in K$ and $\phi \in A(K^\wedge,H)$. Then ϕ is constant on $\{x\} * H$.

3. CHARACTERISATION OF $I(K)$

As mentioned earlier, the general problem of characterising $I(K)$ is far from solved. In the case that G is a locally compact abelian group this was the content of Cohen's well-known paper [4]. Just to see how Cohen's result sits with regard to those of the preceding section, let G denote a locally compact abelian group with dual $G^\wedge$. For any open subgroup Λ of $G^\wedge$, $A(G,\Lambda)$ is a compact subgroup of G and 1_Λ is just the Fourier transform of its Haar measure $\omega_{A(G,\Lambda)}$. Thus $\omega_{A(G,\Lambda)}$ is idempotent. We also observe that if $\hat{\mu}_i = 1_{\Lambda_i}$, where $\Lambda_i \subset G^\wedge$, $i \in \{1,2\}$, and $\chi \in \hat{G}$ then

$$(\chi\mu)^\wedge = 1_{\chi\Lambda} ,\ (\delta_0 - \mu)^\wedge = 1_{\Lambda^c} ,\ (\mu_1 * \mu_2)^\wedge = 1_{\Lambda_1 \cap \Lambda_2} \text{ and}$$
$$(\mu_1 + \mu_2 - \mu_1 * \mu_2)^\wedge = 1_{\Lambda_1 \cup \Lambda_2} ,$$

where δ_0 denotes the point measure at 0 and Λ^c is just the complement of Λ. It follows that if Λ_1, Λ_2 appear as support sets for the Fourier transforms of idempotent measures then so do their translates, complements, finite intersections and finite unions. Hence all members of the ring of sets generated by the open cosets in $G^\wedge$ so appear. Cohen's theorem is that this describes all the idempotent measures. The ring of sets referred to above is usually called the coset ring of $G^\wedge$.

In the above description the basic building blocks are ω_H, where H is a compact subgroup of G, and $\chi\omega_H$ for $\chi \in G^\wedge$. The latter (which include the former!) are known as the elementary idempotents. Section 2 takes care of these for general hypergroups.

Dunkl [6] has extended Cohen's result to hypergroups that are, in some sense, close to locally compact abelian groups. We describe these briefly. Firstly, the centre $Z(K)$ of a hypergroup K is defined by

$$Z(K) = \{x \in K : \varepsilon_x * \varepsilon_{x^-} = \varepsilon_e\} .$$

It is easily shown that for each $x,y \in Z(K)$ there corresponds $z \in Z(K)$ satisfying $\varepsilon_x * \varepsilon_y = \varepsilon_z$, and that $x * y = z$ defines a binary operation on $Z(K)$ making it into a locally compact abelian group; this is just the maximum subgroup of K. It is known (Dunkl [5], Proposition 3.13) that

$$Z(K) = \{x \in K : |\chi(x)| = 1 \quad \text{for all} \quad \chi \in K^\wedge\} .$$

Dunkl [6], Definition 3.1 introduces a class (SP*-) of compact commutative hypergroups having hypergroup duals, which satisfy

$$\{\chi \in K^\wedge : |\chi(x)| = 1\}$$

is finite for all $x \notin Z(K)$. For these hypergroups it turns out that when $\omega_K(Z(K)) = 0$, the natural projection $\pi : M(K) \to M(Z(K))$ is an algebra homomorphism and is bounded, in the sense that $\|\pi(\mu)^\wedge\|_\infty \le \|\hat{\mu}\|_\infty$ for all $\mu \in M(K)$. With these assumptions one can transfer idempotent measures from K to $Z(K)$. It is then the case that $\hat{\mu} - \pi(\mu)^\wedge$ is zero except at finitely many points, and Cohen's theorem gives that

$$S = \{\chi \in Z(K)^\wedge : \pi(\mu)^\wedge(\chi) = 1\}$$

is in the coset ring of $Z(K)^\wedge$, where $\mu \in I(K)$. We then have

$$\{\chi \in K^\wedge : \pi(\mu)^\wedge(\chi) = 1\} = \{\chi \in K^\wedge : \chi|_{Z(K)} \in S\} ,$$

which brings everything back to K, and hence that $S(\mu)$ is in the hypercoset ring of $K^\wedge$ (defined as for the group case, but in terms of open subhypergroups of $K^\wedge$).

The preceding analysis was for the case where the centre of K has Haar measure zero. The case $\omega_K(Z(K)) > 0$ is handled in a similar way.

REFERENCES

[1] Walter R Bloom and Herbert Heyer, *The Fourier transform for probability measures on hypergroups*, Rend. Mat. 2(1982), 315-334.

[2] Walter R Bloom and Herbert Heyer, *Convolution semigroups and resolvent families of measures on hypergroups*, Math. Z. 188(1985), 449-474.

[3] Ajit Kaur Chilana and Kenneth A Ross, *Spectral synthesis in hypergroups*, Pacific J. Math. 76(1978), 313-328.

[4] Paul J Cohen, *On a conjecture of Littlewood and idempotent measures*, Amer. J. Math. 82(1960), 191-212.

[5] Charles F Dunkl, *The measure algebra of a locally compact hypergroup*, Trans. Amer. Math. Soc. 179(1973), 331-348.

[6] Charles F Dunkl, *Structure hypergroups for measure algebras*, Pacific J. Math. 47(1973), 413-425.

[7] Herbert Heyer, *Probability theory on hypergroups : A survey*. Probability Measures on Groups, Proc. Conf., Oberwolfach Math. Res. Inst., Oberwolfach, 1983 pp. 481-550. Lecture Notes in Math., Vol. 1064, Berlin - Heidelberg - New York, Springer, 1984.

[8] Robert I Jewett, *Spaces with an abstract convolution of measures*, Advances in Math. 18(1975), 1-101.

LES VARIABLES ALEATOIRES DE LOI STABLE ET LEUR REPRESENTATION SELON P. LEVY

H. Carnal

Résumé: Une représentation due à P. Lévy des lois stables sur $\mathbb{R}$ est étendue aux lois stables sur un groupe de Lie nilpotent, simplement connexe. Le résultat permet d'étudier le domaine d'attraction d'une telle loi.

1. Les cas de $\mathbb{R}$ et de $\mathbb{R}^d$

Dans un article de 1935, P. Lévy [9] construisait de la manière suivante une v.a. réelle X de loi stable pour un exposant $\alpha < 2$:

Soit $\{N_t\}_{t \geq 0}$ un processus de Poisson de paramètre λ, ayant des sauts aux temps $\Gamma_1 < \Gamma_2 < \Gamma_3 \ldots$, et $\{Y_j\}_{j \geq 1}$ une suite de v.a. équidistribuées, à valeurs dans $\{-1, +1\}$, indépendantes entre elles et indépendantes du processus N_t. On suppose que la suite $\{a_j\}_{j \geq 1}$ est donnée par

$$a_j = E(Y_j)E(\Gamma_j^{-1/\alpha}) \sim E(Y_j)\cdot(\lambda/j)^{1/\alpha} \tag{1.1}$$

pour $j \geq j_0$. Alors la série

$$\sum_{j=1}^{\infty}(\Gamma_j^{-1/\alpha}Y_j - a_j) \tag{1.2}$$

convergera p.s. vers une v.a. X de loi stable ($\alpha < 2$) arbitrairement donnée.

La démonstration de la convergence p.s. se résume à une application du théorème des 3 séries, car la suite $\{\Gamma_j\}$ se comporte p.s. comme la suite $\{j/\lambda\}$ et la variance conditionnelle de $\Gamma_j^{-1/\alpha}Y_j$ pour Γ_j donné est de la forme $c\Gamma_j^{-2/\alpha} \sim c_1 j^{-2/\alpha}$. X est donc la limite, pour $t \to \infty$, d'une somme que l'on peut écrire:

$$S_t = \sum_{j=1}^{N_t} \Gamma_j^{-1/\alpha}Y_j - b_t \tag{1.3}$$

où $b_t = \sum_{1 \leq j \leq \lambda t} a_j$ se comporte asymptotiquement comme $t^{1-1/\alpha}$ (pour $\alpha > 1$), comme log t (pour $\alpha=1$) ou comme une constante (pour $\alpha<1$). On remarquera que $N_t - \lambda t = o(t^{1/\alpha})$ (loi du log itéré et $\alpha < 2$) et donc que

$\sum_{j\leq N_t} a_j - b_t = O(t^{-1/\alpha}(N_t - \lambda t)) \to o.$

Dans (1.3), on peut remplacer les Γ_j par des v.a. U_j indépendantes et uniformément distribuées sur $[0,t]$. Si l'on pose $p=P(Y_j=1)$, $q=P(Y_j=-1)$, on a

$$\left.\begin{array}{ll} P(U_j^{-1/\alpha}Y_j > x) = pP(U_j < x^{-\alpha}) = px^{-\alpha}/t & (x > t^{-1/\alpha}) \\ P(U_j^{-1/\alpha}Y_j < x) = qP(U_j < |x|^{-\alpha}) = q|x|^{-\alpha}/t & (x < -t^{-1/\alpha}) \end{array}\right\} \quad (1.4)$$

Dès lors, S_t aura une loi de Poisson composée dont on calcule la fonction caractéristique par

$$\log E(e^{iuS_t}) = -iub_t + \lambda q \int_{-\infty}^{-t^{-1/\alpha}} (e^{iux}-1)M'(x)dx - \lambda p \int_{t^{-1/\alpha}}^{\infty} (e^{iux}-1)M'(x)dx \quad (1.5)$$

avec $M(x)=|x|^{-\alpha}$. On peut choisir λ et p de manière que λp et λq prennent des valeurs non-négatives c_+, c_- arbitraires et l'on remarque que b_t se comporte lorsque $t \to \infty$, comme

$$\lambda q \int_{-\infty}^{-t^{-1/\alpha}} \frac{x}{1+x^2} M'(x)dx - \lambda p \int_{t^{-1/\alpha}}^{\infty} \frac{x}{1+x^2} M'(x)dx.$$

En passant à la limite dans (1.5) on obtient alors:

$$\log E(e^{iux}) = i\gamma u + c_- \int_{-\infty}^{0-} \psi(u,x)M'(x)dx - c_+ \int_{0+}^{\infty} \psi(u,x)M'(x)dx \quad (1.6)$$

$$\text{avec } \psi(u,x) = e^{iux}-1-\frac{iux}{1+x^2},$$

γ pouvant être choisi arbitrairement si l'on ajuste a_1.

Une représentation analogue a été donnée par R. Le Page [7] pour des v.a. de loi stable sur $\mathbb{R}^d$ en l'absence d'une composante normale. La stabilité considérée est celle de Sharpe [10] ou de Hudson et Mason

[6], définie par rapport à un semi-groupe d'automorphismes de la forme

$$\tau_t x = t^A x = \exp(\log t . A) . x \quad (t>0) \tag{1.7}$$

avec $Sp(A) \subset \{\lambda : \text{Re } \lambda > 1/2\}$.

Un résultat de Siebert [11] permet de trouver une matrice définie positive T telle que tout $x \in \mathbb{R}^d \setminus \{0\}$ s'écrive d'une manière unique sous la forme $t^A . y$, $t>0$, $y \in \mathbb{R}^d$ et $\|T_y\| = 1$ (nous écrirons Q pour l'ellipsoïde donné par les deux dernières conditions). Il existe alors une mesure positive finie sur Q, que nous écrirons $\lambda\nu$ ($\lambda > 0$, ν loi de probabilité sur Q), telle que la mesure de Lévy apparaissant dans la formule de Lévy-Chintchine pour la loi stable en question ait la forme

$$\eta(dx) = \lambda\nu(dy) . t^{-2} dt. \tag{1.8}$$

La fonction caractéristique de la v.a. X que l'on entend construire sera donc donnée par

$$\log E(e^{i<u,X>}) = i<u,\gamma> + \lambda \int_Q \int_0^\infty \psi(u, t^A y)\nu(dy) t^{-2} dt \tag{1.9}$$

$$\text{avec } \psi(u,x) = e^{i<u,x>} - 1 - i \frac{<u,x>}{1+\|x\|^2}$$

En utilisant le raisonnement de tout à l'heure, on s'aperçoit que l'on peut prendre pour X la limite p.s. de la série

$$\sum_{j=1}^\infty (\Gamma_j^{-A} Y_j - a_j), \tag{1.10}$$

où Γ_j est l'instant du j-ième saut de N_t, $\{Y_j\}_{j\geq 1}$ est une suite de v.a. indépendantes de loi ν sur Q et où $a_j = E(\Gamma^{-A} Y_j)$ pour j suffisamment grand.

2. Construction sur le groupe de Heisenberg

Le groupe de Heisenberg H_n sera identifié avec $\mathbb{R}^n \times \mathbb{R}^n \times \mathbb{R}$, dont nous utiliserons la structure additive (avec l'opération +). Le produit scalaire de $\mathbb{R}^n$ est noté <, > et le produit dans H_n est défini par

$$(x', x'', x''')(y', y'', y''') = (x'+y', x''+y'', x'''+\tfrac{1}{2}[x,y])$$

$$[x,y] = \langle x', y''\rangle - \langle x'', y'\rangle.$$

Les automorphismes requis par les lois stables ont été étudiés par Drisch et Gallardo [1]. Ce sont des transformations linéaires de $\mathbb{R}^{2n+1}$, compatibles avec [,]. La non-commutativité empêche de passer directement, comme nous l'avions fait au § 1, des temps Γ_j définis par le processus de Poisson aux v.a. U_j à valeurs dans $[0,t]$. Il faut encore, pour garantir l'équidistribution, procéder à une permutation aléatoire des Γ_j, ce qui peut s'effectuer globalement à l'aide de v.a. $\{V_j\}$ sur $[0,1]$: la permutation $\sigma \in S_n$ est définie par $V_{\sigma_1} < V_{\sigma_2} < \dots$.

Lemme 2.1: Soit $\{N_t\}_{t\geq 0}$ un processus de Poisson de paramètre λ, $\Gamma_1 < \Gamma_2 < \dots$ les temps où N_t effectue des sauts, $\{Y_j\}_{j\geq 1}$ une suite de v.a. de loi ν dans $\mathbb{R}^{2n+1}$, indépendantes entre elles et indépendantes de N_t, $\{V_j\}_{j\geq 1}$ une suite de v.a. i.i.d., uniformes sur $[0,1]$, indépendantes des processus précédents, $\varepsilon_{ij} = \mathrm{sgn}(V_j - V_i)$ pour $i<j$. Soit A un automorphisme de $\mathbb{R}^{2n+1}$. Alors la v.a.

$$S_t = \sum_{j=1}^{N_t} \Gamma_j^{-A} Y_j + \frac{1}{2} \sum_{1\leq i<j\leq N_t} \varepsilon_{ij}[\Gamma_i^{-A}Y_i, \ \Gamma_j^{-A}Y_j],$$

que l'on considère comme un élément de H_n (il faudrait plutôt interpréter S_t comme un élément de l'algèbre de Lie $\mathcal{H}_n$, mais $\exp: \mathcal{H}_n \to H_n$ est ici l'identité), a une loi de Poisson composée dont l'exposant μ est donné par

$$\int_{H_n} f(x)\mu(dx) = \lambda \int_{H_n} \int_0^t f(u^{-A}y)du \ \nu(dy).$$

Preuve: Si, partant des V_j, nous définissons la permutation aléatoire σ comme ci-dessus, si nous posons $\tau = \sigma^{-1}$ et $U_j = \Gamma_{\tau_j}$ $(1 \leq j \leq N_t)$, ces U_j seront des v.a. indépendantes, uniformément distribuées sur $[0,t]$. Les v.a. $Z_j = \Gamma_{\tau_j}^{-A} Y_{\tau_j}$ seront indépendantes, de loi $\mu/\lambda t$ et leur produit $Z_1 \cdot Z_2 \cdot \dots \cdot Z_{N_t}$ aura une loi de Poisson composée d'exposant μ. Ce produit vaut

$$\sum_{j=1}^{N_t} Z_j + \frac{1}{2} \sum_{k<\ell} [Z_k, Z_\ell] = \sum_{j=1}^{N_t} Z_j + \frac{1}{2} \sum_{\tau_i<\tau_j} [\Gamma_{\tau_i}^{-A} Y_{\tau_i}, \ \Gamma_{\tau_j}^{-A} Y_{\tau_j}]$$

et l'on a, pour $i < j$,

$$\tau_i < \tau_j \Leftrightarrow \sigma^{-1}(i) < \sigma^{-1}(j) \Leftrightarrow V_i < V_j \Leftrightarrow \varepsilon_{ij} = 1.$$

Lemme 2.2: Sous les mêmes hypothèses qu'en 2.1 mais en supposant encore le support de μ borné, $E(Y_j) = 0$ et $Sp(A) \subset \{Re\ \lambda > \frac{1}{2}\}$, on obtient en formant les sommes

$$\sum_{1\leq i<j\leq N} \varepsilon_{ij}[\Gamma_i^{-A}Y_i\ ,\ \Gamma_j^{-A}Y_j] = S_N$$

une suite presque sûrement convergente.

Preuve: Posons $Z_i = \Gamma_i^{-A}Y_i$. La convergence p.s. de ΣZ_i se démontre comme au § 1 . On peut répéter l'opération (en fixant d'abord V_i) pour la convergence p.s. de $\sum_{j>i}\varepsilon_{ij}Z_j$, notée W_i. Il faut alors prouver la convergence p.s. de $\Sigma[Z_i\ ,W_i]$. La somme $\Sigma \| Z_i\|^2$ est p.s. finie, grâce à l'hypothèse sur $S_p(A)$ et au fait que $\Gamma_j \sim j/\lambda$. De même, $\| W_i\|^2$ est p.s. borné, ce qui entraîne la convergence p.s. de $\Sigma |[Z_i\ , W_i]|^2$. On pose alors $\mathcal{F}_i = \sigma(\{\Gamma_j\}_{j\geq 1}, \{V_j\}_{j\geq 1}, \{Y_j\}_{j\geq i})$ et l'on constate que $\{[Z_i\ , W_i], \mathcal{F}_i\}$ est une martingale inverse (puisque $E(Y_i) = 0$). Il suffit alors de traduire en termes de martingales inverses la démonstration du théorème 11.1.3., p.381, de [12] pour aboutir au résultat annoncé.

En utilisant les lemmes qui précèdent, on obtient sans peine une représentation du type (1.10) sous des hypothèses particulières. Nous considérons, pour $f \in \mathfrak{D}_0$ (ensemble des fonctions bornées, deux fois dérivables, nulles à l'origine), la fonctionnelle

$$\Psi(f, t^A y) = \begin{cases} <\mathrm{grad}_0 f,\ t^A y> & 0 < t \leq 1 \\ <\mathrm{grad}_0 f,\ y> & t \geq 1 \end{cases} \tag{2.1}$$

($\mathrm{grad}_0 f$: gradient de f à l'origine) et le générateur infinitésimal (G.I.)

$$\mathcal{A} f = \lambda \int_Q \nu(dy) \int_0^\infty \frac{dt}{t^2} (f(t^A y) - \Psi(f, t^A y)) \tag{2.2}$$

($\lambda>0$, t^A comme en (1.7), Q selon [11], $\nu \in M_1(Q)$). Nous appellerons

$\{\mu_t\}_{t\geq 0}$ le semi-groupe de mesures de probabilité ayant $\mathcal{A}$ pour G.I.

Théorème 2.3. Si les processus $\{N_t\}$, $\{Y_j\}$ et $\{V_j\}$ ainsi que les v.a. ε_{ij} sont donnés de la même manière qu'au lemme 2.1, la loi ν de Y_j étant concentrée sur Q, et si $E(Y_j) = 0$, la v.a.

$$X = \sum_{j=1}^{\infty} \Gamma_j^{-A} Y_j + \frac{1}{2} \sum_{i=1}^{\infty} \sum_{j>i} \varepsilon_{ij} [\Gamma_i^{-A} Y_i, \Gamma_i^{-A} Y_j]$$

sera p.s. définie (chacune des sommes convergeant p.s.) et elle aura la loi μ_1 correspondant au G.I. donné par (2.2).

Preuve: Il suffit d'appliquer le lemme 2.1, de remplacer u par $u^{-1}=s$ dans l'intégrale, de constater que, grâce à la forme (2.1) de Ψ et à l'hypothèse $\int y\nu(dy) = 0$, on peut retrancher du G.I. un terme de la forme

$$\int_{t^{-1}}^{\infty} \frac{ds}{s^2} \int_Q \nu(dy) \Psi(f, s^A y),$$

puis de laisser tendre t vers ∞ en tenant compte des propriétés de convergence démontrées au lemme 2.2.

On remarque que l'on peut définir un processus $s \to X_s^t$ $(0\leq s\leq 1)$:

$$X_s^t = \sum_{V_j\leq s,\ j\leq N_t} Z_j + \frac{1}{2} \sum_{V_i,\ V_j\leq s,\ i<j\leq N_t} \varepsilon_{ij}[Z_i, Z_j] \tag{2.3}$$

$(Z_i = \Gamma_i^{-A} Y_i)$ et que celui-ci, ayant des accroissements indépendants et stationnaires, correspond au G.I.

$$\mathcal{A}_t f = \lambda \int_Q \nu(dy) \int_{t^{-1}}^{\infty} \frac{ds}{s^2} f(s^A y) \to \mathcal{A} f \tag{2.4}$$

Si l'on remplace $E(Y_j)=0$ par $E(Y_j)=a$ et que l'on ajoute à $\mathcal{A}$ une forme primitive $\langle \xi, \operatorname{grad}_0 f \rangle$, il faut corriger $\mathcal{A}_t$ par un terme de la forme $\langle \beta_t, \operatorname{grad}_0 f \rangle$, avec

$$\beta_t = \beta_1 - [\lambda \int_{t^{-1}}^{1} \frac{ds}{s^2} s^{-A}]a = \gamma - \sum_{j_0\leq j\leq N_t} E(\Gamma_j^{-A})a + o(1) \tag{2.5}$$

pour un $j_0>1$ et un $\gamma \in \mathcal{H}_n$ appropriés et pour $t \to \infty$ (cf. (1.3) et les remarques ultérieures). On construit un processus de G.I. $\tilde{\mathcal{A}} f = \mathcal{A}_t f + \langle \beta_t, \operatorname{grad}_0 f \rangle$ en appliquant la formule de Lie-Trotter ([2], lemme 2.6) aux processus X_s^t et $Y_s^t = \exp(s\beta_t)$ et l'on trouve

$$\tilde{X}_s^t = X_s^t + Y_s^t + \sum_{V_j \leq s,\ j \leq N_t} [(\tfrac{s}{2} - V_j)Z_j\ ,\ \beta_t]. \tag{2.6}$$

En effet, si $V_j = v$, il faut additionner des crochets $\pm \frac{1}{2}\,[dX_v^t\ ,\ dY_u^t]$ (+ si $v \leq u \leq s$, - si $0 \leq u \leq v$) et l'on obtient au bilan $\frac{1}{2}\,[dX_v^t\ ,\ (s-v-v)\beta_t] = [(\frac{s}{2} - v)Z_j\ ,\ \beta_t]$. En posant s=1 dans (2.6), puis en laissant tendre t vers ∞, on obtient une v.a. $X = \lim_{t\to\infty} \tilde{X}_1^t$, dont la loi aura la forme stable requise, c.à.d. qu'elle sera la loi μ_1 du semi-groupe déterminé par le G.I. $\mathcal{A}f + \langle \xi, \mathrm{grad}_o f \rangle$, $\mathcal{A}$ selon (2.2). Après avoir introduit les abréviations $a_j = E(\Gamma_j^{-A})a$ $(j \geq j_o)$ et $W_j = Z_j - a_j$, nous pourrons utiliser (2.5) pour calculer:

$$X = \sum_{j\geq 1} W_j + \frac{1}{2} \sum_{i\geq 1} \sum_{j>i} \varepsilon_{ij}\ [W_i\ ,\ W_j] + \gamma + \sum_{j\geq 1} [(\tfrac{1}{2} - V_j)Z_j\ ,\ \gamma] + R \tag{2.7}$$

$$R = \frac{1}{2} \sum_{i\geq 1} \sum_{j>i} \varepsilon_{ij}\{[a_i\ ,\ W_j]+[W_i\ ,\ a_j]\}+\frac{1}{2} \sum_{i\geq 1} \sum_{j>i} [Z_i\ ,\ (\varepsilon_{ij}-1+2V_i)a_j] \tag{2.8}$$

On peut démontrer la convergence p.s. de toutes les sommes intervenant dans (2.7) et (2.8). Nous nous bornerons ici au second terme de (2.8), auquel nous pouvons appliquer les mêmes considérations que dans la démonstration du lemme 2.2.

Nous avons en effet $E(\varepsilon_{ij} \mid V_i) = 1 - 2V_i$, donc la somme

$$S_i = \sum_{j>i} (\varepsilon_{ij} - 1 + 2V_i)\, a_j$$

existera p.s., conditionnellement à V_i, et $E(\|S_i\|^2)$ tendra vers 0 lorsque $i \to \infty$ puisque $\|a_j\|^2 \sim c.j^{-1-\delta}$. On écrira alors $[Z_i\ ,\ S_i] = [W_i\ ,\ S_i] + [a_i\ ,\ S_i]$ et on pourra traiter le premier terme comme dans la démonstration déjà évoquée. Pour le second, on posera

$$S_i = \sum_{j>i} (1-2V_j)\, a_j + \sum_{j>i} (\varepsilon_{ij} - (1-2V_j) - (1-2V_i))\, a_j = S_i' + S_i''$$

et $\mathcal{F}_i = \sigma(V_i\ ,\ V_{i+1}\ ,\ \ldots)$. $(S_i'', \mathcal{F}_i)$ est une martingale inverse, donc aussi $([a_i\ ,\ S_i''], \mathcal{F}_i)$ et il suffit de vérifier $\sum E\,[a_i\ ,\ S_i'']^2 < \infty$ pour démontrer la convergence de $\sum [a_i\ ,\ S_i'']$. Le cas de $(S_i\ ,\ \mathcal{F}_{i+1})$ est analogue.

3. Les groupes de Lie nilpotents, simplement connexes

Le cadre naturel pour l'étude des lois stables est celui des groupes de Lie nilpotents et simplement connexes (cf. [3], [4] et [5]). Comme dans le cas particulier du groupe de Heisenberg, on peut identifier le groupe G avec son algèbre de Lie $\mathfrak{G}$ et les automorphismes t^A ont encore la forme (1.7), tout en satisfaisant $[t^A x, t^A y] = t^A[x, y]$. Les G.I. des semi-groupes stables peuvent s'écrire $\mathcal{A}f + \langle \mathrm{grad}_o f, \xi \rangle$, $\mathcal{A}f$ comme sous (2.2), $f \in \mathfrak{D}_o$.

Dès lors, rien ne s'oppose en principe à la construction décrite au §2: On introduit le processus de Poisson N_t, les v.a. i.i.d. Y_j de loi ν sur une section Q convenablement choisie et les v.a. i.i.d. V_j de loi uniforme sur [0,1]. Comme en (2.3), on définit un processus $s \to X_s^t$ en posant $dX_s^t = \Gamma_j^{-A} Y_j$ si $V_j = s$ et si $\Gamma_j \leq t$ (c.à.d. $j \leq N_t$), puis en effectuant les produits de gauche à droite dans l'ordre donné par les V_j. On peut expliciter ce produit comme au lemme 2.1, en tenant compte des termes qui apparaissent dans la formule de Campbell-Hausdorff (et qui sont en nombre fini puisqu'on se trouve sur un groupe nilpotent), mais nous renoncerons ici à cette encombrante formalité. Pour $\xi = E(V_j) = 0$, il suffira alors de laisser t tendre vers ∞ pour obtenir l'analogue du théorème 2.3. Dans le cas général, il faudra corriger X_s^t par un processus Y_s^t de G.I. $\langle \mathrm{grad}_o f, \beta_t \rangle$: $d\tilde{X}_s^t = dX_s^t + \beta_t\, ds$, β_t comme en (2.5).

4. Domaine d'attraction des lois stables

Soit $\{X_j = T_j^A Y_j\}$ une suite de v.a. i.i.d. sur un groupe G du type considéré au § 3, avec $Y_j \in Q$, $T_j > 0$. Nous supposerons

$$P(T_j > t) = t^{-1}L(t), \quad \mathcal{L}(Y_j | T_j > t) \underset{t \to \infty}{\Longrightarrow} \nu \tag{4.1}$$

où L(t) varie lentement pour $t \to \infty$. Soit $\{\mu_t\}$ un semi-groupe de lois stables déterminé par le G.I. $\mathcal{A}f$ de (2.2), avec $\lambda = 1$.

Théorème 4.1: Sous l'hypothèse (4.1), il existe des suites $t_n \to \infty$, $\beta_n \in \mathfrak{G}$ et $b_n = \exp(-\beta_n) \in G$ telles que

$$S_n = t_n^{-A}\Big(\prod_{j=1}^{n} b_n X_j\Big) \xrightarrow{\mathcal{L}} \mu_1.$$

Preuve: Il suffira pour l'essentiel de transcrire la démonstration de [8]. La suite t_n est déterminée par $nP(T_j/t_n>t) \to t^{-1}$ et la suite β_n par

$$\beta_n = \quad E[\varphi_n(X_j)], \ \varphi_n(X_j) = (T_j \wedge t_n)^A Y_j \tag{4.2}$$

(on identifie ici $X_j \in G$ et $\log X_j \in \mathfrak{G}$). Partant d'un processus de Poisson N_t de paramètre $\lambda=n$, on construit

$$X_s^n = t_n^{-A}(\prod_{j\leq N_s} b_n X_j),$$

processus dont le G.I. vaut $\mathcal{A}_n f = nEf(t_n^{-A}(b_n X_1))$. S_n, la valeur de X_s^n en $s=\Gamma_n$, a approximativement la même loi que X_1^n. On démontre alors que $\mathcal{A}_n f$ peut être approché par

$$n[Ef(t_n^{-A}X_1) - \langle \mathrm{grad}_o f, t_n^{-A}\beta_n\rangle] = \sum_{j=1}^{n} E[f(t_n^{-A}X_j) - \Psi(f, t_n^{-A}X_j)], \tag{4.3}$$

Ψ étant défini comme en (2.1). On ordonne les valeurs de T_j par ordre décroissant: $T_{\sigma 1} \geq T_{\sigma 2} \geq \ldots$ et l'on fixe la valeur de $R_m = T_{\sigma m}$ pour un m qui tendra vers ∞ plus lentement que n. Dans l'expression (4.3), on sépare les j tels que $T_j > R_m$ de ceux correspondant à des $T_j \leq R_m$. La façon dont les t_n ont été choisis implique que, conditionnellement à $R_m = rt_n$, les v.a. t_n/T_{σ_k} $(1\leq k<m)$ se comportent asymptotiquement comme les Γ_j du processus N_t, c.à.d. que, pour $T_j > R_m$, la v.a. T_j/t_n peut être asymptotiquement remplacée par une v.a. de densité $rs^{-2}ds\, 1_{\{s>r\}}$. La condition (4.1) permet alors de remplacer Y_j par une v.a. $\overline{Y}_j$ de loi ν sur Q et l'on obtient approximativement pour la somme sur les j du premier groupe:

$$(m-1)r \int_Q \nu\,(dy) \int_{1/r}^{\infty} \frac{ds}{s^2}\,(f(s^A y) - \Psi(f, s^A y)). \tag{4.4}$$

Si l'on prend m assez grand, t_n/R_m sera de la forme $m+o(m)$ et l'on pourra donc remplacer $(m-1)r = (m-1)\,R_m/t_n$ par 1 et $1/r$ par 0+ dans (4.4). La somme sur les j du second groupe est limitée, grâce à (1.7), par une expression du genre $n\cdot O(r^{1+\Delta}) = nO(m^{-1-\Delta})$ et il suffira de choisir $m=n^{1-\delta}$ (avec $(1-\delta)(1+\Delta)>1$) pour pouvoir négliger ce terme. Cela démontre bien que le G.I. $\mathcal{A}_n f$ (qui donne la loi de S_n) se comporte asymptotiquement comme le G.I. $\mathcal{A}f$ de (2.2).

Références

[1] T. Drisch, L. Gallardo: Stable Laws on the Heisenberg Groups. In: Probability Measures on Groups VII. Lecture Notes Math. 1064. Springer Verlag 1984.

[2] W. Hazod: Stetige Faltungshalbgruppen von Wahrscheinlichkeitsmassen und erzeugende Distributionen. Lecture Notes Math. 595. Springer Verlag 1977.

[3] W. Hazod: Stable Probabilities on locally compact Groups. In: Probability Measures on Groups VI. Lecture Notes Math. 928. Springer Verlag 1982.

[4] W. Hazod: Remarks on (semi-) stable Probabilities. In: Probability Measures on Groups VII. Lecture Notes Math. 1064. Springer Verlag 1984.

[5] W. Hazod: Semi-groupes de convolution demi-stables et décomposables sur les groupes localement compacts. In: Probabilités sur les structures géométriques. Université de Toulouse 1985.

[6] W.N. Hudson, J.D. Mason: Operator-Stable Laws. J. Multivariate Analysis 11, 434-447 (1981).

[7] R. Le Page: Multidimensional infinitely divisible Variables and Processes. In: Probability in Banach Spaces III. Lecture Notes Math. 860. Springer Verlag 1981.

[8] R. Le Page, M. Woodroofe, J. Zinn: Convergence to a stable Distribution via Order Statistics. Ann. Probability 9, 624-632 (1981).

[9] P. Lévy: Propriétés asymptotiques des sommes de variables aléatoires indépendantes ou enchaînées. Journal de Math., tome XIV, fasc. IV (1935).

[10] M. Sharpe: Operator-stable Probability Distributions on Vector Groups. Trans. Amer. Math. Soc. 136, 51-65 (1969).

[11] E. Siebert: Supplements to Operator-stable and Operator semi-stable Laws on Euclidean Spaces. To appear in J. Multivariate Analysis.

[12] Y.S. Chow, H. Teicher: Probability Theory. Springer-Verlag, (1978).

Université de Berne
Sidlerstr. 5

CH-3012 Berne

PARABOLIC SUBGROUPS AND FACTOR COMPACTNESS OF MEASURES ON SEMISIMPLE LIE GROUPS

S.G. DANI and M. McCRUDDEN

Introduction. For any locally compact topological G we write $M(G)$ to denote the topological semigroup of all probability distributions on G, furnished with the weak topology and with convolution as the multiplication

Given $\mu \in M(G)$ we write $\mathfrak{F}(\mu, G)$ for the two-sided factor set of μ, so that

$$\mathfrak{F}(\mu, G) = \{\lambda \in M(G) : \exists \nu \in M(G) \text{ s.t. } \lambda\nu = \mu = \nu\lambda\}.$$

We say that μ is factor-compact in G to mean that $\mathfrak{F}(\mu, G)$ is a compact subset of $M(G)$. When applied to a point mass ε_x $(x \in G)$, this notion is equivalent to the centraliser of x in G being compact. We write $G(\mu)$ for the smallest closed subgroup of G which contains the support of μ.

It is known that if G is a connected semisimple Lie group with finite centre and if $\mu \in M(G)$ such that $G(\mu) = G$, then μ is factor-compact in G ([3], Theorem 1). Our aim here is to prove a considerable generalisation of this result.

Any connected semisimple Lie group G with finite centre contains an important family of subgroups known as the parabolic subgroups of G. The precise definition of these subgroups will be given in the next section and for the moment we simply note that they fall into a finite number of conjugacy classes within G, so in some sense there are not too many of them. We shall prove

Main Theorem. Let G be a connected semisimple Lie group with finite centre and suppose $\mu \in M(G)$ such that μ is not factor-compact in G. Then $G(\mu)$ lies inside a proper parabolic subgroup of G.

At the end of the paper we indicate by examples that, in general, connected semisimple Lie groups with finite centre have lots of subgroups which do not lie inside a proper parabolic subgroup, and so support a rich supply of probability measures which are factor-compact in G. Of course the restriction to finite centre is a natural one, since a group with non-compact centre cannot support any factor-compact measures.

§1. Parabolic subgroups of semisimple Lie groups

Throughout this section G will denote a connected semisimple Lie group with finite centre, and $\mathfrak{g}$ will be its Lie algebra. Our aim here is to recall the definition and properties of the family of subgroups of G known as the parabolic subgroups. For a brief introduction to parabolic subgroups the reader is referred to the article of MacDonald [2], for a fuller discussion with complete proofs the book of Warner [8, Chapter I] is recommended.

Let $G = KAN$ be an Iwasawa decomposition of G, and let A^* denote the character group of A, where a character on A is just a homomorphism of A into $\mathbb{R}^+$, the positive reals. For each $\alpha \in A^*$ we write

$$\mathfrak{g}^\alpha = \{X \in \mathfrak{g} : \forall a \in A, \mathrm{Ad}a(X) = \alpha(a)X\}$$

and we let Φ denote the set of all non-identity $\alpha \in A^*$ such that $\mathfrak{g}^\alpha \neq \{0\}$. The set Φ is called the <u>root system of $\mathfrak{g}$ w.r.t. A</u>, it is a finite set, and we have the direct sum decomposition

$$\mathfrak{g} = \mathfrak{z}(A) + \sum_{\alpha\in\Phi} \mathfrak{g}^\alpha$$

where $\mathfrak{z}(A)$ is $\{X \in \mathfrak{g} : \forall a \in A, \mathrm{Ad}a(X) = X\}$, the centraliser of A in $\mathfrak{g}$ (c.f. [2], page 122).

We note that since A is a vector group, A^* has a natural vector space structure, so we may select a basis Δ for Φ (i.e. the simple roots) and we use Φ^+ and Φ^- respectively to denote the set of positive roots and the set of negative roots w.r.t. the chosen basis Δ. We write

$$A^+ = \{a \in A: \alpha(a) \geqslant 1, \forall \alpha \in \Delta\}$$

and we recall that $G = KA^+K$.

Now let Θ be any subset of Δ and let $\langle\Theta\rangle$ denote the intersection of Φ with the multiplicative subgroup of A^* generated by Θ. We note that each of the sets

$$\sum_{\alpha\in\langle\Theta\rangle} \mathfrak{g}^\alpha \quad , \quad \sum_{\alpha\in\Phi^+-\langle\Theta\rangle} \mathfrak{g}^\alpha \quad , \quad \sum_{\alpha\in\Phi^--\langle\Theta\rangle} \mathfrak{g}^\alpha$$

is a subalgebra of $\mathfrak{g}$ invariant under $\mathrm{Ad}A$, and we write H_Θ, N_Θ, N_Θ^- for the corresponding analytic subgroups of G. We write P_Θ for the subgroup of G generated by $H_\Theta \cup N_\Theta \cup Z(A)$, where $Z(A)$ is the centraliser of A in G, and we note that the Lie algebra of P_Θ is $\mathfrak{z}(A) + \sum_{\alpha\in\Phi^+\cup\langle\Theta\rangle} \mathfrak{g}^\alpha$. The subgroups P_Θ and N_Θ^- are clearly normalised by A.

The subgroup P_Θ is called the <u>standard parabolic subgroup corresponding to Θ</u>. We say that a subgroup P of G is <u>parabolic</u> iff it is a conjugate of some P_Θ, for some $\Theta \subseteq \Delta$. When $\Theta = \phi$ we obtain the so-called <u>minimal parabolic subgroups</u> of G. We note that there are a number of different ways to define the parabolic subgroups of G (see [8], page 55) and that the constructive definition we have given here seems to be the one best-suited to our later purpose. The reader unfamiliar with parabolic subgroups should find illuminating the quite explicit description of the standard parabolic subgroups of $SL(n, \mathbb{R})$ given on page 138 of [2].

We now list some properties of parabolic subgroups required in the next section, and indicate a reference where appropriate.

Proposition 1. Let G be a connected semisimple Lie group with finite centre, let Δ, Θ, P_Θ and N_Θ^- be as above. Then

(i) $N_\Theta^- P_\Theta$ is a neighbourhood of e in G.

(ii) P_Θ and N_Θ^- are closed subgroups of G.

(iii) if H is a subgroup of G such that P_Θ is an open subgroup of H, then $P_\Theta = H$.

Proof (i) This follows at once from the inverse function theorem. Indeed it is true that $N_\Theta^- P_\Theta$ is a dense open subset of G ([8], 1.2.4.10) but we do not need this fact.

(ii) These follow from 1.2.4.8 and 1.2.4.11 of [8].

(iii) By 1.2.1.1 of [8], H must itself be a standard parabolic subgroup of G, and since $\dim H = \dim P_\Theta$, we must have $P_\Theta = H$.

§2. A property of semisimple Lie groups

We maintain the notation of the previous section.

Proposition 2. Let $(a_n)_{n\geq 1}$ be a sequence in A^+, and suppose $(a_n)_{n\geq 1}$ has no convergent subsequence in G. Then there is a subsequence $(b_r)_{r\geq 1}$ of $(a_n)_{n\geq 1}$, and a proper subset Θ of Δ, such that

(a) for every compact set $C \subseteq P_\Theta$, $\bigcup_{r=1}^{\infty} b_r^{-1} C b_r$ is relatively compact.

(b) for every compact set $D \subseteq N_\Theta^-$, $b_r x b_r^{-1} \to e$ uniformly in $x \in D$, as $r \to \infty$.

Proof Since Δ is a basis for the vector space A^*, and $(a_n)_{n\geq 1}$ has no convergent subsequence, there is some $\alpha \in \Delta$ such that $(\alpha(a_n))_{n\geq 1}$ is unbounded. By successively choosing subsequences as necessary, we can find a proper (though possibly empty) subset θ of Δ, and a subsequence $(b_r)_{r\geq 1}$ of $(a_n)_{n\geq 1}$ such that

$$\left.\begin{array}{lll} \forall\, \alpha \in \Theta, & (\alpha(b_r))_{r\geq 1} & \text{is bounded} \\ \forall\, \alpha \in \Delta\backslash\Theta & \alpha(b_r) \to \infty & \text{as } r \to \infty \end{array}\right\} \ldots(*)$$

(a)(i) Suppose $C = \exp B$, where B is a compact subset of $\mathfrak{p}$, the Lie algebra of P_Θ. Since $\mathfrak{p}$ equals $\mathfrak{z}(A) + \sum_{\alpha\in\Phi^+\cup<\Theta>} \mathfrak{g}^\alpha$, equation (*) implies that all eigenvalues of $\{\mathrm{Ad}\, b_r^{-1} : r \geq 1\}$ on $\mathfrak{p}$ are uniformly bounded above, hence $\bigcup_{r=1}^{\infty} \mathrm{Ad}\, b_r^{-1}(B)$ is a bounded subset of $\mathfrak{g}$, and so $\bigcup_{r=1}^{\infty} b_r^{-1} C b_r = \exp(\bigcup_{r=1}^{\infty} \mathrm{Ad}\, b_r^{-1}(B))$ is a relatively compact subset of P_Θ, and so of G (P_Θ is closed by Proposition 1 (ii)).

(ii) Suppose C is a compact subset of P_Θ^0, the connected component of e in P_Θ. We may pick a compact neighbourhood B of $0 \in \mathfrak{p}$ such that $\exp(B)$ is a neighbourhood of e in P_Θ, and then there is some $m \geq 1$ such that $C \subseteq \exp(B)^m$. Then $\bigcup_{r=1}^{\infty} b_r^{-1}Cb_r \subseteq \bigcup_{r=1}^{\infty} (b_r^{-1}\exp(B)b_r)^m$ and the last set is relatively compact by (i) above.
(iii) For arbitrary compact $C \subseteq P_\Theta$ the conclusion (a) follows by (ii) above and the observation (obvious from our definition of P_Θ) that $P_\Theta = Z(A)P_\Theta^0$.
(b) The Lie algebra of N_Θ^- is $\mathfrak{n} = \sum_{\alpha \in \Phi^- - \langle\Theta\rangle} \mathfrak{g}^\alpha$, and clearly all eigenvalues of $\mathrm{Ad}\,b_r$ on $\mathfrak{n}$ tend to zero as $r \to \infty$, by (*). An argument along the lines of (i) and (ii) of the proof of (a) now gives the required result, since the group N_Θ^- is connected.

Definition. For $B \subseteq G$ we write $\mathcal{L}(B, G)$ to denote the set of all y in G such that, for each neighbourhood U of y, there exists a compact set $C \subseteq G$ (C depending on U) such that $U \cap aCa^{-1} \neq \phi$, for all $a \in B$.

$\mathcal{L}(B, G)$ is always a closed subgroup of G and its elementary properties are given in [4], where the idea was first introduced.

Proposition 3. Let $(k_n)_{n\geq 1}$, $(h_n)_{n\geq 1}$, $(a_n)_{n\geq 1}$ be sequences in G, and suppose that $(h_n)_{n\geq 1}$ is bounded and $k_n \to k \in G$ as $n \to \infty$. Then

$$\mathcal{L}((k_n a_n h_n)_{n\geq 1}, G) \subseteq k\mathcal{L}((a_n)_{n\geq 1}, G)k^{-1}.$$

Proof By Proposition 3(iv) of [4] we may assume that $(h_n)_{n\geq 1}$ is the identity sequence. Let $y \in \mathcal{L}((k_n a_n)_{n\geq 1}, G)$ and let U be a compact neighbourhood of e. Since $k_n \to k$, there is a neighbourhood V of e and some $N \geq 1$ such that for all $n \geq N$, $k_n^{-1}yVk_n \subseteq k^{-1}ykU$.

But there exists a compact set D such that

$$\forall n \geq 1, \quad yV \cap k_n a_n D a_n^{-1} k_n^{-1} \neq \phi$$

hence $\forall n \geq N$,

$$k^{-1}ykU \cap a_n D a_n^{-1} \neq \phi.$$

If we write $C = D \cup \bigcup_{r=1}^{N} (a_r^{-1}k^{-1}yk\,U\,a_r)$, then $\forall n \geq 1$,

$$k^{-1}ykU \cap a_n C a_n^{-1} \neq \phi$$

and since C is clearly bounded, we deduce that $k^{-1}yk \in \mathcal{L}((a_n)_{n\geq 1}, G)$.

This brings us to the property of G we wish to establish.

Theorem 4. Let G be a connected semisimple Lie group with finite centre and let B be an unbounded subset of G. Then there is a proper parabolic subgroup P of G

such that $\mathcal{L}(B, G) \subseteq P$.

Proof Let $G = KAN$ be an Iwasawa decomposition of G and let $G = KA^{+}K$ be the corresponding Cartan decomposition, where K is of course a maximal compact subgroup of G. Let $(x_n)_{n\geq 1}$ be a sequence in B with no convergent subsequence and write $x_n = k_n a_n h_n$, with k_n, $h_n \in K$ and $a_n \in A^{+}$. Passing to a subsequence if necessary we may assume that $(k_n)_{n\geq 1}$ is convergent and $(a_n)_{n\geq 1}$ has no convergent subsequence. Since the class of parabolic subgroups is closed under conjugation, and since $B \subseteq C$ implies $\mathcal{L}(C, G) \subseteq \mathcal{L}(B, G)$, it is sufficient, by Proposition 3, to show that $\mathcal{L}((a_n)_{n\geq 1}, G)$ lies inside a proper parabolic subgroup.

By Proposition 2 we may select a subsequence $(b_r)_{r\geq 1}$ of $(a_n)_{n\geq 1}$, and a proper subset Θ of Δ, such that (a) and (b) of Proposition 2 are satisfied. To finish the proof it is sufficient to show that $\mathcal{L}((b_r)_{r\geq 1}, G) = P_\Theta$.

For any $x \in P_\Theta$, $(b_r^{-1}xb_r)_{r\geq 1}$ is relatively compact by (a) of Proposition 2, and this is enough to show that $x \in \mathcal{L}((b_r)_{r\geq 1}, G)$. We conclude that $P_\Theta \subseteq \mathcal{L}((b_r)_{r\geq 1}, G)$. By Proposition 1 (iii) the proof will be complete if we can show that P_Θ is an open subgroup of $H = \mathcal{L}((b_r)_{r\geq 1}, G)$. By Proposition 1 (i), this will follow if we show that $H \cap N_\Theta^- P_\Theta = P_\Theta$, and this in turn will follow if we show that $H \cap N_\Theta^- = \{e\}$ (since $P_\Theta \subseteq H$).

So let $y \in H \cap N_\Theta^-$ and suppose $y \neq e$, then by Proposition 1 (i), we can find a compact neighbourhood E of e in P_Θ, and a compact neighbourhood F of y in N_Θ^-, such that EF is a neighbourhood of y in G, and $e \notin F$. Since $y \in H$ there is a compact set C in G such that $EF \cap b_r C b_r^{-1} \neq \phi$, for all $r \geq 1$. Hence

$$\forall r \geq 1, \quad (b_r^{-1}Eb_r)(b_r^{-1}Fb_r) \cap C \neq \phi$$

which by (a) of Proposition 2 implies there is a compact set M in G such that $b_r^{-1}Fb_r \cap M \neq \phi$, $\forall r \geq 1$. Writing $D = N_\Theta^- \cap M$ we have $F \cap b_r D b_r^{-1} \neq \phi$, $\forall r \geq 1$, and as D is a compact set in N_Θ^- (remember N_Θ^- is closed by Proposition 1 (ii)), we obtain a contradiction to (b) of Proposition 2. The proof is now complete.

§3. The main theorem and consequences

Main Theorem. Let G be a connected semisimple Lie group with finite centre and let $\mu \in M(G)$. Suppose that $\mathcal{F}(\mu, G)$ is not compact. Then the closed subgroup $G(\mu)$ lies inside some proper parabolic subgroup of G.

Proof The given condition implies that there are sequences $(\lambda_n)_{n\geq 1}$ and $(\nu_n)_{n\geq 1}$ in $M(G)$ such that $(\nu_n)_{n\geq 1}$ has no convergent subsequence, and $\nu_n\lambda_n = \mu = \lambda_n\nu_n$, for all $n \geq 1$. By shift compactness there is a sequence $(x_n)_{n\geq 1}$ in G such that $(\nu_n x_n)_{n\geq 1}$ and $(x_n^{-1}\lambda_n)_{n\geq 1}$ are relatively compact. We note that $(x_n)_{n\geq 1}$ has no convergent subsequence, so by Theorem 4 there is a proper parabolic subgroup P of G such that $\mathcal{L}((x_n)_{n\geq 1}, G) \subseteq P$.

But if $x \in \text{supp}(\mu)$ and U is a neighbourhood of x, we have $\mu(U) = \delta > 0$.

Since $(x_n^{-1}\mu x_n)_{n\geq 1}$ equals $(x_n^{-1}\lambda_n \nu_n x_n)_{n\geq 1}$ and so is relatively compact, Prohorov's condition tells us there is a compact set $C \subseteq G$ s.t. $x_n^{-1}\mu x_n(C) > 1 - \delta/2$ for all $n \geq 1$, which implies that $x_n C x_n^{-1} \cap U \neq \phi$ for all $n \geq 1$. So $x \in \mathcal{L}((x_n)_{n\geq 1}, G)$, and hence $\mathrm{supp}(\mu) \subseteq P$, giving the result.

Concluding Remarks. (i) Let G be a connected semisimple Lie group with finite centre, and let $R(\mu, G)$ be the root set of μ in G, so

$$R(\mu, G) = \bigcup_{n=1}^{\infty} \{\nu^k : \nu \in M(G),\ \nu^n = \mu,\ 1 \leq k \leq n\}.$$

If $R(\mu, G)$ is not relatively compact in $M(G)$, then $G(\mu)$ lies inside a proper parabolic subgroup, but cannot itself be a parabolic subgroup (which is equivalent to saying it cannot contain a minimal parabolic subgroup). This follows from our main theorem, Propositions 12 and 13 of [4], and the fact that a parabolic subgroup is self-normalising and has a finite group of components.

(ii) Let G be as in (i), let K be a maximal compact subgroup of G, and suppose $\mu \in M(G)$ is left K-invariant i.e. $\forall k \in K$, $k\mu = \mu$. Then $\mathfrak{J}(\mu, G)$ is compact.

For if not then by our main theorem we have $G(\mu) \subseteq P$ for some proper parabolic subgroup P, and since clearly $K \subseteq G(\mu)$ we obtain $K \subseteq P$, which is impossible. As a consequence we obtain the embedding theorem for infinitely divisible probabilities on a symmetric space of non-compact type, due to Parthasarathy [5], by a method different from that of Parthasarathy, which uses spherical functions on G. The problem of obtaining Parthasarathy's result without spherical functions was first raised with the second-named author by H. Heyer, and was a motivating question for the present work.

(iii) Let G be a connected reductive Lie group, and suppose $\mu \in M(G)$ such that $R(\mu, G)$ is not relatively compact. Then $\mathrm{Ad}(G(\mu))$ is contained in a proper parabolic subgroup of the semisimple group $\mathrm{Ad}G$. This follows from our main theorem and an argument as in the proof of Proposition 8 of [3].

(iv) To appreciate the significance of the main theorem it is important to realise that, in general, semisimple Lie groups have a lot of subgroups not contained in any proper parabolic subgroup. We content ourselves with giving some examples.

(a) Any maximal compact subgroup of any semisimple connected Lie group G with finite centre.

(b) Any Zariski dense subgroup of G, in particular any lattice in a non-compact simple G. This depends on the fact that parabolic subgroups are Zariski closed, and the Borel density theorem ([7], Chapter 5). There also exist discrete subgroups other than lattices which are Zariski dense [1].

(c) Any subgroup of $SL(n, \mathbb{R})$ whose natural action on $\mathbb{R}^n$ is irreducible. Since $SL(2, \mathbb{R})$ has an irreducible action on $\mathbb{R}^n$ for all $n \geq 2$, this gives lots of three dimensional subgroups of $SL(n, \mathbb{R})$ which are not in a proper parabolic subgroup.

(v) The method of proof of our main theorem appears to us to be the natural development of the technique of Parthasarathy introduced in [6].

Acknowledgements. The first-named author would like to thank the University of Erlangen Nurenberg, and Professor D. Kölzow in particular, for hospitality during the autumn of 1985, when this work was done.

Both authors would like to thank Professor H. Heyer, whose kind invitation to them to attend the 1985 Oberwolfach conference made possible the present collaboration.

References

[1] Dani, S.G. A simple proof of Borel's Density Theorem, Math. Zeit 174 (1980), 81-94.

[2] MacDonald, I.G. Algebraic structures of Lie groups, in: Representation Theory of Lie Groups, L.M.S. Lecture Notes 34 (1979).

[3] McCrudden, M. Factors and roots of large measures on connected Lie groups, Math. Zeit. 177 (1981) 315-322.

[4] McCrudden, M. Local tightness of convolution semigroups over locally compact groups, in: Probability measures on groups, Oberwolfach 1981, Lecture Notes in Mathematics 928, Springer, 304-314.

[5] Parthasarathy, K.R. On the embedding of an infinitely divisible distribution in a one-parameter convolution semigroup, Theory of Prob. & Appl. 12 (1967), 373-380.

[6] Parthasarathy, K.R. Infinitely divisible distributions in $SL(k, \mathbb{C})$ or $SL(k, \mathbb{R})$ may be embedded in diadic convolution semigroups, in: Probability measures on groups, Oberwolfach 1978, Lecture Notes in Mathematics 706, Springer, 252-256.

[7] Raghunathan, M.S. Discrete subgroups of Lie groups, Springer (1972).

[8] Warner, G. Harmonic analysis on semisimple Lie groups vol. 1, Springer-Verlag, (1972).

S.G. DANI,
School of Mathematics,
Tata Institute,
Homi Bhabha Road,
Bombay 400005,
INDIA.

M. McCRUDDEN,
Department of Mathematics,
University of Manchester,
Oxford Road,
Manchester, M13 9PL,
ENGLAND.

UNE CARACTERISATION DU TYPE
DE LA LOI DE CAUCHY-HEISENBERG.

Jean-Louis DUNAU et Henri SENATEUR

1. INTRODUCTION

Une des manières d'obtenir la loi de Cauchy dans $\mathbb{R}$ est la suivante : si σ est la probabilité uniforme sur le cercle unité du plan complexe, alors la mesure image de σ par la transformation de Cayley : $z \longmapsto i\,\frac{1-z}{1+z}$ (par cette application $e^{i\theta} \longmapsto tg(\theta/2)$) est la loi de probabilité de densité $\frac{dx}{\pi(1+x^2)}$.

Nous considérons ici une généralisation de cette méthode. Nous appelons *loi de Cauchy-Heisenberg* sur $\mathbb{R} \times \mathbb{C}^n$ la mesure de probabilité image de la probabilité uniforme sur $S_{2n+1} = \{(z_0,z) \in \mathbb{C} \times \mathbb{C}^n \;;\; |z_0|^2 + |z|^2 = 1\}$ par l'application
$(z_0,z) \longmapsto \left(i\,\frac{1-z_0}{1+z_0} - i\,\frac{|z|^2}{|1+z_0|^2}\;,\; i\,\frac{z}{1+z_0}\right)$.
Nous verrons dans la Proposition 4 qu'elle a pour densité :

$$(t,z) \longmapsto \frac{n!\;2^{2n}\;\pi^{-(n+1)}}{((1+|z|^2)^2+t^2)^{n+1}} .$$

Bien entendu, cette mesure n'est pas la plus usuelle généralisation de la loi de Cauchy dans $\mathbb{R}$. On considère le plus souvent la mesure de probabilité sur $\mathbb{R}^n$ de densité :

$$x \longmapsto \frac{\Gamma(\frac{n+1}{2})\,\pi^{-(n+1)/2}}{(1+\|x\|^2)^{(n+1)/2}} \;;$$

c'est la mesure image de la probabilité uniforme sur $\mathbb{S}_n$ (la sphère unité de l'espace euclidien $\mathbb{R}^{n+1}$) par l'application :

$$(x_1,\ldots,x_{n+1}) \longmapsto (x_1/x_{n+1},\ldots,x_n/x_{n+1}) \;;$$

On pourrait l'appeler "loi de Cauchy-projective" de $\mathbb{R}^n$; on peut également considérer la "loi de Cauchy-conforme" de $\mathbb{R}^n$, c'est-à-dire la mesure image de la probabilité uniforme sur $\mathbb{S}_n$ par la projection stéréographique :

$$(x_1,\ldots,x_{n+1}) \longmapsto (x_1/(1-x_{n+1}),\ldots,x_n/(1-x_{n+1})) \;;$$

Pour plus de détails sur ces lois, voir [2] ou Letac [7].

Nous considérons ensuite le groupe d'Heisenberg $\mathbb{R} \times \mathbb{C}^n$, et, en suivant KORANYI [6], sont définis sur $\mathbb{R} \times \mathbb{C}^n$ l'analogue d'une translation, d'une homothétie, d'une transformation unitaire et d'une inversion ; nous obtenons ainsi un groupe $\mathcal{K}$ analogue au groupe des similitudes-translations et un groupe $\mathcal{G}$ analogue au groupe conforme. Si nous appelons type d'une mesure μ sur $\mathbb{R} \times \mathbb{C}^n$ l'ensemble des mesures images de μ par $\mathcal{K}$, nous obtenons la caractérisation suivante (c'est le Théorème 3, qui est le but de cet article): si μ est une mesure de probabilité sans atome sur $\mathbb{R} \times \mathbb{C}^n$, alors μ est du type de la loi de Cauchy-Heisenberg si et seulement si le type de μ est préservé par $\mathcal{G}$. Ce résultat est l'analogue de celui que nous avons démontré pour la loi de Cauchy-conforme dans [3] et de celui démontré pour la loi de Cauchy-projective par Knight et Meyer [5] dont nous avons donné dans [1] une démonstration élémentaire. Chacun de ces résultats constitue une généralisation différente de la caractérisation de Knight [4] de la loi de Cauchy dans $\mathbb{R}$.

Cet article est basé sur le Chapitre 5 de notre thèse de 3ème cycle [2] ; dans les parties 2,3 et 4 qui suivent, nous développons les aspects évoqués plus haut de la géométrie du groupe d'Heisenberg, en nous inspirant largement de KORANYI [6] dont nous conservons la plupart des notations. Dans la partie 5, nous étudions la loi de Cauchy-Heisenberg. Le Théorème 3, qui caractérise le type de cette loi est démontré dans la partie 6.

2. NOTIONS SUR LA GEOMETRIE DU GROUPE D'HEISENBERG.

Si $z = (z_1,\dots,z_n)$ et $z' = (z'_1,\dots,z'_n)$ sont dans $\mathbb{C}^n$, on note

$$\bar{z} = (\bar{z}_1,\dots,\bar{z}_n) \quad , \quad zz' = z_1z'_1 + \dots + z_nz'_n \quad \text{et} \quad |z| = (z\bar{z})^{1/2}$$

Nous munissons $\mathbb{C} \times \mathbb{C}^n$ de la loi de groupe (non abélien)

$$(z_0,z)(z'_0,z') = (z_0 + z'_0 + 2\,\mathcal{I}m\, z\bar{z}',\ z+z')$$

Alors $\mathbb{R} \times \mathbb{C}^n$ est un sous-groupe, que nous appellerons groupe d'Heisenberg.

Nous allons définir sur $\mathbb{R} \times \mathbb{C}^n$ et sur $\mathcal{P} = \{(z_0,z) \in \mathbb{C} \times \mathbb{C}^n \ ;\ \mathcal{I}m\, z_0 > 0\}$ des groupes de bijections $\mathcal{H}$, $\mathcal{A}$, $\mathcal{M}$ respectivement analogues aux translations, homotheties et transformationsorthogonales de $\mathbb{R}^n$. $\mathcal{H}$ est l'ensemble des translations à gauche $h_{t,z}$ $(t \in \mathbb{R}\ ,\ z \in \mathbb{C})$:

$$(z_0',z') \longmapsto (t,z)(z_0', z')$$

$\mathcal{A}$ est l'ensemble des applications a_s (s réel > 0) :

$$(z_0,z) \longmapsto (s^2 z_0, sz)$$

$\mathcal{M}$ est l'ensemble des applications m_u (u transformation unitaire de $\mathbb{C}^n$) :

$$(z_0,z) \longmapsto (z_0, u(z)) .$$

Il est clair que tout élément de $\mathcal{H}$, $\mathcal{A}$ ou $\mathcal{M}$ applique $\mathbb{R} \times \mathbb{C}^n$ (resp. $\mathcal{P}$) sur lui-même ; ainsi chacun des ensembles $\mathcal{H}$, $\mathcal{A}$, $\mathcal{M}$ pourra être considéré comme un groupe de bijections de $\mathbb{R} \times \mathbb{C}^n$ (resp. $\mathcal{P}$) sur lui-même. Si l'on compactifie $\mathbb{R} \times \mathbb{C}^n$ par un point à l'infini, $\mathcal{H}$, $\mathcal{A}$ et $\mathcal{M}$ deviennent alors naturellement des groupes de bijections de $(\mathbb{R} \times \mathbb{C}^n) \cup \{\infty\}$.

Nous allons maintenant définir l'analogue d'une inversion de $\mathbb{R}^n$.

Définition 1 : *si (z_0,z) est dans $\mathbb{C} \times \mathbb{C}^n$ et si $z_0 \neq -i|z|^2$, nous posons :*

$$q(z_0,z) = \left(\frac{-\bar{z}_0}{|z_0+i|z|^2|^2}, \frac{z}{iz_0-|z|^2}\right) .$$

Alors,

a) q applique $\mathcal{P}$ dans lui-même

b) q induit l'application de $\mathbb{R} \times \mathbb{C}^n$ privé de $(0,0)$ dans lui-même :

$$(t,z) \longmapsto \left(\frac{-t}{|t|^2+|z|^4}, \frac{z}{it-|z|^2}\right)$$

c) en posant de plus $q(0,0) = \infty$ et $q(\infty) = (0,0)$ on définit une application de $(\mathbb{R} \times \mathbb{C}^n) \cup \{\infty\}$ dans lui-même.

On peut démontrer que q définie comme au a) ou comme au c) est une involution ; cette propriété sera rendue plus évidente par la Proposition 1.

Définition 2 : *nous désignons par $\mathcal{G}$ le sous groupe de bijections de $\mathcal{P}$ (ou de $(\mathbb{R} \times \mathbb{C}^n) \cup \{\infty\}$) engendré par $\mathcal{H} \cup \mathcal{A} \cup \mathcal{M} \cup \{q\}$.*

$\mathcal{G}$ est l'analogue du groupe conforme pour $\mathbb{R}^n$.

Si $a_s \in \mathcal{A}$ et $h_{t,z} \in \mathcal{H}$, en utilisant le fait que a_s est un automorphisme du groupe $\mathbb{C} \times \mathbb{C}^n$, on obtient que

$$a_s\, h_{t,z}\, a_s^{-1} = h_{a_s(t,z)} \in \mathcal{H} \;;$$

on en déduit que $\mathcal{H}\mathcal{A} = \mathcal{A}\mathcal{H}$; ainsi $\mathcal{A}\mathcal{H}$ est le sous groupe de $\mathcal{G}$ engendré par $\mathcal{A} \cup \mathcal{H}$. Par un raisonnement similaire on démontre que le sous groupe de $\mathcal{G}$ engendré par $\mathcal{M} \cup \mathcal{A} \cup \mathcal{H}$ est $\mathcal{M}\mathcal{A}\mathcal{H}$.

Nous considérons maintenant les deux sous-ensembles suivants de $\mathbb{C} \times \mathbb{C}^n$:

$$D = \{(\zeta_0,\zeta) \in \mathbb{C} \times \mathbb{C}^n \;;\; |\zeta|^2 < \mathfrak{Im}\, \zeta_0\}$$

$$\partial D = \{(\zeta_0,\zeta) \in \mathbb{C} \times \mathbb{C}^n \;;\; |\zeta|^2 = \mathfrak{Im}\, \zeta_0\} .$$

Nous allons définir une bijection α entre $\mathbb{R} \times \mathbb{C}^n$ et ∂D (resp. entre $\mathcal{P}$ et D) ; puisque $\mathcal{G}$ opère sur $(\mathbb{R} \times \mathbb{C}^n) \cup \{\infty\}$ (resp. sur $\mathcal{P}$) on en déduira que $\alpha\mathcal{G}\alpha^{-1}$ opère sur $\partial D \cup \{\infty\}$ (resp. sur D).

Proposition 1 : *soit α l'application de $\mathbb{C} \times \mathbb{C}^n$ dans lui-même :*

$(z_0,z) \longmapsto (z_0 + i|z|^2,\, z)$. *Alors*

a) α induit une bijection de $\mathbb{R} \times \mathbb{C}^n$ (resp. $\mathcal{P}$) dans ∂D (resp. D)

b) soit f dans $\mathcal{G}$; alors $\hat{f} = \alpha f \alpha^{-1}$ est une bijection de $\partial D \cup \{\infty\}$ (resp. de D) ; plus précisément :

si $f = h_{t,z} \in \mathcal{H}$ *alors* $\hat{f} : (\zeta_0,\zeta) \longmapsto (t+\zeta_0+2i\bar{z}\zeta + i|z|^2,\, z+\zeta)$

si $f = a_s \in \mathcal{A}$ *alors* $\hat{f} : (\zeta_0,\zeta) \longmapsto (s^2\zeta_0,\, s\,\zeta)$

si $f = m_u \in \mathcal{M}$ *alors* $\hat{f} : (\zeta_0,\zeta) \longrightarrow (\zeta_0,\, u(\zeta))$

et $\hat{q} = \alpha\, q\, \alpha^{-1} : (\zeta_0,\zeta) \longmapsto (-1/\zeta_0,\, -i\zeta/\zeta_0)$.

Notons que la bijection induite par α de $\mathbb{R} \times \mathbb{C}^n$ sur ∂D a pour application inverse : $(\zeta_0,\zeta) \longmapsto (\mathrm{Re}\,\zeta_0,\zeta)$ et que la bijection induite par α de $\mathcal{P}$ sur D a pour application inverse :

$$(\zeta_0,\zeta) \longmapsto (\zeta_0 - i|\zeta|^2,\zeta).$$

Nous allons maintenant établir une bijection entre ∂D et la sphère unité de $\mathbb{C}^{n+1}$ et entre D et l'intérieur de cette sphère.

Définition 3 : *notons* $S_{2n+1} = \{(z_0,z) \in \mathbb{C} \times \mathbb{C}^n \; ; \; |z_0|^2 + |z|^2 = 1\}$

et $\mathfrak{D} = \{(z_0,z) \in \mathbb{C} \times \mathbb{C}^n \; ; \; |z_0|^2 + |z|^2 < 1\}$.

Nous désignons par θ *la bijection de* $(\mathbb{C} \times \mathbb{C}^n) \setminus (\{-1\}\times\mathbb{C}^n)$ *dans* $(\mathbb{C}\times\mathbb{C}^n)\setminus(\{-i\}\times\mathbb{C}^n)$:

$$(z_0,z) \longmapsto \frac{i}{1+z_0}(1 - z_0, z) .$$

L'application θ est appelée *transformation de Cayley* de $\mathbb{C} \times \mathbb{C}^n$. Il est facile de voir que la restriction de θ à $\mathfrak{D}$ induit une bijection de $\mathfrak{D}$ sur D et que la restriction de θ à S_{2n+1} induit une bijection de S_{2n+1} privé de $\{-1,0\}$ sur ∂D ; en posant $\theta(-1,0) = \infty$ on obtient une bijection de S_{2n+1} sur $\partial D \cup \{\infty\}$.

3. L'ISOMORPHISME ENTRE $\mathcal{G}$ ET UN SOUS-GROUPE DU GROUPE PROJECTIF COMPLEXE $\mathbb{P}GL(n+2)$.

Nous allons maintenant donner une traduction matricielle de l'action de $\mathcal{G}$; nous nous plaçons pour cela dans une structure d'espace projectif.

Nous considérons l'*espace projectif complexe* $\mathbb{P}_{n+1}$ qui est le quotient de $\mathbb{C} \times \mathbb{C}^n \times \mathbb{C}$ privé de (0,0,0) par la relation d'équivalence $x \sim y$ si et seulement si il existe un complexe λ non nul tel que $y = \lambda x$. Si V est un sous-espace vectoriel de $\mathbb{C} \times \mathbb{C}^n \times \mathbb{C}$, nous notons $\mathbb{P}(V)$ la variété linéaire projective correspondante. A tout endomorphisme f de $\mathbb{C} \times \mathbb{C}^n \times \mathbb{C}$, de noyau N, il correspond, par passage au quotient, une application $\tilde{f}$ de $\mathbb{P}_{n+1}\setminus\mathbb{P}(N)$ dans $\mathbb{P}_{n+1}$, dite application linéaire projective de $\mathbb{P}_{n+1}$; si N se réduit à (0,0,0), c'est-à-dire si f est dans GL(n+2) (le groupe linéaire de l'espace vectoriel complexe $\mathbb{C}^{n+2}$), alors on dit que $\tilde{f}$ est une *projectivité* de $\mathbb{P}_{n+1}$; si f et g sont dans GL(n+2) et s'il existe un complexe non nul λ tel que $g = \lambda f$, alors $\tilde{g} = \tilde{f}$; l'ensemble des projectivités de $\mathbb{P}_{n+1}$ forme le *groupe projectif* $\mathbb{P}GL(n+2)$; si K est un sous-groupe de GL(n+2), on notera $\mathbb{P}K$ le sous-groupe correspondant de $\mathbb{P}GL(n+2)$. Soit π l'application injective de $\mathbb{C}^{n+1}$ dans $\mathbb{P}_{n+1}$ qui à $(z_0,\ldots,z_n)$ de $\mathbb{C}^{n+1}$ associe le point de coordonnées homogènes $(z_0,\ldots,z_n,1)$ de $\mathbb{P}_{n+1}$; si A est une partie de $\mathbb{C}^{n+1}$, nous noterons $\tilde{A}$ l'ensemble $\pi(A)$; l'identification de A et de $\tilde{A}$ permet de plonger A dans $\mathbb{P}_{n+1}$.

Considérons la matrice (n+2, n+2) :

$$T = \begin{pmatrix} -i & 0 & i \\ 0 & iI & 0 \\ 1 & 0 & 1 \end{pmatrix}$$

où I représente la matrice unité (n,n) ; nous avons la relation $\tilde{T}\,\pi = \pi\,\theta$ d'où l'on déduit que la projectivité $\tilde{T}$ induit une bijection de $\mathfrak{D}$ sur $\tilde{D}$ et une bijection de $\widetilde{S_{2n+1}\setminus\{-1,0\}}$ sur $\widetilde{\partial D}$; de plus, au point de coordonnées homogènes (-1,0,1), $\tilde{T}$ fait correspondre le point de coordonnées homogènes (1,0,0) que l'on prend pour point à l'infini de $\widetilde{\partial D}$; $\tilde{T}$ induit ainsi une bijection de $\widetilde{S}_{2n+1}$ sur $\widetilde{\partial D} \cup \{\infty\}$. Si l'on définit $\pi(\infty) = (1,0,0)$, alors π met en bijection $\partial D \cup \{\infty\}$ et $\widetilde{\partial D} \cup \{\infty\}$.

Définition 4 : *Soit ϕ la forme hermitienne sur $\mathbb{C} \times \mathbb{C}^n \times \mathbb{C}$:*

$$(z_0, z, z_{n+1}) \longmapsto |z_0|^2 + |z|^2 - |z_{n+1}|^2 .$$

Nous désignons par $U(n+1,1)$ le groupe unitaire de ϕ .

U(n+1,1) est le groupe des matrices f de GL(n+2) qui préservent la forme ϕ ; c'est l'ensemble des matrices f vérifiant ${}^t\bar{f}\,\Lambda\, f = \Lambda$ où Λ est la matrice de GL(n+2) :

$$\Lambda = \begin{pmatrix} 1 & 0 & 0 \\ 0 & I & 0 \\ 0 & 0 & -1 \end{pmatrix}$$

Nous désignons par SU(n+1,1) le sous-groupe des f de U(n+1,1) tels que det f = 1.

Si $\Gamma = \{ z \in \mathbb{C}^{n+2} ; \phi(z) = 0 \} = \{z \in \mathbb{C}^{n+2} ; {}^t\bar{z}\,\Lambda\, z = 0\}$, alors $\mathbb{P}\Gamma$ est le sous-ensemble de $\mathbb{P}_{n+1}$ des points de coordonnées homogènes $(z_0,z,1)$ vérifiant $|z_0|^2 + |z|^2 - 1 = 0$; nous avons donc $\mathbb{P}\Gamma = \widetilde{S}_{2n+1}$; l'application π induit ainsi une bijection de S_{2n+1} sur $\mathbb{P}\Gamma$.

Proposition 2 : *Notons $G = T.SU(n+1,1).T^{-1}$. Alors $\mathbb{P}G$ est le sous-groupe des projectivités de $\mathbb{P}_{n+1}$ qui préservent à la fois $\widetilde{\partial D} \cup \{\infty\}$ et $\tilde{D}$.*

Démonstration : notons $C = \{(z_0,z,z_{n+1}) \in \mathbb{C} \times \mathbb{C}^n \times \mathbb{C} ; |z_0|^2 + |z|^2 < |z_{n+1}|^2 \}$; alors $\tilde{C} = \mathfrak{D}$. Montrons d'abord que $\mathbb{P}SU(n+1,1)$ est le sous-groupe des projectivités de $\mathbb{P}_{n+1}$ qui préservent à la fois $\mathbb{P}\Gamma$ et $\tilde{C}$; soit donc $\tilde{g}$ une projectivité de $\mathbb{P}_{n+1}$ qui préserve $P\Gamma$ et $\tilde{C}$; alors les formes hermitiennes définies par Λ et ${}^t\bar{g}\,\Lambda\, g$ ont les mêmes zéros ; leurs parties réelles étant des formes quadratiques de signature (2n+2,2) ayant aussi les mêmes zéros, on sait alors qu'elles sont proportionnelles ; les formes hermitiennes définies par Λ et ${}^t\bar{g}\,\Lambda\, g$ sont donc elles aussi proportionnelles et la condition $({}^t\bar{Z}\,\Lambda\, Z < 0 \Longrightarrow {}^t\bar{Z}\,{}^t\bar{g}\,\Lambda\, gZ < 0)$ impose au coefficient de proportionnalité d'être positif, et donc g est dans PSU(n+1,1).

On termine la démonstration de la Proposition en remarquant que $\tilde{h}$ préserve $\partial\tilde{D} \cup \{\infty\}$ si et seulement si $\tilde{T}^{-1}\,\tilde{h}\,\tilde{T}$ préserve $\tilde{S}_{2n+1} = \mathbb{P}\Gamma$ et que $\tilde{h}$ préserve $\tilde{D}$ si et seulement si $\tilde{T}^{-1}\,\tilde{h}\,\tilde{T}$ préserve $\tilde{\mathcal{D}} = \tilde{C}$.

Nous passons maintenant au théorème qui démontre que le groupe $\mathcal{G}$ est isomorphe à un sous-groupe de projectivités de $\mathbb{P}_{n+1}$.

Théorème 1 : *si f est dans G et si $\tilde{f}$ est sa projectivité associée, alors $\alpha^{-1}\,\pi^{-1}\,\tilde{f}\,\pi\alpha$ est dans $\mathcal{G}$.*

L'application $\tilde{f} \longmapsto \alpha^{-1}\,\pi^{-1}\,\tilde{f}\,\pi\alpha$ est un isomorphisme de groupe de $\mathbb{P}G$ sur $\mathcal{G}$.

En particulier, il existe trois sous-groupes H, A, M de G tels que : $\mathbb{P}H$, $\mathbb{P}A$, $\mathbb{P}M$ sont respectivement isomorphes à $\mathcal{H}, \mathcal{A}, \mathcal{M}$; MAH est un sous-groupe de G ; $\mathbb{P}MAH$ est le stabilisateur du point à l'infini (1,0,0) de $\partial\tilde{D} \cup \{\infty\}$.

Nous effectuons la démonstration en trois étapes a) b) c).

a) à tout élément $h_{t,z}$ de $\mathcal{H}$ nous associons la projectivité $\tilde{H}_{t,z}$ où $H_{t,z}$ est la matrice suivante

$$H_{t,z} = \begin{pmatrix} 1 & 2i\bar{z} & t+i|z|^2 \\ 0 & I & z \\ 0 & 0 & 1 \end{pmatrix}$$

On a alors $\tilde{H}_{t,z}\,\pi = \pi\,\alpha\, h_{t,z}\,\alpha^{-1}$ (égalité dans D ou dans $\partial D \cup \{\infty\}$). On note H l'ensemble de toutes les matrices $H_{t,z}$ où (t,z) est dans $\mathbb{R} \times \mathbb{C}^n$; il est facile de voir que H est un groupe ; de plus, tout élément de $\mathbb{P}H$ préserve $\partial\tilde{D} \cup \{\infty\}$ et $\tilde{D}$ et donc, d'après la Proposition 2, on a $\mathbb{P}H \subset \mathbb{P}G$ d'où l'on déduit facilement $H \subset G$. L'application $\tilde{H}_{t,z} \longrightarrow h_{t,z} = \alpha^{-1}\,\pi^{-1}\,\tilde{H}_{t,z}\,\pi\,\alpha$ est un isomorphisme du groupe $\mathbb{P}H$ sur le groupe $\mathcal{H}$.

On procède de même pour $\mathcal{A}$ et $\mathcal{M}$. Si A_s désigne la matrice :

$$\begin{pmatrix} s & 0 & 0 \\ 0 & I & 0 \\ 0 & 0 & 1/s \end{pmatrix}$$

et si A est l'ensemble des matrices A_s pour $s > 0$, alors A est un sous-groupe commutatif de G et l'application $\tilde{A}_s \longmapsto a_s = \alpha^{-1}\,\pi^{-1}\,\tilde{A}_s\,\pi\,\alpha$ est un isomorphisme du groupe $\mathbb{P}A$ sur le groupe $\mathcal{A}$. Si $M_{\varepsilon,u}$ désigne la matrice :

$$M_{\varepsilon,u} = \begin{vmatrix} \varepsilon & 0 & 0 \\ 0 & u & 0 \\ 0 & 0 & \varepsilon \end{vmatrix}$$

et M l'ensemble de toutes les matrices $M_{\varepsilon,u}$ avec u matrice (n,n) unitaire et ε complexe tel que $\varepsilon^2.\det u = 1$, alors M est un sous-groupe de G et l'application : $\tilde{M}_{\varepsilon,u} \longrightarrow m_u = \alpha^{-1}\pi^{-1}\tilde{M}_{\varepsilon,u}\pi\alpha$ est un isomorphe du groupe $\mathbb{P}M$ sur le groupe $\mathcal{M}$.

On a évidemment AM = MA, donc AM est le sous-groupe de G engendré par $A \cup M$. On voit facilement que MAH est le sous-groupe de G engendré par $M \cup A \cup H$.

b) Nous montrons maintenant que le stabilisateur du point à l'infini de $\widetilde{\partial D} \cup \{\infty\}$ est $\mathbb{P}MAH$; comme ce point à l'infini est le point de coordonnées homogènes (1,0,0), il est équivalent de montrer que, pour g dans G, g appartient à MAH si et seulement si il existe k complexe tel que g(1,0,0) = (k,0,0).

Dans le sens direct, cette assertion est immédiate. Il s'agit donc de démontrer la réciproque. Pour cela, on remarque que $g \in G$ si et seulement si $T\,g\,T^{-1} \in SU(n+1,1)$, soit $\det g = 1$ et ${}^t\bar{g}\,J\,g = J$ où $J = {}^t\bar{T}^{-1}\,\Lambda\,T^{-1} = \begin{pmatrix} 0 & 0 & -i/2 \\ 0 & I & 0 \\ i/2 & 0 & 0 \end{pmatrix}$

Si g(1,0,0) = (k,0,0), $\det g = 1$ et ${}^t\bar{g}\,J\,g = J$, alors, après un calcul que nous ne détaillons pas, on obtient qu'il existe u unitaire, $s > 0$, ε tel que $\varepsilon^2 \det u = 1$, z dans $\mathbb{C}^n$ et t réel de sorte que

$$g = \begin{pmatrix} \varepsilon s & 2i\varepsilon s\,{}^t\bar{z} & \varepsilon st+i\varepsilon s|z|^2 \\ 0 & u & uz \\ 0 & 0 & \varepsilon/s \end{pmatrix}$$

ce qui signifie que g est dans MAH.

c) Si ε est un nombre complexe tel que $\varepsilon^{n+2} = -1$, soit Q_ε la matrice :

$$Q_\varepsilon = \varepsilon \begin{pmatrix} 0 & 0 & -i \\ 0 & I & 0 \\ i & 0 & 0 \end{pmatrix}$$

notons $\tilde{Q}$ pour $\tilde{Q}_\varepsilon$; alors, en raisonnant comme au a), on obtient que $q = \alpha^{-1}\pi^{-1}\tilde{Q}\pi\alpha$ et que Q_ε est dans G.

Ainsi, en reprenant les résultats du a), l'application $\tilde{f} \longmapsto \alpha^{-1}\pi^{-1}\tilde{f}\pi\alpha$ est un isomorphisme du sous-groupe de $\mathbb{P}G$ engendré par $\mathbb{P}MAH \cup \{\tilde{Q}\}$ sur le groupe $\mathcal{G}$.

Il reste à démontrer que le sous-groupe de $\mathbb{P}G$ engendré par $\mathbb{P}MAH \cup \{\tilde{Q}\}$ est $\mathbb{P}G$ tout entier.

Considérons $\tilde{g}$ dans $\mathbb{P}G$ et supposons que $\tilde{g}$ n'appartienne pas à $\mathbb{P}MAH$. Alors $\tilde{g}(\infty) \neq \infty$ et $\tilde{g}(\infty) \in \widetilde{\partial D}$ (puisque $\tilde{g}$ préserve $\widetilde{\partial D} \cup \{\infty\}$) ; ainsi il existe (ζ_0,ζ) dans ∂D tel que $\tilde{g}(\infty) = (\zeta_0,\zeta,1)$; mais il existe (t,z) dans $\mathbb{R} \times \mathbb{C}^n$ tel que $\alpha\, h_{t,z}\, \alpha^{-1}(\zeta_0,\zeta) = (0,0)$; on a alors $\tilde{H}_{t,z}(\zeta_0,\zeta,1) = (0,0,1)$ d'où $(\tilde{Q}\, \widetilde{H_{t,z}}\, \tilde{g})(\infty) = \infty$; ainsi $\tilde{Q}\, \tilde{H}_{t,z}\, \tilde{g}$ est dans $\mathbb{P}MAH$ et donc $\tilde{g}$ est dans le sous-groupe engendré par $\mathbb{P}MAH$ et $\tilde{Q}$. Ceci termine la démonstration du théorème.

4. L'ACTION DE G SUR LES LOIS DE PROBABILITES DANS $\mathbb{R}^{2n+1}$.

Si f est un élément de GL(n+2) tel que $\tilde{f}$ soit dans $\mathbb{P}G$, par l'isomorphisme décrit au théorème 1, il correspond à $\tilde{f}$ un élément $\dot{f}$ de $\mathcal{G}$; il est alors naturel d'identifier f et $\dot{f}$; ainsi, si μ est une mesure sur $(\mathbb{R} \times \mathbb{C}^n) \cup \{\infty\}$, nous noterons $f\mu$ la mesure image de μ par $\dot{f}$. Nous allons étendre cette définition aux éléments qui sont limites simples de suites d'éléments du genre de f.

Proposition 3 : *nous définissons les ensembles suivants :*

$V(n+2) = \{ f \in GL(n+2) \ ; \ \exists\, k > 0 \quad {}^t\bar{f}\, \Lambda\, f = k\, \Lambda \}$

$V_1(n+2)$ *l'adhérence de* $V(n+2)$ *dans l'ensemble des matrices* $(n+2,n+2)$

$G_0 = T\, V(n+2) T^{-1}$

$G_1 = T\, V_1(n+2) T^{-1}$.

Soit μ *une mesure de probabilité sans atome sur* $(\mathbb{R} \times \mathbb{C}^n) \cup \{\infty\}$, *et soit* f *dans* G_1, f *non nulle ; si* $\tilde{f}$ *est l'application linéaire projective associée à* f, *alors* $\dot{f} = \alpha^{-1}\, \pi^{-1}\, \tilde{f}\, \pi\, \alpha$ *est définie* μ-*pp et on peut donc définir la mesure image de* μ *par* $\dot{f}$ *que nous notons* $f\mu$. *Alors :*

(*i*) *si* f *est dans* G_0, *alors* $f\mu$ *est sans atome.*

(*ii*) *si* f *n'est pas dans* G_0, *c'est-à-dire si* $\det f = 0$, *alors* $f\mu$ *est une mesure de Dirac sur* $(\mathbb{R} \times \mathbb{C}^n) \cup \{\infty\}$.

(*iii*) *si* f *est dans* G_1, *si* λ *est complexe non nul, et si* $f = \lambda g$, *alors* $f\mu = g\mu$.

(*iv*) *si* (f_n) *est une suite dans* G_0 *qui converge simplement vers* f *dans* G_1, *alors* $(f_n\mu)$ *converge étroitement vers* $f\mu$.

Démonstration.

Rappelons que $\Lambda = \begin{pmatrix} 1 & 0 & 0 \\ 0 & I & 0 \\ 0 & 0 & -1 \end{pmatrix}$; ainsi

$\mathbb{P}V(n+2) = \mathbb{P}SU(n+1,1)$ et $\mathbb{P}G_0 = \mathbb{P}G$. Le (i) est alors immédiat : si f est dans G_0, alors det $f \neq 0$, la projectivité $\tilde{f}$ est dans $\mathbb{P}G$ et, d'après le Théorème 1, $\dot{f} = \alpha^{-1}\pi^{-1}\tilde{f}\pi\alpha$ est dans $\mathcal{G}$, donc est une bijection de $(\mathbb{R}\times\mathbb{C}^n)\cup\{\infty\}$; on définit $f\mu$ comme la mesure image de μ par $\dot{f}$ et il n'est d'ailleurs pas nécessaire de supposer μ sans atome ; si μ est sans atome, il en est de même de $f\mu$. Pour démontrer la suite de la Proposition nous utilisons trois lemmes. Rappelons que $\Gamma = \{z \in \mathbb{C}^{n+2} ; {}^t\bar{z}\,\Lambda\, z=0\}$.

Lemme 1 : *si x et y sont dans Γ et satisfont à* ${}^t\bar{x}\,\Lambda\, y = 0$ *, alors x et y sont proportionnels.*

Ce résultat s'obtient facilement en utilisant le cas d'égalité de l'inégalité de Schwarz.

Lemme 2 : *si f est dans* $V_1(n+2)$*, alors* ${}^t\bar{f}$ *est dans* $V_1(n+2)$.

Considérons f dans V(n+2) ; il existe un réel $k > 0$ tel que ${}^t\bar{f}\,\Lambda\, f = k\,\Lambda$; puisque $\Lambda^2 = I$, on a ${}^t\bar{f}\,\Lambda\, f\,\Lambda = kI$ d'où l'on déduit $f\Lambda\,{}^t\bar{f}\,\Lambda = kI$ et $f\Lambda\,{}^t\bar{f} = k\Lambda$: ${}^t\bar{f}$ est dans V(n+2). On étend aisément le résultat à $V_1(n+2)$.

Lemme 3 : *Soit f dans* $V_1(n+2)$*,* $f\neq 0$*, tel que det f = 0 ; alors :*

α) $f(\mathbb{C}^{n+2})$ *est un sous-espace vectoriel de dimension 1 inclus dans Γ.*

β) si N est le noyau de f, alors N ∩ Γ est inclus dans un sous-espace vectoriel de dimension 1.

Si f est dans $V_1(n+2)$, il existe un réel $k \geq 0$ tel que ${}^t\bar{f}\,\Lambda\, f = k\,\Lambda$; si, de plus, det f = 0, c'est-à-dire $f \notin V(n+2)$, c'est que k=0 et donc ${}^t\bar{f}\,\Lambda\, f = 0$. Alors pour tout z de C^{n+2}, ${}^t\bar{z}\,{}^t\bar{f}\,\Lambda\, fz = 0$ et donc $f(\mathbb{C}^{n+2}) \subset \Gamma$. Ensuite, si a et b sont dans $f(\mathbb{C}^{n+2})$, alors a,b et a+b sont dans Γ d'où l'on déduit ${}^t\bar{a}\,\Lambda\, b + {}^t\bar{b}\,\Lambda\, a = 0$;en considérant a, ib et a+ib on obtient ${}^t\bar{a}\,\Lambda\, b = 0$ et donc d'après le lemme 1 a et b sont proportionnels et on a obtenu (α). De plus, d'après le lemme 2, ${}^t\bar{f}$ est dans $V_1(n+2)$ et, d'après (α), il existe y dans Γ qui engendre ${}^t\bar{f}(\mathbb{C}^{n+2})$. Posons $x = \Lambda y$; alors $y = \Lambda x$ et x est dans Γ . Soit maintenant $z \in N$; alors $\bar{f}(\bar{z}) = 0$ d'où ${}^t\bar{z}\,\Lambda\, x = {}^t\bar{z}\, y = 0$; si l'on suppose de plus que z est dans Γ, alors, d'après le lemme 1, z est proportionnel à x et on obtient le (β) du lemme 3.

Nous passons maintenant à la démonstration de la partie (ii) de la Proposition. Nous allons ainsi montrer qu'il est possible, si μ est sans atome et si g appartient à G_1 de définir l'image de μ par $\alpha^{-1}\ \pi^{-1}\ \tilde{g}\ \pi\ \alpha = \dot{g}$. Nous allons en effet montrer que $\dot{g}$ est définie μ-presque partout.

Soit donc $g \in G_1 \backslash G_0$, g non nulle. Allors $f = T^{-1}\ g\ T$ est dans $V_1(n+2) \backslash V(n+2)$ et det f = det g = 0. Le noyau N de f est différent de $\{0\}$ et de $\mathbb{C}^{n+2}$ et, d'après le lemme 3(α), l'espace image de f est inclus dans Γ et est de dimension 1 ; donc l'application linéaire projective associée à f est définie sur $\mathbb{P}_{n+1} \setminus \mathbb{P}(N)$ et son image est réduite à un point de $\mathbb{P}\Gamma$. La restriction $\tilde{f}$ de cette application à $\mathbb{P}\Gamma$ est définie sur $\mathbb{P}\Gamma \setminus \mathbb{P}(N \cap \Gamma)$; or, d'après le lemme 3 (β), $N \cap \Gamma$ est inclus dans un sous-espace de dimension 1 ; ainsi $\tilde{f}$ est définie dans $\mathbb{P}\Gamma$ privé d'un point et son image est un point de $\mathbb{P}\Gamma$. Puisque $\tilde{T}$ établit une bijection de $\widetilde{S_{2n+1}} = \mathbb{P}\Gamma$ sur $\widetilde{\partial D} \cup \{\infty\}$, $\tilde{g} = \tilde{T}\ \tilde{f}\ \tilde{T}^{-1}$ est une application linéaire projective définie sur $\widetilde{\partial D} \cup \{\infty\}$ privé d'un point et ayant pour image un point de $\widetilde{\partial D} \cup \{\infty\}$. Alors $\pi^{-1}\ \tilde{g}\ \pi$ est définie sur $\partial D \cup \{\infty\}$ privé d'un point et son image est un point de $\partial D \cup \{\infty\}$; $\dot{g} = \alpha^{-1}\ \pi^{-1}\ \tilde{g}\ \pi\ \alpha$ est définie sur $(\mathbb{R} \times \mathbb{C}^n) \cup \{\infty\}$ privé d'un point et son image est un point de $(\mathbb{R} \times \mathbb{C}^n) \cup \{\infty\}$. Si la mesure μ est supposée sans atome, alors $\dot{g}$ est définie μ-presque partout et nous pouvons considérer la mesure image de μ par $\dot{g}$ que nous notons $g\mu$; c'est une mesure de Dirac en un point de $(\mathbb{R} \times \mathbb{C}^n) \cup \{\infty\}$.

La partie (iii) résulte immédiatement du fait qu'alors $\tilde{g} = \tilde{f}$. Pour la partie (iv), notons $\dot{f}_n = \alpha^{-1}\ \pi^{-1}\ \tilde{f}_n\ \pi\ \alpha$ et $\dot{f} = \alpha^{-1}\ \pi^{-1}\ \tilde{f}\ \pi\ \alpha$. Si f_n converge simplement vers f, alors $\dot{f}_n$ converge vers $\dot{f}$ μ-presque partout et donc, puisque μ est une probabilité, $f_n\ \mu$ converge étroitement vers $f\mu$. Ceci termine la démonstration de la Proposition 3.

<u>Théorème 2</u> : *Soit μ une mesure de probabilité sur $(\mathbb{R} \times \mathbb{C}^n) \cup \{\infty\}$ et K_μ le sous-groupe de $T.U(n+1,1).T^{-1}$ des f tels que $f\mu = \mu$. Si μ est sans atome, alors K_μ est compact.*

<u>Démonstration</u>. Soit (f_p) une suite dans K_μ . Alors il existe des matrices g_p et h_p dans U(n+2) et une matrice diagonale d_p, d'élément diagonaux $a_j(p)$ vérifiant

$0 < a_1(p) \leq \ldots \leq a_{n+2}(p)$ telles que $f_p = g_p\, d_p\, h_p$ (c'est la décomposition de Cartan).

Il est clair que $a_1(p) \ldots\ldots a_{n+2}(p) = \det d_p = 1$. Considérons maintenant la matrice diagonale $(1/a_{n+2}(p))d_p$; ses éléments diagonaux sont dans $[0,1]$; la compacité de $[0,1]$ et de $U(n+2)$ nous permet alors d'affirmer l'existence d'une suite (p_i) d'entiers telle que, lorsque i tend vers $+\infty$, p_i tend vers $+\infty$, (g_{p_i}) converge vers g dans $U(n+2)$, (h_{p_i}) converge vers h dans $U(n+2)$ et, pour tout j, $a_j(p_i)/a_{n+2}(p_i)$ converge vers un élément α_j de $[0,1]$.

Notons d' la matrice diagonale d'éléments diagonaux $\alpha_1,\ldots,\alpha_{n+1}$, 1 et $f'_p = g_p\, d'_p\, h_p = (1/a_{n+2}(p))f_p$. Alors f'_p est dans G_0 et, quand i tend vers $+\infty$, (f'_{p_i}) converge vers g d' h = f' qui est non nul et appartient à G_1. Alors, d'après la Proposition 3 (iv), $f'_{p_i}\,\mu$ converge étroitement vers $f'\mu$ et, puisque $f'_p\,\mu = \mu$, on a $f'\mu = \mu$. Alors $\det f' \neq 0$, sinon, d'après la Proposition 3(ii), μ possèderait un atome. Ainsi $\det d' \neq 0$, et donc, pour tout j, $0 < \alpha_j$.

$$\text{Alors}\quad \frac{1}{[a_{n+2}(p_i)]^{n+2}} = \frac{a_1(p_i)\ldots a_{n+1}(p_i)a_{n+2}(p_i)}{[a_{n+2}(p_i)]^{n+2}} \longrightarrow \alpha_1 \ldots\ldots \alpha_{n+1} > 0$$

Donc $\alpha = \lim\limits_{i\to\infty} a_{n+2}\,(p_i)$ existe dans $]0,+\infty[$ d'où pour tout j, $\lim\limits_{i\to\infty} a_j(p_i)$ existe dans $]0,+\infty[$. On obtient ainsi que la suite (d_{p_i}) converge vers une matrice diagonale $d = \alpha d'$ de déterminant égal à 1 et que la suite (f_{p_i}) converge vers la matrice $f = gdh = \alpha f'$; d'après la Proposition 3(iii), $f\mu = f'\mu = \mu$ de plus f' est dans G_1 avec $\det f' \neq 0$, d'où f' est dans G_0, f est dans G_0 et comme $|\det f| = |\det gdh| = 1$, f est dans $T^{-1}.U(n+1,1).T$; ainsi f est dans K_μ, ce qui démontre la compacité de K_μ.

5. LA LOI DE CAUCHY-HEISENBERG

Rappelons que $S_{2n+1} = \{(z_0,z)\} \in \mathbb{C} \times \mathbb{C}^n \;;\; |z_0|^2 + |z|^2 = 1\}$; d'après la Proposition 1 et la Définition 3, l'application $\alpha^{-1}\,\theta$:

$$(z_0,z) \longmapsto \left(\frac{2\,\mathfrak{I}m\, z_0}{|z_0+1|^2}\;,\;\frac{iz}{z_0+1}\right)$$

est une bijection de S_{2n+1} privé de $(-1,0)$ sur $\mathbb{R} \times \mathbb{C}^n$. Nous appelons *loi de Cauchy-Heisenberg standard*, notée γ, la mesure de probabilité sur $\mathbb{R} \times \mathbb{C}^n = \mathbb{R} \times \mathbb{R}^{2n}$ qui est l'image par $\alpha^{-1}\,\theta$ de la probabilité uniforme sur S_{2n+1}.

Proposition 4 : *La loi de Cauchy-Heisenberg standard γ est la mesure de probabilité sur $\mathbb{R} \times \mathbb{C}^n = \mathbb{R} \times \mathbb{R}^{2n}$ de densité :*

$$(t,z) \longmapsto \frac{C_n}{((1+|z|^2)^2+t^2)^{n+1}}$$

où pour $z = (z_1,\dots,z_n)$ dans $\mathbb{C}^n$, on a $|z|^2 = |z_1|^2 + \dots + |z_n|^2$ et avec $C_n = n!\ 2^{2n}\ \pi^{-(n+1)}$.

Démonstration : soit $(W_0,\dots,W_n)$ uniformément répartie sur S_{2n+1} et posons $W_k = U_k + i\,V_k$; alors il existe des variables aléatoires indépendantes gaussiennes $N(0,1)$: $X_0,Y_0,\dots,X_n,Y_n$ telles que $U_k = \frac{X_k}{R}$ et $V_k = \frac{Y_k}{R}$ avec $R=(X_0^2+Y_0^2+\dots+X_n^2+Y_n^2)^{1/2}$.

La loi de γ est la loi de $(T,Z) = (T,Z_1,\dots,Z_n)$ avec $T = \frac{2\,\mathfrak{I}_m W_0}{|1+W_0|^2}$ et $Z_k = i\,\frac{W_k}{W_0+1}$.

On a $|Z|^2 - iT = \frac{1-W_0}{1+W_0}$; nous sommes donc amenés à chercher la loi de W_0.

Or $|W_0|^2 = \frac{X_0^2 + Y_0^2}{X_0^2+Y_0^2+\dots+X_n^2+Y_n^2}$ suit une loi de bêta de première espèce de paramètres 1 et n, de densité $n(1-x)^{n-1}\ \mathbb{1}_{]0,1[}(x)$. Comme la loi de $W_0 = U_0 + iV_0$ est invariante par rotation, on en déduit que la loi de (U_0,V_0) a pour densité :

$$(u,v) \longmapsto \frac{n}{\pi}\,(1-u^2-v^2)^{n-1}\ \mathbb{1}_B(u,v)$$

où B est la boule unité de $\mathbb{R}^2$.

Alors, de $|Z|^2- iT = (1-W_0)/(1+W_0)$, on déduit que la densité du couple $(|Z|^2,T)$ est :

$$(x,y) \longrightarrow \frac{n}{\pi}\ \frac{2^{2n}\,x^{n-1}}{((1+x)^2+y^2)^{n+1}}\ \mathbb{1}_{]0,+\infty[\times\mathbb{R}}(x,y)\ .$$

On obtient la valeur annoncée pour la loi de (T,Z) en remarquant que cette loi est identique à celle de (T, r(z)) où r est une rotation quelconque de $\mathbb{R}^{2n}$.

Proposition 5 : *Si (ζ_0,ζ) est dans $\mathcal{D}$, notons $\gamma_{\zeta_0,\zeta}$ la mesure de probabilité sur $\mathbb{R} \times \mathbb{C}^n$ de densité :*

$$(t,z) \longrightarrow \frac{C_n(\mathfrak{I}m\,\zeta_0 - |\zeta|^2)^{n+1}}{((\mathfrak{I}m\,\zeta_0 - |\zeta|^2 + |z-\zeta|^2)^2 + (t-\mathrm{Re}\,\zeta_0 - 2\mathfrak{I}m\,\zeta\,\bar{z})^2)^{n+1}}$$

(avec cette notation $\gamma_{i,0} = \gamma$). Alors

(i) l'ensemble des mesures $\{\gamma_{\zeta_0,\zeta};\ (\zeta_0,\zeta) \in \mathcal{D}\}$ est égal à l'ensemble des mesures images de γ par le groupe $\mathcal{HA}$.

(ii) le groupe $\mathcal{G}$ *préserve globalement* $\{\gamma_{\zeta_0,\zeta}\ ;\ (\zeta_0,\zeta) \in D\}$ *; plus précisément, si* f *est une bijection de* $(\mathbb{R} \times \mathbb{C}^n) \cup \{\infty\}$ *appartenant à* $\mathcal{G}$ *et si* $\hat{f}$ *est la bijection de* D *qu'on peut lui associer suivant la Proposition 1, alors*

$f\,\gamma_{\zeta_0,\zeta} = \gamma_{\hat{f}(\zeta_0,\zeta)}$.

Notons $\mathcal{C} = \{\gamma_{\zeta_0,\zeta}\ ;\ (\zeta_0,\zeta) \in D\}$. Si $a_s \in \mathcal{A}$ et $h_{\tau,\zeta} \in \mathcal{H}$, alors on démontre que l'image de γ par $f = h_{\tau,\zeta}\, a_s$ est $\gamma_{\zeta_0,\zeta} \in \mathcal{C}$ avec $\zeta_0 = \tau + i(s^2+|\zeta|^2)$; remarquons de plus qu'alors $(\zeta_0,\zeta) = \hat{f}(i,0)$; ainsi l'image de γ par $\mathcal{H}\mathcal{A}$ est dans $\mathcal{C}$; réciproquement, tout élément $\gamma_{\zeta_0,\zeta}$ de $\mathcal{C}$ est l'image de γ par $h_{\tau,\zeta}.a_s$ où $s = (\mathcal{I}m\zeta_0-|\zeta|^2)^{1/2}$ et $\tau = \mathcal{R}e\ \zeta_0$; la partie (i) est ainsi démontrée.

En considérant successivement les cas où g appartient à $\mathcal{H}, \mathcal{A}, \mathcal{M}, \{q\}$, nous allons montrer que la mesure image de $\gamma_{\zeta_0,\zeta} \in \mathcal{C}$ par g est $\gamma_{\hat{g}(\zeta_0,\zeta)}$ ce qui démontrera la partie (ii) ; nous utiliserons plusieurs fois la remarque faite plus haut, à savoir que si f est dans $\mathcal{H}\mathcal{A}$ alors $f\,\gamma_{i,0} = \gamma_{\hat{f}(i,0)}$.

Soit d'abord $g = h_{t,z} \in \mathcal{H}$; alors d'après le (i), $\gamma_{\zeta_0,\zeta}$ est l'image de $\gamma_{i,0}$ par un élément $h\ a$ de $\mathcal{H}\mathcal{A}$, et, d'après la remarque, $(\zeta_0,\zeta) = \hat{h}\ \hat{a}(i,0)$; alors $g.\gamma_{\zeta_0,\zeta} = h_{t,z}.h.a\ \gamma_{i,0} = \gamma_{\hat{g}(\zeta_0,\zeta)}$.

Si maintenant $g = a_p \in \mathcal{A}$, on raisonne de même en remarquant que si $h_{t,z} \in \mathcal{H}$ et $a_s \in \mathcal{A}$, alors $a_p\, h_{t,z}\, a_s = h_{p^2t,pz} . a_{ps}$.

Si $g = m_u \in \mathcal{M}$, alors, après calcul, on montre que $g\gamma_{\zeta_0,\zeta} = \gamma_{\zeta_0,u(\zeta)} = \gamma_{\hat{g}(\zeta_0,\zeta)}$.

Il reste le cas où $g = q$. Nous remarquons d'abord que l'application $\theta^{-1}\alpha\ q\ \alpha^{-1}\ \theta$ de S_{2n+1} dans lui-même transforme (z_0,z) en $(-z_0,-z)$ et donc préserve σ ; puisque, par définition, $\gamma = \alpha^{-1}\ \theta\ \sigma$, on en déduit que $q\gamma = \gamma$. On introduit alors, comme KORANYI [6] §3, la fonction ρ suivante, définie sur $\mathbb{C} \times \mathbb{C}^n \times \mathbb{C} \times \mathbb{C}^n$:

$$\rho(\zeta_0,\zeta,\eta_0,\eta) = i(\bar{\eta}_0 - \zeta_0) - 2\zeta\bar{\eta}$$

Alors la densité de $\gamma_{\zeta_0,\zeta}$ est

$$(t,z) \longmapsto \frac{c_n(\mathcal{I}m\ \zeta_0 - |\zeta|^2)^{n+1}}{|\rho(\zeta_0,\zeta,t+i|z|^2,z)|^{2(n+1)}}$$

et donc la densité de $\gamma_{\zeta_0,\zeta}$ par rapport à γ est :

$$(t,z) \longmapsto \theta(\zeta_0,\zeta,t,z) = (\mathcal{I}m\, \zeta_0 - |\zeta|^2)^{n+1} \left| \frac{\rho(i,o,t+i|z|^2,z)}{\rho(\zeta_0,\zeta,t+i|z|^2,z)} \right|^{2(n+1)}$$

Alors la densité de $q\,\gamma_{\zeta_0,\zeta}$ par rapport à $q\gamma = \gamma$ est $\theta(\zeta_0,\zeta,q(t,z))$; on termine la démonstration de la Proposition en démontrant que $\theta(\zeta_0,\zeta,q(t,z)) = \theta(\hat{q}(\zeta_0,\zeta),t,z)$.

5. UNE CARACTERISATION DU TYPE DE γ .

Le type d'une loi de probabilité sur $\mathbb{R}^n$ est rarement défini de manière claire ; c'est pourquoi nous en donnons la définition précise qui suit ; pour plus de détails sur la notion de type voir [2] I.3 ou Letac [7].

Définition 5 : *soit* $\mathcal{K}$ *un sous-groupe de* $\mathcal{G}$ *(les éléments de* $\mathcal{K}$ *sont donc des bijections de* $(\mathbb{R} \times \mathbb{C}^n) \cup \{\infty\}$ *) qui préserve le point à l'infini de* $(\mathbb{R} \times \mathbb{C}^n) \cup \{\infty\}$.

Si μ *est une mesure de probabilité sur* $(\mathbb{R} \times \mathbb{C}^n) \cup \{\infty\}$ *, on appelle* $\mathcal{K}$*-type de* μ *l'ensemble des mesures images de* μ *par* $\mathcal{K}$.

Nous appelons type de Cauchy-Heisenberg le $\mathcal{AH}$*-type de* γ .

De la Proposition 5, il résulte immédiatement que le type de Cauchy-Heisenberg est l'ensemble des mesures $\gamma_{\zeta_0,\zeta}$ où $(\zeta_0,\zeta) \in D$; c'est aussi le $\mathcal{MAH}$-type de γ ; nous donnons maintenant notre caractérisation du type de Cauchy-Heisenberg.

Théorème 3 : *Soit* μ *une mesure de probabilité sur* $\mathbb{R} \times \mathbb{C}^n$*, sans atome. Si* $\mathcal{K} = \mathcal{AH}$ *ou si* $\mathcal{K} = \mathcal{MAH}$*, les propriétés suivantes sont équivalentes :*

i) μ *est du type de Cauchy-Heisenberg.*

ii) pour tout f *de* $\mathcal{G}$ *,* $f\mu$ *est du* $\mathcal{K}$*-type de* μ .

Démonstration : l'implication (i) $\Longrightarrow$ (ii) résulte immédiatement de la Proposition 5. Nous allons démontrer la réciproque pour le cas $\mathcal{K} = \mathcal{MAH}$. La probabilité μ sera supposée définie sur $(\mathbb{R} \times \mathbb{C}^n) \cup \{\infty\}$ et vérifiant $\mu(\infty) = 0$. A l'aide de la Proposition 3, l'assertion (ii) se traduit de la manière suivante :

" pour tout f de G_0, il existe g dans MAH tel que $f\mu = g\mu$ ".

Nous noterons $U(n+1) \times U(1)$ le sous-groupe des matrices f de $GL(n+2)$ telles qu'il existe f_1 dans $U(n+1)$ et un nombre complexe ε de module 1 vérifiant :

$$f = \begin{pmatrix} f_1 & 0 \\ 0 & \varepsilon \end{pmatrix}$$

Rappelons que les sous-groupes compacts maximaux de $U(n+1,1)$ sont les conjugués de $U(n+1) \times U(1)$. Si $S = T.(U(n+1) \times U(1)).T^{-1}$, on voit facilement que $S\gamma = \gamma$ (en effet si $g = T f T^{-1}$ est dans S, alors $g\gamma$ est la mesure image de γ par $\alpha^{-1} \theta f_1 \theta^{-1} \alpha$; or $f_1 \sigma = \sigma$ et γ est l'image de σ par $\alpha^{-1} \theta$) et que S opère transitivement sur $(\mathbb{R} \times \mathbb{C}^n) \cup \{\infty\}$ (en effet $U(n+1)$ opère transitivement sur S_{2n+1}).

Soit donc μ vérifiant (ii). Alors, d'après le Théorème 2, $T^{-1} K_\mu T$ est un sous-groupe compact de $U(n+1,1)$ et donc il existe h dans $U(n+1,1)$ tel que $T^{-1} K_\mu T$ soit un sous-groupe compact de $h^{-1} (U(n+1) \times U(1))h$; on voit que $Th T^{-1}$ est dans G_0 et on considère la mesure $\nu = Th T^{-1} \mu$; alors ν est du $\mathcal{K}$-type de μ, ν vérifie (ii) et K_ν est un sous-groupe de S.

On a de plus : $S = K_\nu .(S \cap MAH)$. En effet si f est dans $S \subset G_0$, comme ν vérifie (ii), il existe g dans MAH tel que $f\nu = g\nu$; alors $g^{-1} f$ est dans K_ν, donc dans S, et g est dans S. Alors $f = gg^{-1} f$ est dans $(S \cap MAH).K_\nu$; on en déduit $S=(S \cap MAH).K_\nu$.

Soit z le point à l'infini de $(\mathbb{R} \times \mathbb{C}^n) \cup \{\infty\}$. Alors, puisque S opère transitivement sur $(\mathbb{R} \times \mathbb{C}^n) \cup \{\infty\}$ et que $MAH\, z = z$, on obtient :

$$(\mathbb{R} \times \mathbb{C}^n) \cup \{\infty\} = Sz = K_\nu.(S \cap MAH).z = K_\nu.z .$$

Ainsi le groupe compact K_ν opère transitivement sur $(\mathbb{R} \times \mathbb{C}^n) \cup \{\infty\}$ qui est donc un espace homogène pour K_ν. Alors ν et γ sont deux mesures de probabilité sur $(\mathbb{R} \times \mathbb{C}^n) \cup \{\infty\}$ invariantes par K_ν (γ est invariante par K_ν car $K_\nu \subset S$ et $S\gamma = \gamma$) ; on en déduit que $\nu = \gamma$ et donc que μ est du type de Cauchy-Heisenberg.

REFERENCES

[1] DUNAU, J.L. et SENATEUR, H. An elementary proof of the Knight-Meyer characterization of the Cauchy distribution. *J. Multivar. Anal.* (à paraître).

[2] DUNAU, J.L. et SENATEUR, H. Sur différentes lois de Cauchy dans $\mathbb{R}^n$, caractérisées comme invariantes par certains groupes de transformations. *Thèse de 3ème cycle.* Université Paul Sabatier, Toulouse, 1985.

[3] DUNAU, J.L. et SENATEUR, H. Une caractérisation du type de la loi de Cauchy-conforme (*à paraître*).

[4] KNIGHT, F.B. A characterization of the Cauchy type. *Proc. Amer. Math. Soc.* 55, 130-135 (1976).

[5] KNIGHT, F.B. et MEYER, P.A. Une caractérisation de la loi de Cauchy. *Z. Wahrsch. Verw. Gebiete* 34, 129-134 (1976).

[6] KORANYI, A. Geometric aspects of analysis on the Heisenberg group. *Topics in modern harmonic analysis* (*Proc. semin. Turin/Milan 1982 - L. De Michele et F. Ricci Ed.*) Vol. 1, 209-258. Ist. Naz. Alta Mat. Francesco Severi, Roma, 1983.

[7] LETAC, G. Seul le groupe des similitudes-inversions préserve le type de la loi de Cauchy-conforme de R^n pour $n > 1$, *J. Func. Anal.* (à paraître).

Jean-Louis DUNAU
INSA Toulouse

Henri SENATEUR
U.E.R. M.I.G.

Laboratoire de Statistique et Probabilités
U.A.-C.N.R.S. n° 745
Université Paul Sabatier
118, route de Narbonne
31062 Toulouse, France.

LEVY-SCHOENBERG KERNELS ON RIEMANNIAN SYMMETRIC SPACES OF NONCOMPACT TYPE

B.-J. Falkowski
Universität der Bundeswehr München
Fakultät für Informatik
Institut für Mathematik
Werner-Heisenberg-Weg 39
D-8014 Neubiberg

INTRODUCTION

In [7] Gangolli studied so-called Lévy-Schoenberg kernels on certain homogeneous spaces. Here we consider a noncompact, connected, semi-simple Lie group with finite centre and a maximal compact subgroup K of G. In this case the homogeneous space G/K is a Riemannian symmetric space of noncompact type. Recalling Gangolli's theory we show in section 1 that the description of Lévy-Schoenberg kernels may be reduced to an analysis of infinitely divisible positive definite functions which are bi-invariant with respect to K, cf. [7].

In section 2 we exploit an abstract Lévy-Khintchin formula for the logarithm of infinitely divisible positive functions in terms of 1-cohomology. This requires a result on the number of <u>irreducible</u> unitary representations which admit a non-trivial 1-cohomology (2.2). Via a direct integral decomposition we obtain a more concrete Lévy-Khintchin formula (2.3) and in particular a very precise description of the "Gaussian Part": We show that nonnegative solutions of the functional equation

$$\int_K \Psi(g_1 k g_2)\,dk = \Psi(g_1) + \Psi(g_2)$$

(dk = normalized Haar measure on K)

do not adequately describe the Gaussian Part. This answers a question posed by Gangolli in [7] in a constructive fashion, cf. (2.4). Finally, we compute the Gaussian Part for G = SU(n;1) (this is the example not treated in [5], [7]).

Note: Representation in this paper will always mean a continuous unitary representation. Inner products are, by abuse of notation, always denoted by $\langle\cdot,\cdot\rangle$ and their induced norms by $\|\cdot\|$. For definitions and notation concerning 1-cohomology and infinitely divisible positive definite functions we refer the reader to [3].

1. GANGOLLI's THEORY [7]

We start by discussing

A) Example: Let us consider a continuous function

$$f : \mathbb{R}^d \times \mathbb{R}^d \to \mathbb{R} \text{ (a kernel) given by}$$

$$f(a,b) := \frac{1}{2} [\| a\|^\alpha + \| b\|^\alpha - \| a-b\|^\alpha] \text{ where } 0 \le \alpha \le 2 .$$

Then it is known that f is positive definite, i.e.

$$\sum_{i=1}^{n} \sum_{j=1}^{n} \alpha_i\alpha_j f(a_i,a_j) \ge 0 \quad \forall\alpha_1,\ldots,\alpha_n \in \mathbb{R} ; \forall a_1,\ldots,a_n \in \mathbb{R}^d .$$

Hence, there exists a centred Gaussian Stochastic Process $\{\xi(a) : a \in \mathbb{R}^d\}$ having f as covariance, i.e. such that $E(\xi(a)\xi(b)) = f(a,b)$. Therefore it is clearly of interest to study such kernels.

Note now that $\mathbb{R}^d$ may be viewed as G/K where $G = SO(d) \circledS \mathbb{R}^d$ and $K = SO(d)$. Hence, we may consider $\mathbb{R}^d$ as a homogeneous space with G-action given by $g_1 : gK \to g_1gK$. This idea may be generalized as follows:

B) Generalization: Let G be a separable topological group, K a closed subgroup. Then we explain the first concept appearing in the title of this paper in

(1.1) Definition: A continuous function $f : G/K \times G/K \to \mathbb{R}$ is called a Lévy-Schoenberg (L.-S.) kernel, if

(i) $f(a,b) = f(b,a) \qquad \forall a,b \in G/K$

(ii) $\exists e \in G/K$ such that $f(a,e) = 0 \qquad \forall a \in G/K$

(iii) $r(a,b) := f(a,a) + f(b,b) - 2f(a,b)$ is invariant under G, i.e.
$r(a,b) = r(ga,gb) \qquad \forall g \in G$

(iv) f is positive definite

We note here that f can be recovered from r (r is the "polarization" of f) and that in example A) above r is given by $r(a,b) = \| a-b\|^\alpha$. Moreover, the characteristic abstract features of example A) have been captured in (1.1).

Part (iv) in (1.1) is most difficult to deal with in general. The following lemma is of crucial importance in this context:

(1.2) Lemma: Suppose r is a kernel (continuous) on G/K such that $r(a,b) = r(b,a)$ for all $a,b \in G/K$, and suppose further that there exists a point e such that $r(e,e) = 0$. Let

$$f(a,b) := \frac{1}{2}\left[r(a,e) + r(b,e) - r(a,b)\right] \quad \text{(cf. example A))} \ ,$$

then f is positive definite iff

$$\Theta_t(a,b) := \exp\left[-\, tr(a,b)\right] \text{ is positive definite} \quad \forall t \geq 0 \ .$$

(1.3) Corollary: If f is a L.-S. kernel on G/K, then its polarized kernel r satisfies:

(i) $r(a,b) = r(b,a)$ $\quad \forall a,b \in G/K$

(ii) $r(a,a) = 0$ $\quad \forall a \in G/K$

(iii) $r(ga,gb) = r(a,b)$ $\quad \forall a,b \in G/K \ ; \ \forall g \in G$

(iv) $\Theta_t(a,b) := \exp\left[-\, tr(a,b)\right]$ is positive definite $\quad \forall t \geq 0$

Conversely: If r is any kernel satisfying (i) - (iv), and if for some point $e \in G/K$ f is defined by

$$f(a,b) := \frac{1}{2}\left[r(a,e) + r(b,e) - r(a,b)\right] ,$$

then f is a L.-S. kernel.

Note that e may be taken as the identity coset, and this will be done from now on.

C) Connection to Infinitely Divisible Positive Definite (I.D.P.) Functions: At this stage the second technical term in the title of the paper enters: From now on we let G = connected, noncompact, semi-simple Lie group with finite centre, K = maximal compact subgroup (unless otherwise specified). That is to say we are dealing with a Riemannian symmetric space of noncompact type G/K.

Suppose that we are given a kernel r on G/K satisfying (1.3) (i) - (iv). Let $\Theta(a,b) := \exp\left[-\, r(a,b)\right]$, then $\Theta(ga,gb) = \Theta(a,b)$ $\quad \forall g \in G \ ; \ \forall a,b \in G/K$. Hence, we can lift Θ to a function φ on G by setting

$$\varphi(g) := \Theta(gK,eK)$$

$$\Theta(g_1K,g_2K) = \varphi(g_2^{-1}g_1) \ .$$

The connection between L.-S. kernels and I.D.P. functions is then established by

(1.4) Theorem: Let φ be as described above. Then we have

a) φ is a continuous, normalized I.D.P. function satisfying
$\varphi(k_1 g k_2) = \varphi(g) \quad \forall k_1, k_2 \in K \; ; \; \forall g \in G.$

b) The kernel r defined on G/K by $r(g_1K, g_2K) := \Psi(g_2^{-1} g_1) = -\log \varphi(g_2^{-1} g_2)$ satisfies (1.3) (i) - (iv), if φ is a continuous normalized I.D.P. function satisfying $\varphi(k_1 g k_2) = \varphi(g) \quad \forall k_1, k_2 \in K \; ; \; \forall g \in G.$

In [7] Gangolli proceeds to describe all L.-S. kernels by means of a Lévy-Khintchin (L.-K.) formula. This formula is quite explicit with the exception of the so-called Gaussian Part. Recall example A) and the classical L.-K. formula on $\mathbb{R}^d \cong SO(d) \circledS \mathbb{R}^d / SO(d)$: The Gaussian Part is given by the kernel (up to scalars)

$$r(a,b) = \| a-b \|^2 .$$

The corresponding function Ψ on $SO(d) \circledS \mathbb{R}^d$ may be described as the (up to scalars unique!) nonnegative solution of the functional equation

$$\int_K \Psi(g_1 k g_2)\,dk = \Psi(g_1) + \Psi(g_2) \quad \text{———} \quad (*) .$$

Here K = SO(d), and dk denotes the normalized Haar measure on K. This functional equation occurs in many different contexts, cf. [5], [6], [8]. The interesting fact in our case is, however, that if G/K is a Riemannian symmetric space of noncompact type, then the same functional equation is satisfied by the Gaussian Part of the L.-K. formula describing the logarithm of I.D.P. functions (normalized and biinvariant with respect to K). Thus, Gangolli in [7] asked the question whether any nonnegative solution of (*) might serve as a Gaussian Part in the L.-K. formula.

In [5] it was shown (by a nonconstructive method) that this is not the case for the group Sp(n;1). Thus, the search for an adequate description of the Gaussian Part motivates the considerations of the following section.

2. CONSEQUENCES OF AN ABSTRACT L.-K. FORMULA

We wish to recall an abstract L.-K. formula in terms of cohomology.

(2.1) Theorem (Abstract L.-K. Formula): Let φ be as in (1.4) a). Then there exists a 1-cocycle δ associated with a unitary representation U of G in a Hilbert space H such that

(i) $\{\delta(g) : g \in G\}$ is total in H

(ii) $\mathrm{Im} \langle \delta(g_1), \delta(g_2) \rangle \; 0 \quad \forall g_1, g_2 \in G$

(iii) $\log \varphi(g) = -\frac{1}{2} \| \delta(g) \|^2$

(iv) The pair (δ, U) is unique up to unitary equivalence

(v) $\delta(k) = 0 \qquad \forall k \in K$

Conversely: Every pair (U, δ) satisfying (i), (ii), (v) determines a φ as in (1.4) a) via (iii) above.

Proof: The theorem is an immediate consequence of theorem (1.1) in [4]: We just need to remember that we are dealing with realvalued functions here. An application of lemma (2.2), chapter V in [2] completes the proof.

$\boxtimes$

Although (2.1) provides a L.-K. formula which is rather nice from an aesthetical point of view, it doesn't help much in practical terms. Thus, we prepare the way for a direct integral decomposition by proving

(2.2) Theorem: Let G = KAN be the Iwasawa decomposition where dim A = n. Then there exist at most 2n pairwise inequivalent irreducible representations of G admitting a nontrivial cohomology.

Proof: Suppose that there are 2n+1 pairwise inequivalent irreducible representations $U_1, \dots, U_{2n+1}$ with associated nontrivial 1-cocycles $\delta_1, \dots, \delta_{2n+1}$. Without loss of generality assume $\delta_i(k) = 0$ for $1 \le i \le 2n+1$ and for all $k \in K$. Set

$$\Psi_i(g) := -\frac{1}{2} \| \gamma_i(g) \|^2$$

$$\text{where } \gamma_i(g) := \begin{cases} \delta_i(g) \text{ if } \mathrm{Im} \langle \delta_i(g_1), \delta_i(g_2) \rangle = 0 \quad \forall g_1, g_2 \in G \\ \delta_i(g) \oplus \overline{\delta}_i(g) \text{ otherwise (cf. lemma (2.2), chapter V in [2])} \end{cases} .$$

This way we obtain at least n+1 distinct functions Ψ_i satisfying

$$\int_K \Psi(g_1 k g_2)\,dk = \Psi(g_1) + \Psi(g_2) \quad \text{(cf. [2] p. 90)} .$$

But the functions satisfying this equation form a vector space of dimension n over $\mathbb{R}$, cf. [6]. Hence, we have

$$\sum_{i=1}^{N} \lambda_i \Psi_i(g) \equiv 0 \text{ where } N \geq n+1 \text{ and}$$

$\lambda_i \neq 0$ for at least two indices (nontriviality of the 1-cocycles!).

Now suppose without loss of generality

$$\lambda_1, \ldots, \lambda_m > 0 \; ; \; \lambda_{m+1}, \ldots, \lambda_k < 0 \; ; \; \lambda_{k+1} = \ldots = \lambda_N = 0 .$$

Then

$$\sum_{i=1}^{m} \lambda_i \Psi_i(g) = \sum_{i=m+1}^{k} - \lambda_i \Psi_i(g) .$$

Set

$$\alpha_i := \begin{cases} \sqrt{\lambda_i} & 1 \leq i \leq m \\ \sqrt{-\lambda_i} & m+1 \leq i \leq k \end{cases} , \text{ then}$$

$$\sum_{i=1}^{m} \alpha_i^2 \Psi_i(g) = \sum_{i=m+1}^{k} \alpha_i^2 \Psi_i(g) \Rightarrow$$

$$-\frac{1}{2} \| \alpha_1 \gamma_1(g) \oplus \ldots \oplus \alpha_m \gamma_m(g) \|^2 = -\frac{1}{2} \| \alpha_{m+1} \gamma_{m+1}(g) \oplus \ldots \oplus \alpha_k \gamma_k(g) \|^2$$

where γ_i defined as above is associated with

$$V_i := \begin{cases} U_i & \text{if } \operatorname{Im} \langle \delta_i(g_1), \delta_i(g_2) \rangle = 0 \quad \forall g_1, g_2 \in G \\ U_i \oplus \overline{U}_i & \text{otherwise} \end{cases}$$

But this implies, cf. (2.1) above, that a subrepresentation of $V_1 \oplus \ldots \oplus V_m$ is equivalent to a subrepresentation of $V_{m+1} \oplus \ldots \oplus V_k$. According to Cor. 3, p. 17 in [10] these two representations are, however, disjoint. Thus, a contradiction follows by Cor. 1, p. 16 in [10].

⊠

(2.2) together with a result due to Pinczon/Simon, cf. [12], gives an upper bound on the number of nontrivial cocycles associated with irreducible representations (modulo coboundaries). Now note that we are dealing with a type I group and hence, a direct integral decomposition is available, cf. [10], [11]. So, suppose an arbitrary representation V and an associated 1-cocycle Δ satisfying $\Delta(k) = 0 \quad \forall k \in K$ are given. Then we may find a measure space (Ω,μ) such that

$$V(g) = \int_{\Omega}^{\oplus} U_g^{\omega} d\mu(\omega)$$

in the sense of direct integrals.

$$\Delta(g) = \int_{\Omega}^{\oplus} \delta^{\omega}(g) d\mu(\omega)$$

Here U^{ω} is an irreducible representation of G in a Hilbert space H^{ω}, and δ^{ω} is a 1-cocycle associated with $U^{\omega} \quad \forall \omega \in \Omega$ and satisfies $\delta^{\omega}(k) = 0 \quad \forall k \in K$.

Suppose that $\delta_1,\ldots,\delta_M$ are the nontrivial 1-cocycles (modulo coboundaries) associated with $U_1,\ldots,U_M$ (irreducible). Then, if $\Omega_i := \{\omega \in \Omega : U^{\omega} = U_i\}$ for $1 \leq i \leq M$, it follows from the results in [10] that all the Ω_i must be measurable. Hence, we may write Δ as a direct sum

$$\Delta(g) = \bigoplus_{i=1}^{M} \int_{\Omega_i}^{\oplus} c_i(\omega)\delta_i(g) d\mu(\omega) + \int_{\Omega\setminus\left(\bigcup_{i=1}^{M}\Omega_i\right)}^{\oplus} [U_g^{\omega}v^{\omega} - v^{\omega}] d\mu(\omega)$$

where

a) $c_i \in L^2(\Omega_i, d\mu) \qquad 1 \leq i \leq M$

b) $U_k^{\omega}v^{\omega} = v^{\omega} \qquad \forall k \in K, \ \forall \omega \in \Omega\setminus\left(\bigcup_{i=1}^{M}\Omega_i\right)$

c) M is finite

Note that the U_i's are not necessarily all distinct, and that almost everywhere statements may be avoided by the trick given in [11].

We finally set $c_i := \int_{\Omega_i} |c_i(\omega)|^2 \, d\mu(\omega)$ and apply (2.1) to obtain

<u>(2.3) Theorem (L.-K. Formula):</u> Let φ be as in (1.4) a). Then there exists a Δ (as described above) such as

$$\log \varphi(g) = -\frac{1}{2} \| \Delta(g) \|^2$$

$$= -\frac{1}{2} \left[\sum_{i=1}^{M} c_i^2 \| \delta_i(g) \|^2 + \int_{\Omega\setminus\left(\bigcup_{i=1}^{M}\Omega_i\right)} \| U_g^{\omega}v^{\omega} - v^{\omega} \|^2 \, d\mu(\omega) \right]$$

where all the relevant terms have been specified above. Moreover, if $\{\Delta(g) : g \in G\}$ is total in the space where the associated representation V acts, then Δ is, up to unitary equivalence, uniquely determined by φ.

Conversely: Every Δ (as described above) gives rise to an I.D.P. function φ, bi-invariant with respect to K, via the formula

$$\varphi(g) = \exp\left[-\frac{1}{2}\,\|\Delta(g)\|^2\right].$$

The reader should compare (2.3) with theorem (3.31), p. 167 in [7]. Normalizing the measure μ appropriately we obtain perfect agreement, and it becomes clear that our description of the Gaussian Part is just

$$-\frac{1}{2}\sum_{i=1}^{M} c_i^2\,\|\delta_i(g)\|^2 .$$

Thus, it is seen that this may be zero ($Sp(n;1)$, $SL(n;\mathbb{R})$ for $n \geq 3$ are known to admit only trivial 1-cohomologies!) and hence, we have obtained a more precise description of the Gaussian Part, since a nonnegative solution of equation (*), i.e.

$$\int_K \Psi(g_1 k g_2)\,dk = \Psi(g_1) + \Psi(g_2)$$

can always be constructed. This may be seen as follows:

Let $G = KAN$ as before, and let MAN, where M is the centralizer of A in K, be a minimal parabolic subgroup. Consider $B := KAN/MAN$. Then K acts transitively on this homogeneous space, and in fact there exists a unique K-invariant probability measure m on B, cf. [6]. Let $m^g(E) := m(gE)$ (E: Borel set) and set

$$\sigma(g,x) := \frac{dm^g}{dm}(x) \qquad \text{(Radon-Nikodym derivative)}.$$

Then we have

<u>(2.4) Theorem:</u> Let σ be as above, then

$$\Psi_\sigma(g) := \int_B \log \sigma(g,x)\,dm(x)$$

is a nonnegative solution of equation (*).

<u>Proof:</u> Ψ_σ is an A-spherical function in the sense of [6] and hence, cf. (6.2) in [6], a solution of equation (*). Moreover,

$$\Psi_\sigma(g) \geq \log \int_B \sigma(g,x)\,dm(x) = 0 .$$

Also Ψ_σ can't be identically zero, cf. [6], p. 417.

⊠

<u>Remark 1:</u> It seems that the L.-K. formula given in (2.3) is best possible, if one doesn't have explicit knowledge of the 1-cohomology of the group or of the irreducible class I representations. It should also be noted that one obtains a more

general L.-K. formula (where biinvariance of the I.D.P. functions is not demanded) from our formula quite easily: The abstract formula (2.1) holds, if condition (v) is dropped. Moreover, the most general 1-cocycle differs from one satisfying (2.1) (v) only by a coboundary.

Remark 2: The bound given in (2.2) is achieved for $G = SU(n;1)$: In this case $\dim A = 1$, and there are two nontrivial 1-cocycles associated with irreducible representations, cf. [1].

Remark 3: If equation (*) has a unique solution (up to scalars) and a nontrivial 1-cocycle is known to exist (as in the case of the classical split rank one groups with the exception of $Sp(n;1)$), it may be convenient to explicitly calculate the Gaussian Part of the L.-K. formuly by realizing the Ψ_σ described in (2.4).

Example: We calculate the Gaussian Part of the L.-K. formula for $SU(n;1)$. Here

$$B = KAN/MAN = \left\{ \underline{x} = (x_o,\ldots,x_{n-1}) \in \mathbb{C}^n : \sum_{i=o}^{n-1} |x_i|^2 = 1 \right\} .$$

Note that $SU(n;1)$ acts by fractional linear transformations on B in this case, and thus the action may be extended to the whole of $\mathbb{C}^n$. Then a nonnegative solution of equation (*) is given by

$$f(g) = - \int_B \log \left\{ \frac{1 - \| g^{-1}(\underline{O}) \|^2}{|1 - \langle g^{-1}(\underline{O}), \underline{x} \rangle|^2} \right\} dm(x) .$$

See [9], p.17, and [13], p. 474 for calculations concerning the Radon-Nikodym derivative. Note that for $SU(1;1)$ the integration may be performed, and f is explicitly given by

$$f(g) = - \log \left[1 - \| g^{-1}(\underline{O}) \|^2 \right] .$$

Acknowledgement: The author wishes to thank Fr. I. Schon for her excellent typing.

References

[1] Delorme, P.: 1-Cohomologie des Représentations Unitaires des Groupes de Lie Semi-Simple et Résolubles. Produits Tensoriels Continus de Représentations, Bull. Soc. Math. France, 105, (1977), p. 281-336

[2] Erven, J., Falkowski, B.-J.: Low Order Cohomology and Applications, Lecture Notes in Mathematics, Vol. 877, Springer-Verlag (1981)

[3] Erven, J., Falkowski,B.-J.: Continuous Cohomology, Infinitely Divisible Positive Definite Functions, and Continuous Tensor Products for SU(1;1), in Probability Measures on Groups (Ed. H. Heyer), Lectures Notes in Mathematics, Vol. 928, Springer-Verlag (1982), p. 76-89

[4] Falkowski, B.-J.: An Analogue of the Lévy-Khintchin Formula on SL(2;$\mathbb{C}$), in Probability Measures on Groups VII (Ed. H. Heyer), Lecture Notes in Mathematics, Vol. 1064, Springer-Verlag (1984), p. 80-85

[5] Faraut, J., Harzallah, K.: Distances Hilbertiennes Invariantes sur un Espace Homogène, Ann. L'Inst. Fourier (Grenoble) XXIV (1974), p. 171-217

[6] Fürstenberg, H.: Noncommuting Random Products, TAMS, 108, (1963), p. 377-428

[7] Gangolli, R.: Positive Definite Kernels on Homogeneous Spaces and Certain Stochastic Processes Related to Lévy's Brownian Motion of Several Parameters, Ann. Inst. H. Poincaré, Vol. III, n° 2, (1967), p. 121-225

[8] Heyer, H.: Convolution Semigroups of Probability Measures on Gelfand Pairs, Expo. Math. 1 (1983), p. 3-45

[9] Lipsman, R.L.: Group Representations, Lecture Notes in Mathematics, Vol. 388, Springer-Verlag (1974)

[10] Mackey, G.W.: The Theory of Unitary Group Representations, Chicago Lectures in Mathematics, University of Chicago Press, (1976)

[11] Parthasarathy, K.R., Schmidt, K.: Positive Definite Kernels, Continuous Tensor Products, and Central Limit Theorems of Probability Theory, Lecture Notes in Mathematics, Vol. 272, Springer-Verlag, Berlin (1972)

[12] Pinczon, G., Simon, E.: On the 1-Cohomology of Lie Groups, Letters in Math. Phys. 1, (1975), p. 83-91

[13] Warner, G.: Harmonic Analysis on Semi-Simple Lie Groups I, Grundlehren der math. Wissenschaften, Bd. 188, Springer-Verlag (1972)

EXEMPLES D'HYPERGROUPES TRANSIENTS

Léonard Gallardo
(Nancy)

Summary : An hypergroup K is said transient if every random walk with spread out law on K is transient. We study here two examples. For polynomial hypergroups on $\mathbb{N}$ and Chebli-Trimèche hypergroups on $\mathbb{R}_+$, we give an intrinsic criterion involving the Plancherel measure of K to decide if the hypergroup is transient or not. These results can be compared to dimensional or growth criteria in the case of groups.

1. Introduction

Soit K un hypergroupe de Jewett commutatif. Pour les notions générales concernant les hypergroupes, nous renvoyons le lecteur à l'article de Jewett ([8]) et au Survey de H. Heyer ([6]) dont nous adoptons les notations (à la différence près que nous écrirons δ_x pour la mesure de Dirac en x et m pour la mesure de Haar de K). Dans cet article nous supposerons que la topologie de K est toujours à base dénombrable.

1.1- Soit $\mu \in M^1(K)$ une mesure de probabilité sur K. On appelle marche aléatoire de loi μ sur K toute chaîne de Markov homogène avec K pour espace d'états et avec un noyau markovien de la forme

$$P(x,A) = \delta_x * \mu(A) \qquad (x \in K,\ A \subset K) .$$

Nous avons montré dans [4] que si μ est adaptée (i.e. le sous hypergroupe de K engendré par le support de μ est K tout entier) et étalée (i.e. μ^{*p} est non singulière par rapport à m pour un $p \in \mathbb{N}^*$), la marche aléatoire de loi μ vérifie le principe de "dichotomie" :

- ou bien tout état $x \in K$ est récurrent (et la marche aléatoire est Harris récurrente par rapport à m)
- ou bien tout état $x \in K$ est transient (et le potentiel de tout compact est une fonction bornée).

Dans le premier (resp. le second) cas on dit que la marche aléatoire est récurrente (resp. transiente) et on dit que l'hypergroupe K est transient si toute marche aléatoire de loi μ adaptée et étalée est transiente. Toutes les mesures de probabilité qui interviendront dans la suite de cet article seront supposées adaptées et étalées. Notons que si K est discret, toute $\mu \in M^1(K)$ est étalée car m charge tous les points de K. L'hypothèse d'étalement n'est donc pas restrictive dans ce cas.

1.2- Dans [4] nous avons montré (voir aussi [5]) qu'une marche aléatoire symétrique (i.e. de loi $\mu = \mu^-$) est transiente si et seulement si

$$\frac{1}{1-\hat{\mu}} \in L^1_{\ell oc}(\hat{K}, \pi) ,$$

où $\hat{K}$ désigne le dual de K, π est la mesure de Plancherel et $\hat{\mu}$ est la transformée de Fourier de μ. Notons que μ étant supposée adaptée, on a $\hat{\mu}(\chi) = 1$ uniquement lorsque $\chi = \mathbb{1}$ (le caractère identiquement égal à 1) ; il suffit donc de tester l'intégrabilité de $(1-\hat{\mu})^{-1}$ au voisinage de $\mathbb{1}$. On a également montré que les critères de Baldi Lohoué et Peyrière ([1]) restaient valables dans le cas des hypergroupes :

1.3- Si la marche aléatoire de loi $\nu = \frac{1}{2}(\mu + \mu^-)$ est transiente, la marche aléatoire de loi μ est aussi transiente.
Ainsi K est transient si et seulement si toute marche aléatoire de loi symétrique est transiente. On a aussi le second résultat :

1.4- Soit $\nu \in M^1(K)$ symétrique sur K et $\phi : K \to [0,1]$ une fonction symétrique (i.e. $\phi^- = \phi$) et strictement positive sur un ouvert de K. Soit c une constante telle que $(\frac{1}{c}\phi) \cdot \nu = \nu_1 \in M^1(K)$. Alors si la marche aléatoire de loi ν_1 est transiente, la marche de loi ν l'est aussi.

Le but de cet exposé est de montrer que pour deux classes d'hypergroupes (les hypergroupes polynomiaux sur $\mathbb{N}$ et les hypergroupes de Chebli-Trimèche associés à un opérateur de Sturm Liouville sur $\mathbb{R}_+$), on peut tester s'ils sont ou non transients uniquement sur la mesure de Plancherel π. Ceci pourra être comparé au cas classique des groupes abéliens (ou des groupes de Lie) où on dispose du critère de la dimension (ou de la croissance) pour décider de la transience.

2. Hypergroupes polynomiaux

2.1- Soit $(P_n(x))_{n \in \mathbb{N}}$ une suite de polynômes orthogonaux sur $[-1,1]$ par rapport à une mesure positive $\pi(dx)$ et tels que $d°P_n = n$ et $P_n(1) = 1$ pour tout entier n, $P_0 \equiv 1$ et $P_1(x) = x$. On fait l'hypothèse que les P_n vérifient la propriété de linéarisation à coefficients non négatifs i.e. pour tous m et $n \in \mathbb{N}$, on a

$$P_m(x)P_n(x) = \sum_r C(m,n,r)P_r(x)$$

avec $C(m,n,r) \geq 0$ pour tout $r \in \mathbb{N}$ (notons que $C(m,n,r) = 0$ si $r \notin [|m-n|, m+n]$). Avec ces notations on a alors le résultat suivant dû à R. Lasser ([9]) :

2.2- L'ensemble $\mathbb{N}$ des entiers naturels avec 0 comme unité, l'identité comme involution et avec une convolution des mesures définie par

$$\delta_n * \delta_m = \sum_{r=|m-n|}^{m+n} C(m,n,r)\delta_r \qquad (m,n \in \mathbb{N}) ,$$

est un hypergroupe commutatif dont la mesure de Haar est donnée par

$$m(\{n\}) = \left(\int_{-1}^{1} |P_n(x)|^2 \pi(dx)\right)^{-1} \qquad (n \in \mathbb{N}) .$$

De plus à tout caractère $\chi \in \hat{\mathbb{N}}$, on peut associer un unique nombre complexe z tel que

$$\chi(n) = P_n(z) \qquad (\forall n \in \mathbb{N}) ,$$

de sorte que l'on a $\hat{\mathbb{N}} = \{z \in \mathbb{C} ; |P_n(z)| \leq 1 \text{ pour tout } n \in \mathbb{N}\}$ et la mesure de Plancherel a son support dans $[-1,1]$ et s'identifie à $\pi(dx)$.

2.3- **Définition**. L'ensemble $\mathbb{N}$ muni d'une structure d'hypergroupe associée à une famille de polynômes orthogonaux à la manière de 2.2, est appelé *hypergroupe polynomial*.

Nous attribuons le résultat qui suit et qui caractérise les hypergroupes polynomiaux, à A.L. Schwartz car tous les ingrédients se trouvent déjà (aux différences de vocabulaire près) dans son article ([10]). Ce fait a également été remarqué par W.R. Bloom et S. Selvanathan ([2]) :

2.4- Proposition. Une structure d'hypergroupe commutatif sur $\mathbb{N}$ est un hypergroupe polynomial si et seulement si les quatre conditions suivantes sont satisfaites par la convolution :

(i) $\delta_0 * \mu = \mu \qquad (\forall \mu \in M^1(\mathbb{N}))$

(ii) $(\delta_n * \delta_n)(0) > 0 \quad (\forall n \in \mathbb{N})$

(iii) $(\delta_n * \delta_1)(k) = 0$ si $|n-k| > 1 \quad (\forall n,k \in \mathbb{N})$

(iv) $(\delta_n * \delta_1)(n+1) > 0 \quad (\forall n \in \mathbb{N})$

La convolution et les polynômes sont alors liés par $\delta_n = P_n(\delta_1)$.

Démonstration : La condition est clairement nécessaire. Réciproquement si une structure d'hypergroupe commutatif vérifie (i) ... (iv), il est clair que 0 est l'unité et que l'involution est l'identité d'après (ii). Il suffit alors d'appliquer le Théorème 1 de [10] pour compléter le résultat □

On s'intéresse maintenant à la transience des marches aléatoires sur un hypergroupe polynomial :

2.5- Définition. On appelle marche aléatoire simple sur l'hypergroupe $\mathbb{N}$, la marche aléatoire de loi $\mu := \delta_1$.

2.6- Théorème. Les trois conditions suivantes sont équivalentes pour un hypergroupe polynomial :

(i) l'hypergroupe polynomial $\mathbb{N}$ est transient.

(ii) la marche aléatoire simple est transiente.

(iii) $\int_{-1}^{1} \frac{\pi(dx)}{1-x} < +\infty$.

Démonstration : (i) $\Longrightarrow$ (ii) est clair. Montrons que (ii) $\Longrightarrow$ (i) : Soit ν une probabilité adaptée sur $\mathbb{N}$. Supposons que ν est à support fini. L'involution de $\mathbb{N}$ étant l'identité, ν est symétrique et on a

$$\mathbb{N} = \bigcup_{k \in \mathbb{N}} \operatorname{supp} \nu^k .$$

Il existe ainsi $k_0 \in \mathbb{N}$ tel que $\operatorname{supp} \nu^{k_0} \ni 1$. Soit alors φ l'application de $\mathbb{N}$ dans $\{0,1\}$ définie par

$$\varphi(x) = \begin{cases} 1/\nu^{k_0}(\{1\}) & \text{si } x = 1 \\ 0 & \text{si } x \neq 1 \end{cases} .$$

On a évidemment $\varphi \cdot \nu^{k_0} = \delta_1$. Le résultat rappelé en 1.4 nous assure alors que la marche aléatoire de loi ν^{k_0} est transiente si la marche simple est transiente. Mais ceci implique que la marche de loi ν est transiente d'après le résultat suivant (bien connu dans le cas des groupes) :

2.6.1- **Lemme**. Soit K un hypergroupe, $\nu \in M^1(K)$ une probabilité à support compact et $n_o \in \mathbb{N}$ un entier quelconque. Alors les marches aléatoires de loi ν et ν^{n_o} sont toutes les deux transientes ou toutes les deux récurrentes.

Démonstration : Pour toute fonction $f \in C_c^+(K)$, on vérifie sans peine que

$$\sum_{n \geq 0} \nu^n(f) = \sum_{n \geq 0} (\nu^{n_o})^n(g) ,$$

où $g = \left(\sum_{k=1}^{n_o-1} \nu^k \right)^- * f \in C_c^+(K)$. Le résultat en découle □

Fin de la preuve du Théorème : Toute marche aléatoire de loi ν à support compact est transiente. Il en résulte, en choisissant une fonction ϕ à support compact dans 1.4, que toute marche aléatoire sur $\mathbb{N}$ est transiente. Donc (i) $\Longleftrightarrow$ (ii). L'équivalence entre (ii) et (iii) résulte de 1.2 car $\hat{\delta}_1(x) = 1-x$ □

2.7- **Exemple**. Soient $\alpha, \beta \in \mathbb{R}$ des nombres tels que $\alpha \geq \beta > -1$ et $\alpha + \beta + 1 \geq 0$. Les polynômes de Jacobi $(P_n^{\alpha,\beta})_{n \geq 0}$ orthogonaux sur $[-1,+1]$ par rapport à la mesure $\pi(dx) := (1-x)^\alpha(1+x)^\beta$, donnent à $\mathbb{N}$ une structure d'hypergroupe polynomial ([9]) qui est donc transient si et seulement si $\alpha > 0$.

3. Hypergroupes de Chebli-Trimèche

Sur $\mathbb{R}_+$ on considère l'opérateur du second ordre

$$L := -\frac{1}{A(t)} \frac{d}{dt} (A(t) \frac{d}{dt}) ,$$

où $A : [0,+\infty[\to \mathbb{R}$ est une fonction de la forme

$$A(t) := t^{2\alpha+1} C(t) ,$$

où $\alpha > -\frac{1}{2}$ et C est une fonction strictement positive, paire et de classe $\mathscr{C}^\infty$. On suppose que A est croissante, A'/A est décroissante et $\lim_{t \to +\infty} A'(t)/A(t) = 2\rho \geq 0$. Sous ces conditions H. Chebli ([3]) puis K. Trimèche ([11]) ont prouvé le résultat suivant :

3.1- L'équation aux valeurs propres

$$\begin{cases} Lu = su & (s \in \mathbb{C}) \\ u(o) = 1 \ , \ u'(o) = 0 \end{cases}$$

a pour chaque $s \in \mathbb{C}$ une solution φ_s qui est telle que

$$\varphi_s(x)\, \varphi_s(y) = \int_0^{+\infty} \varphi_s(z) W_{x,y}(dz) \quad (x,y \in \mathbb{R}_+) \quad ,$$

où $W_{x,y}(dz)$ est une mesure de probabilité sur $[0,+\infty[$ à support dans $[|x-y|,\ x+y]$.

Si on pose alors

$$\delta_x * \delta_y := W_{x,y} \quad \text{si } x \text{ et } y > 0$$
$$\delta_x * \delta_o := \delta_o * \delta_x := \delta_x \quad \text{si } x \geq 0 \quad ,$$

on définit sur $\mathbb{R}_+ = [0,+\infty[$ avec sa topologie usuelle, une structure d'hypergroupe commutatif (car $W_{x,y} = W_{y,x}$) d'involution l'identité et de mesure de Haar

$$m(dt) = A(t)\, dt \quad .$$

Un tel hypergroupe sera appelé <u>hypergroupe de Chebli-Trimèche</u>.
On prouve facilement que les caractères de l'hypergroupe sont les fonctions $x \to \varphi_s(x)$ pour $s \geq 0$. Ainsi $\hat{\mathbb{R}}_+ \cong \mathbb{R}_+$ et la mesure de Plancherel $\pi(dx)$ associée est portée par l'intervalle $[\rho^2,\ +\infty[$. La transformée de Fourier d'une probabilité $\mu \in M^1(\mathbb{R}_+)$ est donc donnée par

$$(3.1.1) \qquad \hat{\mu}(s) = \int_0^{+\infty} \varphi_s(x)\mu(dx) \qquad (s \in \mathbb{R}_+) \quad .$$

3.2- <u>Définition</u> : La variance de $\mu \in M^1(\mathbb{R}_+)$ est le nombre $V(\mu)$, lorsqu'il existe, donné par

$$V(\mu) := - \frac{d}{ds}\, \hat{\mu}(s)\Big|_{s=0} \quad .$$

3.3- <u>Remarque</u> : Cette notion de variance, à notre connaissance introduite par K. Trimèche ([12]), a été utilisée par plusieurs auteurs ([7]).

3.4- Proposition. On peut calculer la variance de $\mu \in M^1(\mathbb{R}_+)$ lorsqu'elle existe par la formule

$$V(\mu) := \int_0^{+\infty} Q(x)\mu(dx) ,$$

où $Q(x) = \int_0^x \frac{1}{A(t)} \int_0^t A(u)dudt$.

Démonstration : D'après un résultat de K. Trimèche ([13]), on a

$$\varphi_s(x) = \sum_{p=0}^{+\infty} (-1)^p s^p b_p(x) ,$$

où les $b_p(x)$ sont des fonctions telles que pour tout $p \geqq 1$

$$L b_{p+1} = b_p , \quad b_p(0) = 0$$

et $L b_0 = 0$ et $b_0(0) = 1$. On voit alors facilement que

$$Q(x) = b_1(x) ;$$

le résultat en découle □

3.5- Théorème. L'hypergroupe de Chebli-Trimèche $\mathbb{R}_+$ est transient si et seulement si $\int_0^1 \frac{\pi(ds)}{s} < +\infty$.

Démonstration : Si $\mu \in M^1(\mathbb{R}_+)$ est à support compact et charge un ouvert, $V(\mu)$ existe et c'est un nombre strictement positif. Pour une telle mesure on a

$$\hat{\mu}(s) \sim 1 - V(\mu)s \qquad (s \to 0) ,$$

et d'après le critère 1.2 une marche aléatoire de loi μ est transiente si et seulement si la fonction $(1 - \hat{\mu}(s))^{-1} \sim V(\mu)^{-1} s^{-1}$ est intégrable au voisinage de zéro pour la mesure de Plancherel $\pi(ds)$. Mais on peut toujours se ramener à une telle mesure μ d'après le critère 1.4 en prenant une fonction φ à support compact. Toute marche aléatoire est donc transiente si les marches aléatoires à support compact sont transientes. D'où le théorème □

3.6- Remarque. Si $\rho > 0$, l'origine n'est pas dans le support de la mesure de Plancherel. Un tel hypergroupe est donc transient.

4. Remarques finales

Nous avons annoncé dans l'introduction que les critères de transience pouvaient être comparés à des caractéristiques géométriques concernant l'hypergroupe. En effet par exemple :

4.1- Dans le cas des marches aléatoires sur $\mathbb{R}^d$, on sait que pour une loi de probabilité μ centrée, adaptée et ayant un moment d'ordre 2 on a

$$\hat{\mu}(x) \sim 1 - C|x|^2 \quad (x \to 0) .$$

Le critère de transience 1.2 se réduit à considérer l'intégrabilité de

$$\int_{V_o} \frac{dx}{|x|^2} ,$$

sur un voisinage V_o de 0 . Or cette intégrale converge si et seulement si $d \geqq 3$. De même dans le cas d'un groupe de Lie la transience équivaut à dire que la croissance du groupe est polynomiale de degré $\geqq 3$ ou exponentielle ([1]) ;

4.2- Dans [5], nous avons, avec des idées analogues, étudié l'hypergroupe D dual d'une paire symétrique (G,K). On a pu dans ce cas traduire le résultat de transience de manière plus géométrique :

$$D \text{ est transient si et seulement si } \dim p^* \geqq 3 ,$$

où p^* est l'espace tangent en eK à l'espace symétrique G/K .

L'étude du lien entre la transience des hypergroupes et leur degré de croissance est donc tout un programme.

Bibliographie

[1] BALDI, P., LOHOUE, N. et PEYRIERE, J. : Sur la classification des groupes récurrents. Comptes Rendus Acad. Sc. de Paris Série A t. 285 (1977) p. 1103-1104.

[2] BLOOM, W.R. and SELVANATHAN, S. : Hypergroup structures on the set of natural numbers. To appear in Bull. Aust. Math. Soc.

[3] CHEBLI, H. : Positivité des opérateurs de translation généralisée associés à un opérateur de Sturm-Liouville et quelques applica-

tions à l'analyse harmonique. Thèse. Université Louis Pasteur (1974).

[4] GALLARDO, L. et GEBUHRER, O. : Marches aléatoires et hypergroupes. A paraître dans Expositiones Mathematicae.

[5] GALLARDO, L. et GEBUHRER, O. : Un critère usuel de transience pour certains hypergroupes commutatifs. Application au dual d'un espace symétrique. Probabilités sur les structures géométriques. Journées S.M.F. 1984. Publications de l'Université P. Sabatier de Toulouse.

[6] HEYER, H. : Probability theory on hypergroups : a survey. Probability measures on groups VII Oberwolfach - Lecture Notes in Math. N° 1064 (1984) p. 481-550.

[7] HEYER, H. : Convolution semi-groups of probability measures on Gelfand pairs. Expositiones Mathematicae, 1, (1983) p. 3-45.

[8] JEWETT, R.I. : Spaces with an abstract convolution of measures. Advances in Math., 18, (1975) p. 1-101.

[9] LASSER, R. : Orthogonal polynomials and hypergroups. Rendiconti di Matematica (2) Vol. 3, Série VII (1983) p. 185-209.

[10] SCHWARTZ, A.L. : ℓ^1-convolution algbras : representation and factorization. Zeit. für Wahr. und verw. Geb. 41, (1977) p. 161-176.

[11] TRIMECHE, K. : Transformation intégrale de Weyl et théorème de Paley-Wiener associés à un opérateur différentiel singulier sur $[0,+\infty[$. J. Math. Pures et Appl. 60 (1981) p. 51-98.

[12] TRIMECHE, K. : Probabilités indéfiniment divisibles et théorème de la limite centrale pour une convolution généralisée sur la demi droite. Séminaire d'analyse harmonique. Université de Tunis (1976).

[13] TRIMECHE, K. : Convergence des séries de Taylor généralisées. Comptes Rendus Acad. Sc. de Paris t. 281, Série A (1975) p. 1015-1017.

Léonard Gallardo
Université de Nancy II
et
UA N° 750 du CNRS
Mathématiques
B.P. N° 239
54506 VANDOEUVRE les NANCY
FRANCE

QUELQUES PROPRIETES DU NOYAU POTENTIEL D'UNE MARCHE ALEATOIRE SUR LES HYPERGROUPES DE TYPE KUNZE-STEIN.

par Olivier GEBUHRER
Université Louis Pasteur (Strasbourg I)
I.R.M.A.
7, rue René Descartes
67084 STRASBOURG CEDEX

Summary :

For locally compact commutative hypergroups such that the trivial character does not belong to the support of the Plancherel measure (Kunze-Stein type hypergroups) we prove here some results of L^2 - type (some of which were known in particular cases, but our proofs are completely different).

§ 1. Définitions, Notations.

Dans la suite, on notera X un hypergroupe localement compact, à base dénombrable commutatif ; nous nous placerons dans l'axiomatique de Jewett [Je] . On notera σ la mesure de Haar de X et ϖ sa mesure de Plancherel. La transformée de Fourier sur X est notée $\wedge$. Le caractère trivial de X est noté $\mathbb{1}$. On rappelle que $C_{\mathcal{K}}(X)$ désigne l'espace vectoriel complexe des fonctions continues sur X à support compact.

Définition 1.1. :

On dira que l'hypergroupe X est de Godement si, dans $\hat{X}$, on a $\{\mathbb{1}\} \in \operatorname{Supp} \bar{\varpi}$.
On a alors le

THEOREME 1.1. : Soit X un hypergroupe localement compact, à base dénombrable commutatif. Alors X est un hypergroupe de Godement si et seulement s'il vérifie la propriété

(G) : pour la convergence compacte sur X , la fonction   est limite d'une suite $\{\varphi_n\}$ de fonctions continues de type positif à support compact sur X ; en outre, on peut supposer que $\varphi_n = \psi_n * \tilde{\psi}_n$ où $\psi_n \in C_{\mathcal{K}}(X)$ pour tout $n \in \mathbb{N}$ et

$$\sup_n \|\psi_n\|_2 < \infty .$$

Démonstration :

On la trouvera dans [Ga - Ge][(1)]. p. 62. Th. 1.4.

THEOREME 1.2. Soit X un hypergroupe localement compact à base dénombrable, non compact, commutatif. Supposons que pour un $p \in]1,2[$, l'hypergroupe X possède la propriété $(K\text{-}S)_p$: L'espace $L^p_{\mathbb{C}}(X,\sigma)$ convole $L^2_{\mathbb{C}}(X,\sigma)$. Alors le support de la mesure de Plancherel ϖ de X ne contient pas le caractère trivial $\mathbb{1}$.

Démonstration :

Supposons que $\in \mathrm{Supp}\,\varpi$; par le théorème 1.1. rappelé ci-dessus, l'hypergroupe X vérifie la propriété (G) . Soit $\varphi \in C_{\mathcal{K}}(X)$. Soit $\{\varphi_n\}_{n\in\mathbb{N}}$ une suite de $C_{\mathcal{K}}(X)$ telle que i) $\varphi_n = \psi_n * \psi_n$ où pour tout n , $\psi_n \in C_{\mathcal{K}}(X)$

ii) $\mathrm{Sup}\,\|\psi_n\|_2 < +\infty$

iii) $\{\varphi_n\}$ converge vers $\mathbb{1}$ sur X , uniformément sur tout compact.

Alors, par propriété $(K\text{-}S)_p$ nous avons :

$$\|\varphi * \psi_n * \tilde{\psi}_n\|_\infty \le \|\varphi * \psi_n\|_2 \, \|\tilde{\psi}_n\|_2 \le K_p \|\varphi\|_p \, .$$

Mais pour $x \in X$ on a alors

$$|\varphi * \psi_n * \tilde{\psi}_n(x)| \le K_p \|\varphi\|_p$$

et en passant à la limite en n , nous avons $\|\varphi\|_1 \le K_p \|\varphi\|_p$ pour toute $\varphi \in C^+_{\mathcal{K}}(X)$ et ceci implique que X est compact.

Le théorème précédent nous permet de poser la

Définition 1.2. :

On dira que l'hypergroupe X est de type Kunze-Stein si dans $\hat{X}$, on a $\{\mathbb{1}\} \notin \mathrm{Supp}\,\varpi$.

Remarque 1.1. :

Si (G,K) est une paire riemannienne symétrique de type non compact, on sait que l'hypergroupe associé vérifie la propriété de Kunze-Stein $(K\text{-}S)_p$ pour tout $p \in [1,2[$. Il en est de même des hypergroupes de Chébli-Trimèche [Che],[Tri] . La question d'une réciproque du théorème 1.2. est à notre connaissance ouverte.

§ 2. Théorie du renouvellement sur les hypergroupes de type (K-S) .

THEOREME 2.1. Soit X un hypergroupe de type (K-S) . Toute marche aléatoire sur X de loi μ adaptée et étalée est transitoire. (Voir définition 2.1 i) ci-dessous)/

Démonstration : Par application de [Ga-Ge] (2) (Corollaire 2.10), toute marche aléatoire symétrique adaptée sur X est transitoire.
Compte tenu du résultat précédent, on va s'intéresser maintenant au comportement à l'infini des mesures de Radon $\delta_x * V_\mu$ où $V_\mu = \Sigma\mu^{*n}$ est la mesure de Radon potentiel de μ . Ce comportement sera obtenu sous hypothèse d'étalement grâce à des propriétés sensiblement plus générales du noyau V_μ .
Soit μ une probabilité de Radon sur l'hypergroupe X . On appellera T_μ l'opérateur de convolution sur $L^2_{\mathbb{C}}(X)$, $f \to \mu * f$. Nous noterons $\|\mu\|_{CV_2}$ la norme de T_μ et $\rho(\mu)$ son rayon spectral.

THEOREME 2.2. Soit X un hypergroupe de type (K-S) . On suppose que la probabilité μ sur X est adaptée et étalée. Alors $\rho(\mu) < 1$.

Démonstration : elle s'appuie sur deux lemmes. Le premier est dû à C. Berg et J.P.R. Christensen ; on en trouvera la démonstration dans [Fa] ou [Be][(1)].

LEMME 1. Supposons que $\rho(\mu) = 1$. Il existe alors une suite (f_n) d'éléments de $L^2(X,\sigma)$, telle que pour tout $n \in \mathbb{N}$, $f_n \geq 0$, $\|f_n\|_2 = 1$ et que

$$\lim_{n\to\infty} \int_X \|{}_x f_n - f_n\|_2^2 \, d\mu(x) = 0 .$$

Par le théorème de Plancherel, ceci s'écrit

$$\lim_{n\to\infty} \int_{\hat{X}} |\hat{f}_n(\chi)|^2 \, \gamma(\chi) \, d\varpi(\chi) = 0$$

où on a posé

$$\gamma(\chi) = \int_X |1 - \chi(x)|^2 \, d\mu(x) .$$

On a alors le

LEMME 2. Si la probabilité μ est adaptée et étalée on a

$$\inf_{\chi \in \operatorname{Supp} \varpi} \gamma(\chi) > 0 .$$

Démonstration du lemme 2 :

a) Si X est discret, toute probabilité régulière μ sur X est étalée, $\hat{X}$ est compact, et l'assertion du lemme résulte alors trivialement de l'adaptation de μ : γ est >0 sur $\operatorname{supp} \omega$.

b) X n'est pas discret ; alors si μ est étalée, on a

$$\overline{\lim_{\substack{\chi \in \mathrm{Supp}(\varpi) \\ \chi \to \infty}}} |\hat{\mu}(\chi)| < 1$$

Si tel n'était pas le cas, pour une suite (χ_n) dans $\mathrm{supp}\,\varpi$ tendant vers l'infini dans $\hat{X}$, on aurait $\lim_{n\to\infty} |\hat{\mu}(\chi_n)| = 1$. Mais pour tout entier $p > 0$, on peut écrire (étalement de μ) $\mu^{*p} = \mu_{r,p} + \mu_{s,p}$ où $\mu_{r,p}$ est absolument continue par rapport à σ, où $\mu_{s,p}$ est singulière par rapport à σ et $\lim_{p\to\infty} \|\mu_{s,p}\|_1 = 0$. Par suite $\lim_{p\to\infty} \|\hat{\mu}_{s,p}\|_\infty = 0$. Mais pour tout entier $p > 0$, tout caractère χ de $\hat{X}$, on a

$$|\hat{\mu}(\chi)|^p \leq |\hat{\mu}_{r,p}(\chi)| + |\hat{\mu}_{s,p}(\chi)| .$$

De plus, pour p fixé, on a $\lim_{n\to\infty} |\hat{\mu}_{r,p}(\chi_n)| = 0$. On a ainsi clairement une contradiction.

Si maintenant, on avait $\inf_{\chi \in \mathrm{Supp}\,\omega} \gamma(\chi) = 0$, il existerait une suite (χ_n) d'éléments de $\mathrm{supp}(S(\omega)$ dans $\hat{X}$ telle que, $\lim_{n\to\infty} \chi_n(x) = 1$ p.p. μ, sur $\mathrm{supp}(\mu)$. Ainsi $\lim_{n\to\infty} \hat{\mu}(\chi_n) = 1$. Comme μ est adaptée, cela impose que la suite (χ_n) tend vers l'infini sur $\hat{X}$ (ou une suite extraite) et on vient de voir que c'est impossible compte tenu de l'étalement de μ.

Revenons à la démonstration du théorème : on a

$$\int_{\hat{X}} |\hat{f}_n(\chi)|^2 \gamma(\chi)\, d\varpi(\chi) \geq \alpha \int_{\hat{X}} |\hat{f}_n(\chi)|^2\, d\varpi(\chi)$$

où $\alpha = \inf_{\chi \in \mathrm{Supp}(\varpi)} \gamma(\chi)$. Alors $\lim_{n\to\infty} \|\hat{f}_n\|_2^2 = 0$ et cela contredit le fait que $\|f_n\|_2 = 1$ pour tout $n \in \mathbb{N}$. Le théorème est démontré.

<u>Remarque</u> : Sur un groupe abélien localement compact non discret G, le fait que pour toute probabilité étalée μ on a $\overline{\lim}_{\chi\to\infty} |\hat{\mu}(\chi)| < 1$ était connu : voir par exemple Revuz [Re] (Exerc. 5.18 p. 104).

DEFINITIONS 2.1. i) <u>Soit</u> X <u>un hypergroupe commutatif. Nous dirons qu'une marche aléatoire de loi</u> μ <u>sur</u> X <u>est transitoire si la série de terme général</u> μ^{*n} <u>converge vaguement sur</u> X. <u>Sa somme est alors une mesure de Radon appelée potentiel de</u> μ <u>et noté</u> V_μ.

ii) <u>Soit</u> X <u>un hypergroupe de type</u> (K-S). <u>Pour une marche aléatoire transitoire de loi</u> μ <u>sur</u> X, <u>nous dirons que la mesure de Radon</u> V_μ <u>(potentiel de</u> μ) <u>convole</u> $L^2_{\mathbb{C}}(X)$ <u>si l'équation de Poisson</u> (P) : $\varphi - \varphi * \mu = f$ <u>admet une unique solution</u> φ <u>dans</u> $L^2(X)$ <u>pour toute</u> $f \in L^2_{\mathbb{C}}(X)$:

PROPOSITION 2.1. <u>Soit</u> X <u>un hypergroupe commutatif. Si</u> μ <u>est une probabilité sur</u>

X , <u>on a</u> $\|\mu\|_{CV_2} = \sup_{\chi \in S(\varpi)} |\hat{\mu}(\chi)|$.

<u>Démonstration</u> : Immédiate par transformation de Fourier. Voir aussi [Je] .

THEOREME 2.3. <u>Soit</u> X <u>un hypergroupe de type</u> (K-S) . <u>Alors, pour une marche aléatoire de loi</u> μ <u>adaptée et transitoire sur</u> X , <u>la mesure potentiel</u> V_μ <u>convole</u> $L^2_{\mathbb{C}}(X)$ <u>si et seulement si</u> $\rho(\mu) < 1$.

<u>Démonstration</u> : Si $\rho(\mu) < 1$, la série de terme général $\mu^{*n} * f$ converge dans $L^2_{\mathbb{C}}(X)$ pour toute $f \in L^2_{\mathbb{C}}(X)$ et sa somme définit un opérateur borné sur $L^2_{\mathbb{C}}(X)$ (par le théorème de Banach-Steinhaus) qui prolonge évidemment V_μ . Alors V_μ convole évidemment $L^2_{\mathbb{C}}(X)$ car d'une part, par transformée de Fourier on a $(\delta - \mu) * V_\mu(f) = f$ pour toute $f \in L^2_{\mathbb{C}}(X)$ et d'autre part, l'adaptation de μ donne l'unicité.

Réciproquement, si V_μ convole $L^2_{\mathbb{C}}(X)$, et si μ est adaptée, l'opérateur $\varphi \to \varphi * \mu$ qui est borné sur $L^2_{\mathbb{C}}(X)$ est bijectif. Par le théorème de Banach, il admet un inverse continu sur $L^2_{\mathbb{C}}(X)$; de la sorte on a $\varliminf_{\substack{\chi \to \infty \\ \chi \in S(\varpi)}} |1 - \hat{\mu}(\chi)| > 0$.

Mais cela implique que pour $f \in L^2_{\mathbb{C}}(X)$, la série de terme général $\mu^{*n} * f$ converge dans $L^2_{\mathbb{C}}(X)$ vers un opérateur borné sur $L^2_{\mathbb{C}}(X)$ qui n'est autre que l'opérateur précédent, lequel prolonge à $L^2_{\mathbb{C}}(X)$ la mesure de Radon V_μ .

<u>Remarque</u> :

1) D'après ce qu'on vient de voir, si μ est une probabilité adaptée sur l'hypergroupe X de type (K-S) , l'opérateur

$$f \to \int_{\hat{X}} \frac{\hat{f}(\chi)}{1 - \hat{\mu}(\chi)} \, d\varpi(\chi)$$

est bornée sur $L^2_{\mathbb{C}}(X, \sigma)$ si et seulement si

$$\varliminf_{\substack{\chi \to \infty \\ \chi \in S(\varpi)}} |1 - \hat{\mu}(\chi)| > 0 .$$

Nous savons que, pour toutes les marches aléatoires adaptées transitoires (et en particulier, si μ est étalée), cet opérateur est une mesure de Radon (<u>non bornée</u>). Peut-on décrire cet opérateur dans les autres situations ? (A priori, on sait alors seulement que pour $\varphi \in C_{\mathcal{K}}(X)$, $V_\mu \varphi$ est $[\sigma]$ p.p. finie (et dans $L^2_{\mathbb{C}}(X)$) . Par ailleurs, C. Berg nous a fait aimablement observer que, dans le cas d'un espace riemannien symétrique de rang 1, il n'y a pas d'intégrale singulière du type précédent, voir $[Be]^{(2)}$; le problème de leur existence en rang > 1 est semble-t-il ouvert.

2) Dans [9], H. Chébli démontre à l'aide de théorèmes de structure fins que pour certains hypergroupes de type (K-S) sur $]0, +\infty[$, la condition précédente est

satisfaite par tous les semi groupes markoviens d'opérateurs autoadjoints invariants par translations sur $L^2_{\mathbb{C}}(X)$. Nous ne disposons pas ici de ces théorèmes de structure mais nous retrouvons complètement ces résultats en notant que sur un hypergroupe de type (K-S) , toute marche symétrique adaptée est transitoire.

DEFINITION 2.2. Soit X un hypergroupe de type (K-S) . Soit μ une probabilité régulière adaptée sur X . On dira que la marche aléatoire de loi μ sur X est de potentiel nul à l'infini sur X si les mesures de Radon $\delta_x * V_\mu$ convergent vaguement vers 0 sur X lorsque x tend vers l'infini.

COROLLAIRE 2.3. Soit X un hypergroupe de type (K-S) . Soit μ une probabilité régulière sur X . Si μ est adaptée et étalée, la marche aléatoire de loi μ est de potentiel nul à l'infini sur X .

Démonstration : C'est une conséquence facile d'une application simultanée des théorèmes 2.1. $(\rho(\mu) < 1)$ et 2.3. (V_μ convole $L^2(X)$) .
Si on oublie le théorème 2.1., c'est une conséquence du théorème 2.3. car si μ est étalée on a $\overline{\lim}_{\substack{\chi \to \infty \\ \chi \in \mathrm{Supp}(\varpi)}} |\hat{\mu}(\chi)| < 1$ et si μ est adaptée, la proposition 2.1. implique $\|\mu\|_{CV_2} < 1$ donc V_μ convole $L^2(X)$.

Remarque : Nous ne savons pas si toute marche aléatoire sur un hypergroupe de type (K-S) est de potentiel nul à l'infini. En revanche, les hypergroupes discrets de type (K-S) sont de type (I) au sens de Revuz [Re] . Pour ces hypergroupes, en effet, toute probabilité est étalée.

BIBLIOGRAPHIE

[Je] : Robert I. JEWETT ; Spaces with an abstract convolution of measures : Adv. in Math. 18(1975), 1-101.

[Ga-Ge](1) : Léonard GALLARDO et Olivier GEBUHRER ; Marches aléatoires et analyse harmonique sur les hypergroupes localement compacts : Prépublication IRMA Strasbourg (1985)

[Ga-Ge](2) : Léonard GALLARDO et Olivier GEBUHRER ; Marches aléatoires et hypergroupes : à paraître dans "Expositiones Mathematicae" (1987)

[Fa] : Jacques FARAUT ; Moyennabilité et normes d'opérateurs de convolution : Analyse Harmonique sur les groupes de Lie in Séminaire Nancy-Strasbourg 1973-75 Lecture Notes in Math. n° 497.

[Re] : Daniel <u>REVUZ</u> : Markov Chains, North Holland. (1984) 2e Ed.

[Che] : Houcine CHEBLI : Positivité des opérateurs de translations généralisées associées à un opérateur de Sturm-Liouville et quelques applications à l'analyse harmonique. Thèse d'Etat (1974) Université de Strasbourg.

[Tri] : Khélifa TRIMECHE : Transformation intégrale de Weyl et Théorème de Paley Wiener associés à un opérateur différentiel singulier sur $]0,\infty[$; J. Math. Pures et Appliquées 60 (1981) ; p. 51-98.

[Be](1) : Christian BERG : On the relation between amenability of locally compact groups and the norm of convolution operators. Math. Ann. 149-153 (1974).

[Be](2) : Christian BERG : Dirichlet forms on symmetric spaces, Annales de l'Institut Fourier Tome XXIII Fasc. 1 (1973) p. 135-156.

SOBOLEV INEQUALITIES AND RANDOM WALKS

by Peter Gerl (Salzburg)

1. INTRODUCTION

At present there are known several versions of local limit theorems on free groups ([3], [5], [7], [11]); but they all treat only very special probabilities (either radial or supported by the natural generators). In this paper we reduce the validity of local limit theorems for more general probabilities to the existence of "almost" common eigenfunctions for two probabilities (see section 6). This seems to be an important problem. Our approach uses a recent result of Varopoulos ([9]) and the algebraicity of the Green function ([1], [8], [12]). Although we did not (yet) succeed in deriving a general local limit theorem on free groups in this way we explain the method and give some by-products. They show the role of the norming sequence (λ_x) of reversible probabilities (section 5B) and the possibility of deriving a local limit theorem from a Sobolev inequality (section 5A). A discrete weighted Sobolev type inequality is proved in section 4.

2. THE RESULTS OF VAROPOULOS

Let Γ be a connected (countable) infinite graph and $p(x,y)$ the one-step transition probabilities of a Markov chain on Γ, i.e.

$$p: \text{edges of } \Gamma \to \mathbb{R}$$

with

$$p(x,y) = 0 \quad \text{if } xy \text{ is not an edge of } \Gamma$$

$$p(x,y) \geq 0 \quad \text{for all vertices } x,y \text{ of } \Gamma$$

$$\sum_y p(x,y) = 1$$

(this last sum runs over all edges through x). We will always assume that p is <u>irreducible</u>, i.e. to arbitrary vertices x,y of Γ there are vertices $x = x_0, x_1, \ldots, x_n = y$ such that

$$p(x_{i-1},x_i) > 0 \qquad \text{for} \quad i = 1,\ldots,n$$

and that p is <u>reversible</u>, i.e. there is a function

$$\lambda\colon \text{vertices of } \Gamma \to \mathbb{R}$$

with $\lambda_x > 0$ for all vertices x of Γ and

$$\lambda_x\, p(x,y) = \lambda_y\, p(y,x).$$

We then call (Γ,p) a reversible <u>random walk on Γ.</u>

Let further

$$p^n(x,y) = \text{Prob}\,(x \xrightarrow[\text{edges of }\Gamma]{\text{along } n} y)$$

(= n-step transition probabilites),

$$\|f\|_s = \frac{1}{2}\sum_{x\sim y} \lambda_x\, p(x,y)\,|f(x) - f(y)|,$$

$$\|f\|_D^2 = \frac{1}{2}\sum_{x\sim y} \lambda_x\, p(x,y)\,(f(x) - f(y))^2$$

(where $x \sim y$ means that xy is an edge of Γ with endpoints x,y)

and for $r \geq 1$

$$\|f\|_r^r = \sum_x \lambda_x\, |f(x)|^r,$$

where

$$f \in c_o(\Gamma) = \{\, f\colon \text{vertices of } \Gamma \to \mathbb{R} \mid f \text{ has finite support} \,\}.$$

Then consider the following conditions (they depend in general on the reversible probability p):

$(S_k(p))$: There is a constant $C > 0$ such that

$$\|f\|_{\frac{k}{k-1}} \leq C\,\|f\|_s$$

for all $f \in c_o(\Gamma)$

(= <u>Sobolev inequality</u>),

$(D_k(p))$: There is a constant $C > 0$ such that

$$\|f\|_{\frac{2k}{k-2}} \leq C\,\|f\|_D$$

for all $f \in c_o(\Gamma)$

(= Dirichlet inequality).

The main results in [9] are as follows:

If $k > 2$ then

$$(S_k(p)) \Longrightarrow (D_k(p))$$

and

$$(D_k(p)) \Longleftrightarrow \sup_{x,y} \lambda_y^{-1} p^n(x,y) = O(n^{-k/2})$$

$$(\text{for } n \to \infty).$$

3. A REDUCTION

Let (Γ,p) be a reversible random walk on Γ (as in section 2) and let

$$\sigma_p = \limsup_{n\to\infty} (p^n(x,y))^{1/n}$$

be the spectral radius of p; σ_p is independent of x,y and $0 < \sigma_p \leqq 1$. It is known that

$$p^n(x,y) = O(\sigma_p^{\ n})$$

for $n \to \infty$ ([10]) which implies that $p^n(x,y)$ converges exponentially fast to 0 if $\sigma_p < 1$. To get more precise informations also in this case and to be able to apply the results of section 2 we may proceed as follows:

It is shown in [6] that

$$p * f(x) = [\sum_y p(x,y) f(y)] = \sigma_p f(x)$$

has a positive solution f (i.e. $f(x) > 0$ for all vertices x of Γ). Now put

$$q(x,y) = \sigma_p^{-1} \frac{f(y)}{f(x)} p(x,y).$$

Then q is again a one-step transition probability on Γ, $\sigma_q = 1$ and q is reversible,

$$\mu_x q(x,y) = \mu_y q(y,x)$$

with

$$\mu_x = \lambda_x f^2(x) \qquad (>0)$$

for all vertices x of Γ. Therefore (Γ,q) is a reversible random walk on Γ. So if $(S_k(q))$ or $(D_k(q))$ holds we get from what was said in section 2 that

$$\sup_{x,y} \mu_y^{-1} q^n(x,y) = O(n^{-k/2})$$

or rewritten in terms of p that

$$\sup_{x,y} (\lambda_y f(x)f(y))^{-1} p^n(x,y) = O(\sigma_p^n n^{-k/2}).$$

Of course we would like to get such a result under $(S_k(p))$ (or $(D_k(p))$) but unfortunately there is in general no relation between these conditions and the corresponding conditions for q. In some cases however one can derive good results with this method.

4. A SOBOLEV (DIRICHLET) INEQUALITY ON $\mathbb{N}$

We will derive here a discrete weighted Sobolev (Dirichlet) inequality on $\mathbb{N}$; perhaps the result is known but I could not find it in the literature.

Proposition 1: For all $f: \mathbb{N} \to \mathbb{R}$ with finite support we have

(i) $$\left(\sum_n n^a |f_n|^b\right)^{1/b} \leq \sum_n n^a |f_n - f_{n+1}|$$

for $a > 0$ and $b \geq \frac{a+1}{a}$,

(ii) $$\left(\sum_n n^a |f_n|^c\right)^{1/c} \leq (c+2)\sqrt{2}\left(\sum_n n^a |f_n - f_{n+1}|^2\right)^{1/2}$$

for $a > 1$ and $c \geq \frac{2(a+1)}{a-1}$.

Proof: (i) Suppose that the support of f is contained in $\{1,2,\dots,N\}$, so $f(n) = 0$ for $n \geq N+1$. If we write

$$c_n = f_n - f_{n+1}, \qquad d_n = |c_n|$$

then

$$\begin{aligned} f_1 &= c_1 + c_2 + \dots + c_N \\ f_2 &= \phantom{c_1 +{}} c_2 + \dots + c_N \\ &\dots\dots \\ f_N &= \phantom{c_1 + c_2 + \dots + {}} c_N \end{aligned}$$

and we have to show that

$$\Big(\sum_{n\leq N} n^a (d_n + \dots + d_N)^b\Big)^{1/b} \leq \sum_{n\leq N} n^a d_n .$$

But this follows easily from Minkowski's inequality, because

$$\Big(\sum_{n\leq N} (n^{a/b} d_n + \dots + n^{a/b} d_N)^b\Big)^{1/b} \leq$$

$$\leq \sum_{n\leq N} (1^a + 2^a + \dots + n^a)^{1/b} d_n$$

$$\leq \sum_{n\leq N} n^{(a+1)/b} d_n \qquad \leq \sum_{n\leq N} n^a d_n$$

for $\frac{a+1}{b} \leq a$, i.e. $b \geq \frac{a+1}{a}$.

As the example $f_1 = 1$, $f_n = 0$ for $n \neq 1$ shows this inequality is best possible.

(ii) follows from (i) as in [9] (section 3) if we write $c = \frac{2b}{2-b}$. □

Proposition 2: The estimates for the exponents b,c in proposition 1 are best possible, i.e. if there is a constant $K > 0$ such that for all $f: \mathbb{N} \to \mathbb{R}$ with finite support we have

(i)

$$\Big(\sum_n n^a |f_n|^b\Big)^{1/b} \leq K \sum_n n^a |f_n - f_{n+1}|$$

for some $a > 0$, $b > 0$ then $b \geq \frac{a+1}{a}$

(ii)

$$\Big(\sum_n n^a |f_n|^c \Big)^{1/c} \leq K \Big(\sum_n n^a |f_n - f_{n+1}|^2 \Big)^{1/2}$$

for some $a > 1$, $c > 0$ then $c \geq \frac{2(a+1)}{a-1}$.

<u>Proof:</u> (i) For a positive integer N let

$$f_n = 1 \quad \text{if } n = 1,2,\dots,N$$
$$f_n = 0 \quad \text{if } n > N.$$

Since (i) has to hold for this function we get

$$c_1 N^{(a+1)/b} \leq \Big(\sum_{n \leq N} n^a \Big)^{1/b} \overset{(i)}{\leq} K N^a$$

for some positive constant c_1. Since N was arbitrary we obtain the result.

(ii) We observe at first that

$$N^{-(r+s+1)} \sum_{n=1}^{N} n^r (N-n)^s = \frac{1}{N} \sum_{n=1}^{N} \Big(\frac{n}{N}\Big)^r \Big(1 - \frac{n}{N}\Big)^s \xrightarrow[(N\to\infty)]{} \beta \neq 0,$$

since the last sum is a Riemann sum for the Beta function $B(r+1,s+1)$. Now let N be a positive integer and define

$$f_n = (N-n)^2 \quad \text{for } n = 1,2,\dots,N$$
$$f_n = 0 \quad \text{for } n \geq N.$$

Since (ii) has to hold for this function we get

$$c_1 N^{(a+2c+1)/c} \leq \Big(\sum_{n \leq N} n^a (N-n)^{2c} \Big)^{1/c} \leq$$

$$\leq 2K \Big(\sum_{n \leq N} n^a (N-n)^2 \Big)^{1/2} \leq c_2 N^{(a+2+1)/2}$$

for some positive constants c_1, c_2. Since N was arbitrary we get the result. □

5. RANDOM WALKS ON $\mathbb{N}_0$

A) As a very simple but typical example we consider the random walk on $\mathbb{N}_0 = \mathbb{N} \cup \{0\}$ with transition probabilities

$$p(0,1) = 1, \quad p(n,n+1) = 1-r, \quad p(n,n-1) = r$$

for $n = 1,2,\ldots$ and r fixed with $0 < r < \frac{1}{2}$ (see e.g. [4], Ex. 3.3). Then

$$\sigma_p = 2\,(r\,(1-r))^{1/2} < 1.$$

Proceeding as in section 3 we find: A positive solution f of

$$p * f = \sigma_p f$$

is given by

$$f_n = (1+(1-2r)\,n)\,\left(\frac{r}{1-r}\right)^{n/2} \quad (n = 0,1,\ldots).$$

This gives a new transition probability

$$q(0,1) = 1$$

$$q(n,n+1) = \frac{1+(1-2r)(n+1)}{2\,(1+(1-2r)\,n)}$$

$$q(n,n-1) = \frac{1+(1-2r)(n-1)}{2\,(1+(1-2r)\,n)} \quad (n = 1,2,\ldots)$$

which is reversible $(\mu_n q(n,m) = \mu_m q(m,n))$ with

$$\mu_0 = 1, \quad \mu_n = \frac{(1+(1-2r)n)^2}{1-r} \quad (n = 1,2,\ldots).$$

Therefore we can apply proposition 1 for a = 2; this gives $b \geqq 3/2$ and we obtain $(S_3(q))$. The results of [9] stated at the end of section 2 then yield

$$\sup_{x,y} \mu_y^{-1}\, q^n(x,y) = O\,(n^{-3/2})$$

and the exponent 3 cannot be replaced by any larger number by proposition 2 and the last statement of section 2. Now

$$\sup_{x,y} \mu_y^{-1}\, q^{2n}(x,y) = \sup_x \mu_x^{-1}\, q^{2n}(x,x)$$

and an elementary calculation shows that

$$q^{2n}(0,0) \leqq \sup_x \mu_x^{-1}\, q^{2n}(x,x) \leqq q^{2n}(0,0),$$

so we have equality here. Therefore

$$q^{2n}(0,0) = O(n^{-3/2})$$

and the number 3 in the exponent cannot be replaced by any larger one.

So in terms of the probability p we obtain

$$p^{2n}(0,0) = O(\sigma_p^{\,n}\, n^{-3/2}).$$

A direct calculation or better [12] shows that the Green function

$$G(x,y|z) = \sum_{n\geq 0} p^n(x,y)\, z^n$$

is an algebraic function and $z = \sigma_p^{-1}$ is the only singularity of maximal weight on its circle of convergence. Therefore

$$p^n(x,y) \underset{(n\to\infty)}{\sim} c(x,y)\, \sigma_p^{\,n}\, n^{-q}$$

for a nonnegative rational number q. This and what has been said before imply

$$p^n(x,y) \underset{(n\to\infty)}{\sim} c(x,y)\, \sigma_p^{\,n}\, n^{-3/2}$$

(where n has to be even or odd according to the distance of x and y).

So we have derived a local limit theorem (this is of course well known).

B) As another example consider a random walk on $\mathbb{N}_0$ (which is automatically reversible) with transition probabilities

$$p(n,m) = 0 \quad \text{if} \quad |n-m| \geq 2$$

$$\inf_n p(n,n\pm 1) > 0,$$

$$\lambda_n\, p(n,m) = \lambda_m\, p(m,n)$$

(where $\lambda_n > 0$) and $\lambda_0 = 1$

$$c_1 n^a \leq \lambda_n \leq c_2 n^a$$

for some positive constants c_1, c_2, a $(n = 1,2,\ldots)$. Then proposition 1 implies $(S_k(p))$ for $k = a+1$, since $b = \frac{a+1}{a} = \frac{k}{k-1}$. Therefore we conclude from section 2 that

$$\sup_{x,y} \mu_y^{-1} p^n(x,y) = O(n^{-(a+1)/2})$$

and by proposition 2 that the exponent (a+1)/2) is best possible.

This shows in particular that $\sigma_p = 1$ *and that such a random walk is transient if* $a > 1$.

A typical case is

$$p(0,1) = 1$$

$$p(n,n+1) = \frac{1}{2}\left(1 + \frac{c}{n+c}\right) \qquad (c > 0)$$

$$p(n,n-1) = \frac{1}{2}\left(1 - \frac{c}{n+c}\right) \qquad (n = 1,2,\dots).$$

A short calculation shows that

$$\lambda_n = \frac{2c(2c+1)\dots(2c+n-1)}{n!} \cdot \frac{n+c}{c} \sim \frac{1}{c\Gamma(2c)} n^{2c}$$

so that this random walk is transient for $2c > 1$.

In some cases we can argue as in A) to derive a local limit theorem.

6. A COMMENT ON THE ORIGIN OF THAT ARTICLE

The original intention was to prove a local limit theorem for random walks on free groups. Unfortunately I did not succeed (see the problem at the end of this section). But I still think that this method (the one used in section 5) will give good results and therefore I would like to go a little into details:

Let F_d be the free group with $d \geq 2$ generators $a_1,\dots,a_d$ and let p be a group invariant probability ($p(x,y) = p(zx, zy)$) on F_d with either

(i) support of $p = \{a_1,\ldots,a_d,\ a_1^{-1},\ldots,a_d^{-1}\}$

or

(ii) p is radial ($p(x,y)$ depends only on the length of $x^{-1}y$ written as a reduced word in the generators) and its support generates F_d.

Then ([2], [3], [5], [7], [11])

$$p^n(e,x) \underset{(n\to\infty)}{\sim} c(x)\,\sigma_p^n\, n^{-3/2}$$

(here and also later n is always even or odd according to the length of x). Now one would like to have such a local limit theorem for more general (group invariant) probabilities p'and to get such a result one could argue as follows:

Let f be a positive eigenfunction of p to the eigenvalue σ_p (this exists, see e.g. [6]), i.e.

$$p * f = \sigma_p f\ , \qquad f > 0$$

and define

$$q(x,y) = \sigma_p^{-1}\,\frac{f(y)}{f(x)}\,p(x,y)\ .$$

Then q is again a probability on F_d (but no longer group invariant) with $\sigma_q = 1$. If p has one of the properties (i) or (ii) as before then

$$q^n(x,y) \underset{(n\to\infty)}{\sim} d(x,y)\,n^{-3/2}$$

with $d(x,y) = \frac{f(y)}{f(x)}\,c\,(x^{-1}y)$. If p is reversible,

$$\lambda_x\,p(x,y) = \lambda_y\,p(y,x) \qquad (\lambda_x > 0)$$

then q is reversible too,

$$\mu_x\,q(x,y) = \mu_y\,q(y,x) \qquad (\mu_x > 0)$$

with $\mu_x = \lambda_x f^2(x)$. But then we get from [9] (the last statement in section 2) that $(D_3(q))$ holds, i.e.

$$\left(\sum_x \mu_x |g(x)|^6 \right)^{1/6} \leq c \left(\sum_{x\sim y} \mu_x g(x,y) (g(x) - g(y))^2 \right)^{1/2}$$

for some positive constant c and all functions $g: F_d \to \mathbb{R}$ with finite support. This is a Dirichlet inequality on F_d. If this implied $(D_3(q'))$ for some other reversible probability q' with $\sigma_{q'} = 1$,

$$\left(\sum_x \mu'_x |g(x)|^6 \right)^{1/6} \leq c' \left(\sum_{x\sim y} \mu'_x q'(x,y) (g(x) - g(y))^2 \right)^{1/2},$$

then we could infer from [9] that

$$\sup_{x,y} \mu'^{-1}_y q'^n(x,y) = O(n^{-3/2}).$$

So if p' is related to q' as p to q above we would obtain

$$p'^n(x,y) = O(\sigma^n_{p'} n^{-3/2}).$$

If $(D_3(q)) \Longleftrightarrow (D_3(q'))$ then the exponent 3/2 in $n^{-3/2}$ is best possible. Using this and the property that the Green function

$$G(x,y|z) = \sum_{n\geq 0} p^n(x,y) z^n$$

is an algebraic function if p has finite support (see [1], [8] or [12]) we can conclude as in section 5A) that

$$p'^n(e,x) \underset{(n\to\infty)}{\sim} c'(x) \sigma^n_{p'} n^{-3/2} .$$

The open point in this reasoning in the equivalence of $(D_3(q))$ and $(D_3(q'))$. For this it would be very important to know something about the following

<u>PROBLEM:</u> Under what conditions on two probabilities p, p' (on a free group F_d) do there exist positive solutions f, f' of

$$p * f = \sigma_p f$$
$$p' * f' = \sigma_{p'} f'$$

with $c_1 f \leq f' \leq c_2 f$ for some positive constants c_1, c_2 ?

Of course we can assume $\sigma_p = \sigma_{p'}$ if we replace p by $a\delta_e + (1-a)p$ with $a = (1-\sigma_p)^{-1}(\sigma_{p'} - \sigma_p)$ (if $\sigma_{p'} > \sigma_p$).

I conjecture that the conclusion of the problem is correct if p, p' are symmetric with finite supports (which generate the group F_d). Then we would obtain a local limit theorem for symmetric finitely supported probabilities on F_d of the form

$$p^n(x,y) \sim d(x,y)\, \sigma_p^{\,n}\, n^{-3/2}.$$

T. Steger informed me that he can prove such a result using an entirely different approach.

REFERENCES

[1] AOMOTO, K: Spectral theory on a free group and algebraic curves. J. Fac. Sci. Univ. Tokyo, Sect. IA, 31 (1984), 297-317.

[2] GERL, P.: Eine asymptotische Auswertung von Faltungspotenzen in gewissen Gruppen. Sitzungsber. d. österr. Akad. d. Wiss. 186, (1978), 385-396

[3] GERL, P.: Ein Gleichverteilungssatz auf F_2. In "Probability measures on groups". Lecture Notes in Math. 706 (1979), 126-130

[4] GERL, P.: Continued fraction methods for random walks on $\mathbb{N}$ and on trees. In "Probability measures on groups". Lecture Notes in Math. 1064 (1984), 131-146

[5] PICARDELLO, M.A.: Spherical functions and local limit theorems on free groups. Annali Math. Pur. Apl. 33 (1983), 177-191

[6] PRUITT, W.E.: Eigenvalues of non-negative matrices. Ann. Math. Stat. 35 (1966), 1797-1800

[7] SAWYER, S.: Isotropic random walks in a tree. Z. Wahrscheinlichkeitsth. 42 (1978), 279-292

[8] STEGER, T.: Harmonic analysis for an anisotropic random walk on a homogeneous tree. Ph.D. thesis, Washington Univ., St. Louis, 1985

[9] VAROPOULOS, N.Th.: Isoperimetric inequalities and Markov chains. J. Funct. Anal. 63 (1985), 215-239

[10] VERE-JONES, D.: Ergodic properties of nonnegative matrices I. Pac. J. Math. 22 (1967), 361-386

[11] WOESS, W.: Puissances de convolution sur les groupes libres ayant un nombre quelconque de générateurs. Inst. Elie Cartan 7 (1983), 181-190

[12] WOESS, W.: Context-free languages and random walks on groups. Preprint (1986)

Peter Gerl
Institut für Mathematik
Universität Salzburg
Petersbrunnstraße 19
A-5020 Salzburg / Austria

UNIFORM DISTRIBUTION IN SOLVABLE GROUPS

By

K. Gröchenig, V. Losert, H. Rindler
(Wien)

Dedicated to Professor Elmar Thoma on the occasion of his 60th birthday

Abstract: We study the relations of Hartmann and unitary uniform distribution in solvable groups, in particular in semidirect products of Abelian groups. In every nilpotent group these notions of uniform distribution coincide, but, in general, they are different in solvable groups, as is demonstrated by the motion group of the plane. However, we show that Hartmann and unitary uniform distribution coincide in every solvable analytic group whose Lie algebra has no purely imaginary roots. Finally, we give two six-dimensional solvable analytic groups with the same set of roots, such that the concepts of uniform distribution coincide in one group and differ in the other.

Let G be a locally compact (lc.) group and U a unitary continuous (u.c.) representation of G on a Hilbert space H. Denote the orthogonal projection onto the invariant subspace $\{h \in H: U(x)h = h \quad \forall x \in G\}$ by P_U.

Definition: A sequence (x_n) in G is called uniformly distributed with respect to the representation U, if

$$\lim_{N\to\infty} \frac{1}{N} \sum_{n=1}^{N} U(x_n) = P_U \qquad (*)$$

holds in the strong operator topology.
If (*) holds for all finite-dimensional u.c. representations of G, (x_n) is called Hartmann uniformly distributed (short: u.d.); if (*) holds for all u.c. representations of G, (x_n) is called unitary uniformly distributed (u.u.d.)

Obviously, every u.u.d. sequence is u.d. If G is a compact group or if G is l.c. Abelian, more generally if G is a Moore group where every irreducible u.c. representation is finite dimensional, both notions of uniform distribution coincide trivially. Another example where u.d. and u.u.d. sequences coincide is the ax+b-group ([R2], S.9). Apart from these groups nothing is known about the relations of these two notions. It is the aim of this paper to gain deeper insight into this problem.

Remarks: (i) If G is separable then G admits both u.d. and u.u.d. sequences ([R1], [LR], see also [BR],[B],[V],[KN],[R2]). If either G is almost connected or if G is σ-compact and maximally almost periodic then the converse is also true ([GL]).

We shall complete these results in Corollary 3 of Proposition 3, where we obtain a class of non-separable groups admitting u.u.d. sequences.
(ii) In our definition of uniform distribution it suffices to consider irreducible u.c. representations ($P_U = 0$) because any u.c. representation can be written as a direct integral of irreducible ones. This is possible because for Hartmann uniform distribution one takes only finite-dimensional representations, and in the case of unitary uniform distribution the group is σ-compact (see below) hence any cyclic representation of G acts on a separable Hilbert space and has a separable image.
(iii) If G admits a u.u.d. sequence (x_n), we can conclude that G is σ-compact. For a compact neighbourhood V of the identity the subgroup L generated by (x_n) and V is open and σ-compact. But a u.u.d. sequence cannot be contained in a proper open subgroup of G (take the representation $U(x_n)h = h$ with $h(xL') = 1 \iff x \in L'$ on $l^2(G/L')$ - L' generated by (x_n) - and $P_U h \neq h$), thus L must be all of G.
(iv) Let N(G) be the von Neumann kernel of G, i.e. the intersection of the kernels of all finite-dimensional u.c. representations of G. Then (x_n) is u.d. in G if and only if its image $(\dot{x}_n)$ is u.d. in G/N(G).

The first proposition constitutes an important tool to check the uniform distribution with respect to certain infinite-dimensional representations.

PROPOSITION 1: Let G be a lc. group with a closed normal Abelian subgroup A, U an infinite-dimensional irreducible u.c. representation of G on H whose restriction to A is a multiple of a character χ of A, $U|_A = \chi \otimes Id_H$, and (x_n) a u.d. sequence in G. If $(\dot{x}_n)$ is u.u.d. in G/A, then (x_n) is u.d. with respect to U in G. In particular, if A is a central subgroup then equality of Hartmann and unitary uniform distribution in G/A implies that Hartmann and unitary uniform distribution coincide in G.

Proof: Since $U(a) = \chi(a)\, Id_H$ for all $a \in A$ and some character $\chi \in \hat{A}$ its tensor product with the conjugate representation $\bar{U}$ on $\bar{H}$ - $U \otimes \bar{U}$ on $H \otimes \bar{H}$ is the identity on A and $U \otimes \bar{U}$ gives rise to a u.c. representation of G/A.

By assumption, $(\dot{x}_n)$ is u.u.d. in G/A , therefore

$$\frac{1}{N} \sum_{n=1}^{N} U \otimes \bar{U}\,(x_n) \quad \rightarrow \quad P_{U\otimes\bar{U}}$$

the convergence being in the strong operator topology on $H \otimes \bar{H}$.

We claim that $P_{U\otimes\bar{U}} = 0$. $H \otimes \bar{H}$ is isomorphic to the space of Hilbert-Schmidt-operators on H, the norm on $H \otimes \bar{H}$ corresponds to the Hilbert-Schmidt norm $\| E \|^2 = Tr\, E\cdot E^*$ and $U \otimes \bar{U}$ acts by $(U \otimes \bar{U})(x)\, E = U(x) E\, U(x^{-1})$ (see for example [M2]). Then E is invariant if and only if it commutes with all U(x), $x \in G$. Since U is irreducible it follows that $E = c\cdot Id$; but $c\cdot Id$ is Hilbert-Schmidt if and only if $c = 0$. Thus $P_{U\otimes\bar{U}} = 0$.

Therefore we get for all $h \in H$ in the Hilbert-Schmidt norm

$$\| \frac{1}{N} \sum_{n=1}^{N} U(x_n)h \otimes \overline{U(x_n)h} \|^2 \to 0$$

For $E = \frac{1}{N} \sum_{n=1}^{N} U(x_n)h \otimes \overline{U(x_n)h}$ one calculates

$$E \cdot E^* = \frac{1}{N^2} \sum_{n,m=1}^{N} (U(x_m)h | U(x_n)h) \; U(x_n)h \otimes \overline{U(x_m)h}$$

and

$$\| E \|^2 = \operatorname{tr} E \cdot E^* = \frac{1}{N^2} \sum_{n,m=1}^{N} |(U(x_n)h | U(x_m)h)|^2 \geq \| \frac{1}{N} \sum_{n=1}^{N} U(x_n)h \|^2$$

We conclude that

$$\frac{1}{N} \sum_{n=1}^{N} (U \otimes \overline{U})(x_n) \to 0 \qquad \text{on } H \otimes \overline{H}$$

implies that

$$\frac{1}{N} \sum_{n=1}^{N} U(x_n) \to 0 \qquad \text{on } H$$

and (x_n) is u.d. with respect to the representation U.

Because the restriction of an irreducible u.c. representation of G to a central subgroup is always (a multiple of) a character the second statement of the Proposition follows immediately. □

<u>THEOREM 1:</u> In every nilpotent lc. group the notions of Hartmann and unitary uniform distribution coincide.

<u>Proof:</u> This follows from Proposition 1 by induction on the upper central series of the group. □

<u>COROLLARY to Proposition 1:</u> If (x_n) is a u.u.d. sequence in G and U is a <u>projective</u> infinite-dimensional irreducible u.c. representation of G then (x_n) is u.d. with respect to U.

<u>Proof:</u> Uniform distribution with respect to a projective representation is defined as in Definition (*). Now observe that $U \otimes \overline{U}$ is in fact a representation of G and repeat the proof of Proposition 1 word by word. □

<u>Remark:</u> An analogous statement is by no means true for finite-dimensional projective representations. Uniform distribution with respect to finite-dimensional projective representations can be studied as a separate concept.

Next let us describe briefly the general form of a u.c. representation U of a lc. σ-compact group G according to MACKEY theory (cf. [M1],[M2]). Let N be a closed normal subgroup of type I in G with unitary dual $\hat{N}$ and $g\cdot\lambda$, $g \in G$, $\lambda \in \hat{N}$, the usual action (by inner automorphisms) of G on $\hat{N}$. By decomposing the restriction of the representation U to N into irreducible ones, one can associate a finite, quasi-invariant Borel measure μ on $\hat{N}$ to every U.

$$U|_N = \int_{\hat{N}} \lambda d\mu(\lambda) \quad \text{and} \quad \mu_g(E) := \mu(g\cdot E) = 0 \ \forall g \in G,\ E \subseteq \hat{N} \iff \mu(E) = 0$$

For an irreducible u.c. representation U the associated measure μ is ergodic in the sense that every quasi-invariant measure ν which is absolutely continuous with respect to μ is already equivalent to μ.

Then the representation U is equivalent to a representation V with representation space $L^2(\hat{N},\mu,H)$ for some fixed Hilbert space H: $V(g) = Q'(g)W(g)$ where $W(g)\,f(\lambda) = \rho(g,\lambda)\,f(g^{-1}\cdot\lambda)$, $f \in L^2(\hat{N},\mu,H)$, is the "regular" representation of G on $\hat{N}$, $\rho(g,.) = (d\mu_g/d\mu)^{\frac{1}{2}}$ is the Radon-Nikodym derivative of μ_g with respect to μ and $Q'(g)f(\lambda) = Q(g,\lambda)f(\lambda)$ is a "cocycle", i.e. a measurable function of $G \times \hat{N}$ into the unitary operators on H satisfying $Q(g_1g_2,\lambda) = Q(g_1,\lambda)Q(g_2,g_1^{-1}\cdot\lambda)$.
At last define $\tilde{W}$ to be the representation

$$\tilde{W}(\dot{g})\ h(\lambda) = \rho(g,\lambda)\ h(g^{-1}\cdot\lambda)$$

of G/N on $L^2(\hat{N},\mu)$.

After these preparations we can formulate Proposition 2, which is essential for our further investigations.

<u>PROPOSITION 2:</u> Let G be a σ-compact lc. group with a closed normal subgroup of type I. Assume that Hartmann and unitary uniform distribution coincide in G/N. Then a u.d. sequence (x_n) is u.d. with respect to every irreducible u.c. representation U whose associated measure μ has infinite support and is not equivalent to a finite G-invariant measure on $\hat{N}$.

<u>Proof:</u> (i) Under these assumptions on U the representation $\tilde{W}$ of G/N does not contain the identity representation. If there were an $h \in L^2(\hat{N},\mu)$ with $\tilde{W}(\dot{g})\,h = h$ for all $\dot{g} \in G/N$, the new measure μ' – $d\mu' = |h|^2 d\mu$ – would be a finite G-invariant measure on $\hat{N}$, which is absolutely continuous with respect to μ. Because U is irreducible μ' would have to be equivalent to μ - a contradiction to our assumptions. (The invariance of μ' would be the result of the following calculation: for $E \subseteq \hat{N}$ and $\dot{g} \in G/N$ one has

$$\mu'(\dot{g}\cdot E) = \int_{\hat{N}} c_{g\cdot E}(\lambda)\ |h(\lambda)|^2\ d\mu(\lambda) = \qquad (\lambda \to g\cdot\lambda)$$

$$= \int_{\hat{N}} c_E(\lambda)\ |h(g\cdot\lambda)|^2\ \rho(g^{-1},\lambda)^2\ d\mu(\lambda) =$$

$$= |\tilde{W}(\dot{g}^{-1})h(\lambda)|^2 = |h(\lambda)|^2 \qquad \text{(by invariance of h)}$$

$$= \int_{\hat{N}} c_E(\lambda)\ |h(\lambda)|^2\ d\mu(\lambda) = \mu'(E)$$

(ii) Let (x_n) be u.d. in G and $V \cong U$ the representation described above and $f \in L^2(\hat{N},\mu,H)$. Then we estimate

$$\| \frac{1}{M} \sum_{n=1}^{M} V(x_n)f \|^2_{L^2(\hat{N},\mu,H)} =$$

$$= \int_{\hat{N}} \| \frac{1}{M} \sum_{n=1}^{M} Q(x_n,\lambda)\ f(x_n^{-1}\cdot\lambda)\ \rho(x_n,\lambda) \|^2_H\ d\mu(\lambda) \leq$$

$$\leq \int_{\hat{N}} \{ \frac{1}{M} \sum_{n=1}^{M} \| f(x_n^{-1}\cdot\lambda) \|^2_H\ \rho(x_n,\lambda) \}^2\ d\mu(\lambda) =$$

$$= \left\| \frac{1}{M} \sum_{n=1}^{M} \tilde{W}(\dot{x}_n) \|f\| \right\|^2_{L^2(\hat{N},\mu,H)} \longrightarrow 0$$

Because the image of a u.d. sequence under a group homomorphism is again u.d. and because Hartmann and unitary uniform distribution coincide in G/N this last expression converges to the projection $P_{\tilde{W}} \|f\|$, which is zero by (i). □

We now apply Proposition 2 to semidirect products of Abelian groups where the abstract condition of above gains a more visualizable interpretation in terms of orbits and we obtain sufficient conditions for Hartmann and uniform distribution to coincide.

<u>PROPOSITION 3:</u> Let $G = L \times_\beta M$ be a semidirect product of two σ-compact Abelian groups L and M with separable M. If there does not exist an M-invariant finite measure on $\hat{L}$ with infinite support, then any u.d. sequence is u.u.d in G.

<u>Proof:</u> Because both concepts of uniform distribution coincide in M it suffices to take into consideration those irreducible u.c. representations of G whose associated

measure μ is finite and M-invariant (these are the only representations of G which are not covered by Proposition 2). By our assumption the support of such a μ has to be finite. But then U is finite dimensional and nothing is to be proved. □

If the semidirect product $G = L \times_\beta M$ is regular, i.e. if G has a countable base for its topology and if there are countable many M-invariant Borel sets in $\hat{L}$ that separate any two orbits $O_\chi := \{ \beta_h^* \chi \ (=\chi \circ \beta_h) : h \in M\}$ where $\chi \in \hat{L}$ then one can give a simple sufficient condition for Hartmann and unitary uniform distribution to coincide.

<u>COROLLARY 1:</u> If G is additionally a regular semidirect product and if for all $\chi \in \hat{L}$ the orbit O_χ is either finite or non-compact then u.d. and u.u.d. sequences coincide.

<u>Proof:</u> In this case the measure μ associated to an irreducible representation U is concentrated on one orbit O_χ and every orbit O_χ is homeomorphic to a certain quotient group M/M_χ of M. Therefore μ is equivalent to the image of the Haar measure on M/M_χ. By our assumption it is either finitely supported or infinite (for the non-compact orbits) so that Proposition 3 is applicable. □

<u>COROLLARY 2:</u> Assume that the semidirect product $G = L \times_\beta M$ is regular and that all non-trivial M-orbits on $\hat{L}$ are infinite and non-compact. Then N(G) is isomorphic to L and $(x_n) = (y_n, z_n)$ is u.u.d. in G if and only if (z_n) is u.d. in M. In this case u.u.d. sequences can be contained in proper closed subgroups of G.
(Compare [R2], Thm.9 and [LR])

The <u>proof</u> follows from Corollary 1 and a theorem of WILCOX [W] which describes the von Neumann kernel of a semidirect product $K \times_\beta H$ as $N(G) = S \times_\beta N(H)$ where S is the intersection of the kernels of all finite-dimensional u.c. representations of K with finite H-orbit.

<u>COROLLARY 3:</u> In a semidirect product $G = K \times_\beta D$ of a compact Abelian group K and a discrete countable Abelian group D the concepts of Hartmann and unitary uniform distribution coincide. In particular, if K is chosen to be non-separable one obtains a non-separable group G admitting u.u.d. sequences.

<u>Proof:</u> A non-zero D-invariant measure on $\hat{K}$ with infinite support is necessarily infinite because $\hat{K}$ is discrete and every point in the support of the measure must have the same mass. Now apply Proposition 3. □

EXAMPLE 1 : A concrete group to which these assumptions apply is the following: $G = K \times_\beta \mathbb{Z}$ where $K = \mathbb{Z}_2^{\mathbb{Z}\times I}$ – card I > card $\mathbb{R}$ – is compact and $\mathbb{Z}$ acts as a shift operator, i.e. $\beta(z)(x_{m,i})_{m\in\mathbb{Z},i\in I} = (x_{m+z,i})$. Since the von Neumann kernel of this group is K every sequence $(0,h_n)$ where (h_n) is u.d. in Z is u.u.d. in G. Thus we have obtained the announced example of a non-separable group admitting u.u.d. sequences.

EXAMPLE 2 : The assumptions of Proposition 3 are fulfilled in the ax+b-group $G_0 = \mathbb{R} \times_\beta \mathbb{R}$ with $\beta(x)(y) = e^x y$ (this was already proved in [R2]). Proposition 3 gives another proof that in the Heisenberg group $H = \mathbb{R}^2 \times_\beta \mathbb{R}$ with multiplication $(x,y,z,)\cdot(u,v,w) = (x+u,y+v+zu,z+w)$ u.d. and u.u.d. sequences coincide.

More generally, let $G = \mathbb{R}^n \times_\beta \mathbb{R}$ where $\mathbb{R}$ acts by an nxn-matrix A in the following way: $\beta(t)\,\underline{x} = \exp(tA)\,\underline{x}$ $(t\in\mathbb{R},\ \underline{x}\in\mathbb{R}^n)$. If A does not have an imaginary eigenvalue one easily checks the assumptions of Corollary 2 to hold and therefore every u.d. sequence is u.u.d.

We come to the question to what extent the converse of Proposition 3 is true.

PROPOSITION 4 : Let $G = L \times_\beta M$ be as in Proposition 3 and assume that all non-trivial M-orbits on $\hat{L}$ are infinite and that there exists a finite M-invariant measure μ different from the Dirac measure δ_0 on $\hat{L}$. Then there are u.d. sequences (x_n) in G which are not u.u.d.

Proof: Take any u.d. sequence (z_n) in M, then $(0,z_n)$ is u.d. in G because $N(G) = L$ (by Corollary 2). Consider the u.c. representation U of G

$$U(y,z)\, f(\chi) = \chi(y)\, f(\beta_z^*\chi) \qquad \text{for } f\in L^2(\hat{L},\mu)$$

If $(0,z_n)$ were u.u.d. then we would get for the constant function $1 \in L^2(\hat{L},\mu)$

$$1 = \frac{1}{N}\sum_{n=1}^{N} U(0,z_n)\,1 \quad\rightarrow\quad P_U 1 \quad ,$$

i.e. the constant 1 would have to be U-invariant, especially $\chi(y) = 1$ for all $y \in L$ and μ-almost all $\chi\in\hat{L}$. But this is impossible for $\mu \neq \delta_0$, the Dirac measure, so $(0,z_n)$ cannot be u.u.d. □

EXAMPLE 3 : Consider the groups $D = \mathbb{R}^2 \times_\beta \mathbb{R}$ and $G_1 = \mathbb{R}^2 \times_\beta \mathbb{Z}$, where $\beta(x)$ acts as a rotation by an angle $x \in \mathbb{R}$ and $x\theta$, $x\in\mathbb{Z}$ and θ irrational, in the second case. The non-trivial orbits (resp. their closures) are circles, the Lebesgue measure on a circle is invariant under rotations and Proposition 4 applies.

Another group with differing concepts of uniform distribution is the Mautner group $\mathbb{C}^2 \times_\beta \mathbb{R}$ with $\beta(r)(z_1,z_2) = (e^{ir}z_1, e^{i\alpha r}z_2)$, $r \in \mathbb{R}$, $z_i \in \mathbb{C}$ and irrational α. There is an abundance of finite $\mathbb{R}$-invariant measures on $\hat{\mathbb{C}}^2$ and again Proposition 4 is applicable. This shows that the type of the group is not important for this problem.

Combining the last two propositions we obtain the following criterion:

<u>PROPOSITION 5:</u> Let $G = L \times_\beta M$ be as in Proposition 3 and 4. If all non-trivial M-orbits on $\hat{L}$ are infinite then Hartmann and unitary uniform distribution coincide if and only if there do not exist any finite M-invariant measures $\mu \neq \delta_0$ on $\hat{L}$.

<u>Remark:</u> The condition on the orbits is necessary. The discrete Heisenberg group $\mathbb{Z}^2 \times_\beta \mathbb{Z}$ admits finite $\mathbb{Z}$-invariant measures on $\mathbb{T}^2$ with infinite support (even the Haar measure of $\mathbb{T}^2$ has this property) nevertheless every u.d. sequence is u.u.d. by Theorem 1. For general semidirect products no criterion like Proposition 5 is known to us.

In our last theorem we give a sufficient condition when u.d. and u.u.d. sequences coincide in connected solvable Lie groups G. Let $\mathfrak{g}$ be the Lie algebra of G. By Lie's theorem one can triangulate the adjoint representation <u>ad</u> of $\mathfrak{g}$ on its complexification $\mathfrak{g}_{\mathbb{C}}$. Then the diagonal entries of <u>ad</u> are complex-valued linear functionals λ_i on $\mathfrak{g}$. They are independent of the choice of the triangulation and are called the <u>roots</u> of $\mathfrak{g}$.

<u>THEOREM 2:</u> Let G be a connected solvable Lie group with Lie algebra $\mathfrak{g}$. If $\mathfrak{g}$ has no non-trivial purely imaginary roots then Hartmann and unitary uniform distribution coincide in G. In particular, this applies to all groups of exponential type.

<u>Proof:</u> We proceed by induction over the dimension of $\mathfrak{g}$. The theorem is trivial for dim $\mathfrak{g} = 1$ (and moreover, the examples belonging to Corollary 2 - Ex.2 - show that this is true for dim $\mathfrak{g} \leq 3$).

Now let dim $\mathfrak{g} \geq 2$. By Lie's theorem there is an Abelian ideal $\mathfrak{w} \subseteq \mathfrak{g}$ of dimension 1 or 2 which is the eigenvector of a root $\lambda \in \mathfrak{g}^*$. dim $\mathfrak{w} = 1$ if λ is a real root and dim $\mathfrak{w} = 2$ for a complex-valued root.

Let W be the analytic subgroup of $\mathfrak{w}$ in G. Because $\mathfrak{g}/\mathfrak{w}$ cannot have a purely imaginary root we may apply the induction hypothesis to G/W and conclude that

Hartmann and unitary uniform distribution coincide in G/W.

We now calculate the orbits of G on $\hat{W}$.

If dim $\mathfrak{w}$ = 1 and $\lambda \equiv 0$ then W is a central subgroup of G and Proposition 1 tells us that u.d. and u.u.d. sequences coincide in G.

In the other cases the automorphisms $\exp X \exp V \exp(-X) = \exp(e^{\operatorname{ad} X} V)$, $X \in \mathfrak{g}$, $V \in \mathfrak{w}$, do not leave fixed any discrete subgroup of W, therefore we may identify W with $\mathbb{R}$ or $\mathbb{R}^2$ and via exp also with $\mathfrak{w}$. Then the orbits of G on W and $\hat{W}$ are the same and we have only to compute $e^{\operatorname{ad} X} V$.

If dim $\mathfrak{w}$ = 1 and $\lambda \not\equiv 0$ then $[X,V] = \lambda(x) V$, $X \in \mathfrak{g}$, $V \in \mathfrak{w}$, and $e^{\operatorname{ad} X} V = e^{\lambda(X)} V$. Thus the orbits are $\{0\}$, $\mathbb{R}^+$ and $\mathbb{R}^-$.

If dim $\mathfrak{w}$ = 2 there is a basis V_1, V_2 of $\mathfrak{w}$ such that

$$[X,V_1] = \lambda_1(X)V_1 - \lambda_2(X)V_2 \quad \text{and} \quad [X,V_2] = \lambda_2(X)V_1 + \lambda_1(X)V_2$$

for all $X \in \mathfrak{g}$, where the root $\lambda \in \mathfrak{g}^*$ is given by $\lambda = \lambda_1 + i\lambda_2$ with real-valued functionals λ_1 and λ_2. Observe that $\lambda_1, \lambda_2 \not\equiv 0$ by our assumption. Then for $w \in \mathbb{R}^2$

$$e^{\operatorname{ad} X} w = e^{\lambda_1(X)} \begin{pmatrix} \cos\lambda_2(X) & -\sin\lambda_2(X) \\ \sin\lambda_2(X) & \cos\lambda_2(X) \end{pmatrix} w$$

Keeping $w \in \mathbb{R}^2$ fixed and varying $X \in \mathfrak{g}$ one obtains the G-orbits on $\hat{W}$: if λ_1 and λ_2 are linearly independent one obtains $\{0\}$ and $\mathbb{R}^2 \setminus \{0\}$ as orbits, for linearly dependent λ_i the orbits are spirals. In any case the non-trivial orbits are unbounded and locally closed (i.e. they are open in their closure).

Now consider the restriction of an irreducible u.c. representation U of G to W. The decomposition of U into characters of W gives rise to a finite G-ergodic measure μ on $\hat{W}$. As a consequence of the local closedness of all orbits, μ is in fact concentrated on an orbit ([G]). The case that μ is a point measure is settled by Proposition 1. Since the unbounded orbits are homeomorphic to $\mathbb{R}$ or $\mathbb{R}^2$ where $\mathbb{R}(\mathbb{R}^2)$ is to be understood as a quotient group of G modulo the stabilizing subgroup of the orbit and since consequently every G-invariant measure on the orbit corresponds to the Haar measure (cf. [M1] and the proof of Corollary 1) there can be no finite G-invariant measure with infinite support on $\hat{W}$. Now in this situation Proposition 2 yields that every u.d. sequence in G is u.d. with respect to such an U. The combination of these facts finishes the proof. □

EXAMPLE 4: Example 3 might suggest that even the converse of this theorem is true. However, consider the Lie algebra $\mathfrak{g}$, defined by the non-vanishing brackets of a basis $\{e_0,\dots,e_5\}$:

$$[e_1,e_2] = \epsilon_4, \quad [e_1,e_3] = e_5, \quad [e_0,e_1] = e_1, \quad [e_0,e_2] = -e_2 - e_3$$

$$[e_c,e_3] = e_2 - e_3, \quad [e_0,e_4] = -e_5, \quad [e_0,e_5] = e_4$$

$\mathfrak{g}$ is solvable and its associated simply connected solvable Lie group is the semi-direct product $G = \mathbb{C}^2 \times_\beta G_0$ where G_0 is the ax+b-group and the action β is

$$\beta_{(s,t)}(z_1,z_2) = (e^{(-1+i)t} z_1, \; e^{it} z_2 + s\, e^{(-1+i)t} z_1)$$

with $(s,t) \in G_0$ and $z_1,z_2 \in \mathbb{C}$. The action of an element $(z_1,z_2,s,t) \in G$ on a character $\chi_{a,b}(z_1,z_2) = e^{i\,\mathrm{Re}(\bar{a}z_1 + \bar{b}z_2)}$, $a,b \in \mathbb{C}$ is then given by

$$\beta^*_{(s,t)^{-1}} \chi_{a,b} = \chi_{(e^{(-1-i)t}(a+sb),\, e^{-it}b)}$$

One shows without difficulty that all non-trivial orbits $O_{a,b} = \{\beta^*_{(s,t)}\chi_{a,b},\ (s,t) \in G_0\}$ are unbounded and locally closed. Then the same argument as in the preceding theorem allows to apply Propcsition 2 and together with Example 2 one concludes that Hartmann and unitary uniform cistribution coincide in G. Nevertheless, $\mathfrak{g}$ has the purely imaginary root $\lambda(\Sigma\alpha_i e_i) = i\,\alpha_0$, which even belongs to the common eigenvector $e_4 + i\,e_5$.

The situation changes completely if one replaces G by $G^* = \mathbb{C}^2 \times_{\beta^*} G_0$ where β^* is the dual action of β defined as above. Now the action of G on a character $\chi_{a,b}$ is β and suddenly there appear the compact orbits $O^*_{0,z}$. A slight generalization of Proposition 4 (one dces not need that M is Abelian, it is only necessary that the notions of u.d. coincide in M) shows that Hartmann and unitary uniform distribution differ in G*. Its Lie algebra $\mathfrak{g}^*$, however, with the brackets $[e_0,e_1] = e_1$, $[e_0,e_2] = e_2 - e_3$, $[e_0,e_3] = e_2 + e_3$, $[e_0,e_4] = -e_5$, $[e_0,e_5] = e_4$, $[e_1,e_4] = -e_2$ and $[e_1,e_5] = -e_3$, possesses the same roots as $\mathfrak{g}$.

References

[B] Benzinger, L.: Uniformly distributed sequences in locally compact groups I. Trans.Amer.Math.Soc. 188, 149-165(1974).

[BR] Berg, I.D., Rajagopalan,M., Rubel, L.A.: Uniform distribution in locally compact Abelian groups. Trans.Amer.Math.Soc. 133, 435-446(1968).

[G] Glimm, J.: Locally compact transformation groups. Trans.Amer.Math.Soc. 101, 124-138(1961).

[GL] Gröchenig, K., Losert, V., Rindler, H.: Separabilität, Gleichverteilung und Fastperiodizität. Anz.d.Österr.Akad.Wissensch., math.-naturwiss. Klasse 121,117-119(1984).

[KN] Kuipers, L., Niederreiter, H.: Uniform Distribution of Sequences. New York: John Wiley & Sons.1974.

[LR] Losert, V., Rindler, H.: Uniform distribution and the mean ergodic theorem. Inventiones Math. 50,65-74(1978).

[M1] Mackey, G.W.: Unitary representations of group extensions I. Acta Math.99. 265-311(1958).

[M2] Mackey, G.W.: The Theory of Unitary Group Represention. Chicago Lectures in Math., Chicago, London.1976.

[P] Parthasarathy, K.R.: Multipliers on Locally Compact Groups. Lecture Notes in Math.93, Berlin- Heidelberg-New York:Springer.1969.

[R1] Rindler, H.: Uniform distribution on locally compact groups. Proc.Amer.Math. Soc. 57,130-132(1976).

[R2] Rindler, H.: Gleichverteilte Folgen in lokalkompakten Gruppen. Mh. Math.82, 207-235(1976).

[V] Veech, W.A.: Some questions of uniform distribution. Ann. of Math. 94,125-138(1971).

[W] Wilcox, T.W.: On the structure of maximally almost periodic groups. Math. Scand. 23,221-232(1968).

K. Gröchenig, V. Losert, H. Rindler
Mathematisches Institut der Universität
Strudlhofgasse 4
1090 Wien
Austria

Absolute Continuity and Singularity of Distributions of Dependent Observations: Gaussian and Exchangeable Measures

Arnold Janssen
Universität GSH Siegen,
FB Mathematatik
Hölderlinstr. 3, D-59 Siegen 21

Abstract. A general version of Kakutani's dichotomy theorem concerning absolute continuity and singularity of measures on infinite product spaces is proved. Based on this result it is possible to deduce certain dichotomy properties for product measures, Gaussian measures and exchangeable measures.
On the other hand the same arguments can be used to characterize those exchangeable distributions which are presentable as a mixture of product measures in de Finetti's sense. Furthermore it is proved that each exchangeable probability measure can be decomposed in a part which is presentable and a second part being mutually singular with respect to all presentable distributions.

1. Introduction

The present paper is devoted to dichotomy results concerning absolute continuity and singularity of infinite sequences of dependent random variables. The statistical background can be described as follows.
The model consists of a sequence of possibly dependent observations $X_n : (\Omega, \mathfrak{A}) \to \mathbb{R}$ which are distributed according to probability measures P or Q on Ω. Assume that the distributions of a finite number of observations are absolutely continuous

$$(X_1, \ldots, X_n)(P) \ll (X_1, \ldots, X_n)(Q). \tag{1.1}$$

Then the question arises whether the distributions of the infinite vector $X = (X_1, X_2, \ldots)$ on $\mathbb{R}^\infty$ fulfil the subsequent dichotomy:

$$X(P) \ll X(Q) \quad \text{or} \tag{1.2}$$

$$X(P) \perp X(Q). \tag{1.3}$$

The dichotomy has the following statistical meaning. Absolute continuity (1.2) holds if and only if $(X_1, \ldots, X_n)(P)$ is contiguous with respect to $(X_1, \ldots, X_n)(Q)$, i.e.: for all sequences $A_n \in \mathcal{L}^n$

$$Q(\{(X_1, \ldots, X_n) \in A_n\}) \to 0 \text{ implies } P((\{X_1, \ldots, X_n) \in A_n\}) \to 0. \tag{1.4}$$

Note that contiguity is a standard assumption in asymptotic statistics. On the other hand (1.3) is equivalent to the existence of a consistent sequence φ_n of tests for P against Q where φ_n is based on $(X_1, \ldots, X_n)$, i.e.

$$\max\{\int \varphi_n \, dP, 1- \int \varphi_n \, dQ\} \to 0, \quad n \to \infty. \tag{1.5}$$

It is also possible to interpret (1.3) in terms of statistical experiments. Note that the sequence of binary experiments

$$E_n = \{ \mathbb{R}^n, \mathcal{L}^n, (X_1, \ldots, X_n)(P), (X_1, \ldots, X_n)(Q)\}$$

tends weakly to $\{ \mathbb{R}^\infty, \mathcal{L}^\infty, X(P), X(Q)\}$ in the sense of LeCam [15]. Then E_n tends weakly to the total informative experiment (depending on mutually singular measures) if and only if (1.3) is valid.
If the statistical model consists of independent observations X_n with respect to P and Q the dichotomy (1.2), (1.3) is covered by the famous theorem of Kakutani [14]. If X(P) and X(Q) are Gaussian measures then $X(P) \approx X(Q)$ or $X(P) \perp X(Q)$ follows. This dichotomy is known to be the theorem of Hājek and Feldman, cf. Chatterji and Mandrekar [2] and references therein. In general, martingal convergence arguments can be applied to compare X(P) and X(Q). It is well-known that

$$X(P) \perp X(Q) \text{ iff } \lim_{n\to\infty} \int \left(\frac{d(X_1,\ldots,X_n)(P)}{d(X_1,\ldots,X_n)(Q)}\right)^{1/2} d(X_1,\ldots,X_n)(Q)=0. \tag{1.6}$$

In view of further results for the martingal approach cf. Engelbert and Sirjaev [7]. Conditions in terms of conditional expectations can be found in Kabanov, Lipcer and Sirjaev [13].

It is the aim of this present paper to compare product measures X(P) with distributions X(Q) belonging to the following classes of probability measures: Gaussian measures, exchangeable measures and infinite convolution products of probability measures.
Let us mention the meaning of presentable exchangeable measures Q. Consider the following model. First a distribution ϕ is selected by a prior distribution μ lying on the set of probability measures. Then the sample is choosen according to the countable product measure ϕ^∞, i.e.

$$Q = \int \phi^\infty \, d\mu(\phi). \tag{1.7}$$

2. Preliminaries

In this section the notation is introduced. Let X(P) denote the image measure of a measure P with respect to a measurable function X. Suppose that μ and ν are measures on a measurable space $(S, \mathfrak{F})$. Then μ is said to be absolutely continuous with respect to ν if $\nu(A)$ implies $\mu(A) = 0$, $A \in \mathfrak{F}$, write $\mu \ll \nu$. Define $\mu \approx \nu$ iff $\mu \ll \nu$ and $\nu \ll \mu$. The measures μ and ν are called to be mutually singular if there exists a set $A \in \mathfrak{F}$ such that $\mu(A) = 0$ and $\nu(A^c) = 0$ where A^c denotes the complement of A. Let $P_{|A}$ denote the restriction of P on a measurable set A. The countable product of a set S is defined by S^∞. Let P^n be the product of n copies of the measure P and let ε_x and λ be the one point measure at x, the Lebesgue measure respectively. $\|\mu\|$ denotes the total mass of a finite measure μ.

3. A version of Kakutani's theorem for dependent random variables

In the present part Kakutani's theorem [14] concerning absolute continuity and singularity of product measures is generalized. It turns out that under suitable assumptions the same dichotomy (1.2), (1.3) carries over if X(Q) is not to far away from the product of certain marginal distributions. Note that (1.6) uniquely determines the behaviour of the distributions μ and ν in all cases if the dichotomy $\mu \ll \nu$ or $\mu \perp \nu$ is proved.
Let $(S_n, \mathfrak{F}_n)$ be measurable spaces,

$$(S_o, \mathfrak{F}) := \left(\prod_{n=1}^{\infty} S_n, \bigotimes_{n=1}^{\infty} \mathfrak{F}_n \right) \text{ and define}$$

$$p_n : S_o \to S_1 \times \ldots \times S_n, \quad \Pi_n : S_o \to \prod_{k=n+1}^{\infty} S_k$$

to be the canonical projections on the coordinates of S_o.

<u>(3.1) Theorem</u>

Let μ and ν be probability measures on $(S_o, \mathfrak{F})$.

a) Let $\mu = \bigotimes_{n=1}^{\infty} \mu_n$ be a product measure

of probability measures μ_n on $(S_n, \mathfrak{F}_n)$. Suppose that the following conditions are satisfied for an infinite number of positive integers k:

(i) $p_k(\mu) \ll p_k(\nu)$.

(ii) $p_k(\nu) \otimes \Pi_k(\nu) \ll \nu$.

Then either $\mu \ll \nu$ or $\mu \perp \nu$.

b) Let μ be arbitrary and k a positive integer such that

(iii) $\mu \ll p_k(\mu) \otimes \Pi_k(\mu)$

and (i) is satisfied for k.
Then

(iv) $p_k(\nu) \otimes \Pi_k(\nu) \perp \nu$

implies $\mu \perp \nu$.

<u>Proof:</u> a) Let $\mu = \mu_a + \mu_b$ denote Lebesgue's decomposition of μ with respect to ν such that $\mu_a \ll \nu$ and $\mu_b \perp \nu$. Suppose that (i) and (ii) are satisfied for k. Define

$$\rho_k = \mu_1 \otimes \ldots \otimes \mu_k, \tag{3.1}$$

$$\Pi_k(\mu) = \beta_k^{(1)} + \beta_k^{(2)} \tag{3.2}$$

such that $\beta_k^{(1)} \ll \Pi_k(\nu)$, $\beta_k^{(2)} \perp \Pi_k(\nu)$.
Then

$$\rho_k \otimes \beta_k^{(1)} \ll p_k(\nu) \otimes \Pi_k(\nu) \ll \nu \tag{3.3}$$

follows and

$$\rho_k \otimes \beta_k^{(2)} \perp \nu \tag{3.4}$$

since $\Pi_k(\rho_k \otimes \beta_k^{(2)}) \perp \Pi_k(\nu)$. Therefore

$$\mu_a = \rho_k \otimes \beta_k^{(1)} \quad \text{and} \quad \|\mu_a\| = \|\beta_k^{(1)}\|. \tag{3.5}$$

Let $A \in p_m^{-1}(\mathfrak{F}_1 \otimes \ldots \otimes \mathfrak{F}_m)$. Choose $k \geq m$ such that the assumptions are fulfilled. Then by (3.5)

$$\mu_a(A) = \mu(A)\,\|\mu_a\|. \tag{3.6}$$

Note that (3.6) holds on $\bigcup_{m\in\mathbb{N}} p_m^{-1}(\mathfrak{F}_1 \otimes \ldots \otimes \mathfrak{F}_m)$.

Therefore $\mu_a = \|\mu_a\|\mu$ and $\|\mu_a\| = 0$ or $\mu = \mu_a$ follows.

b) Choose a measurable set $A_k \subset \prod_{j=1}^{\infty} S_{k+j}$

such that

$$\Pi_k(\mu)_{|A_k} \ll \Pi_k(\nu) \text{ and } \Pi_k(\mu)_{|A_k^c} \perp \Pi_k(\nu). \tag{3.7}$$

Define $B_k = S_1 \times \ldots \times S_k \times A_k$. Then by (iii)

$$\mu_{|B_k} \ll (p_k(\mu) \otimes \Pi_k(\mu))_{|B_k} \ll p_k(\nu) \otimes \Pi_k(\nu) \tag{3.8}$$

follows. Hence assumption (iv) implies

$$\mu_{|B_k} \perp \nu. \tag{3.9}$$

On the other hand the second part of (3.7) proves that

$$\mu_{|B_k^c} \perp \nu \tag{3.10}$$

and the result follows. □

(3.2) Remarks

a) The assumption (ii) can not be canceled out. Choose two different equivalent probability measures P and Q on $\mathbb{R}$, $\mu = P^\infty$. Then there exists a set $A \in \mathcal{L}^\infty$ such that $0 < \mu(A) < 1$. Define $\nu = {}^1/_2(Q^\infty + \mu_{|A/\mu(A)})$

Then (3.1) (i) is satisfied but neither $\mu \ll \nu$ nor $\mu \perp \nu$.

b) If we choose $A = \mathbb{R}^\infty$ in case of the example above then $p_k(\mu) \approx p_k(\nu)$ but μ and ν are not equivalent. This example shows that the sign "<<" in theorem (3.1) can not be replaced by "≈".

Theorem (3.1) includes Kakutani's theorem [14]. Moreover the result can be used to compare product measures μ with various probability measures. First note that the theorem can be applied to Gaussian measures ν on $\mathbb{R}^\infty$. Recall that a probability measure ν on $\mathbb{R}^\infty$ is said to be Gaussian if $p_n(\nu)$ is a normal or degenerate distribution on $\mathbb{R}^n$ for each $n \in \mathbb{N}$.

(3.3) Examples

a) Let ν be a Gaussian measure on $\mathbb{R}^\infty$. Then either $p_n(\nu) \otimes \Pi_n(\nu) \approx \nu$ or $p_n(\nu) \otimes \Pi_n(\nu) \perp \nu$ by the dichotomy theorem of Hájek and Feldman, for references see Chatterji and Mandrekar [2], IV. In case of (3.1) (i) the dichotomy $\mu = \bigotimes_{i=1}^{\infty} \mu_i \perp \nu$ or $\mu \ll \nu$ follows. Stronger results were proved in the papers of Fernique [8], [9] by completely different arguments. For related results compare with Chatterji and Ramaswamy [3] and Ramaswamy [16].

b) Let ν_1, ν_2 be Borel probability measures on G^∞ where G is a second countable locally compact group. If ν_1, ν_2 satisfies (3.1) (ii) both then also the convolution product $\nu_1 * \nu_2$.

c) A probability measure ν satisfying (3.1) (iv) for one k is mutually singular with respect to all product measures. This result is obviously also true if condition (i) is violated.
The assumption (3.1) (ii) can be checked for an infinite convolution product ν.

(3.4) Theorem

Let $X_n : (\Omega, \mathfrak{A}, P) \to \mathbb{R}$ be independent random variables such that $X = \sum_{n=1}^{\infty} X_n a_n$ is weakly convergent in $\mathbb{R}^{\infty}$, $a_n \in \mathbb{R}^{\infty}$. Suppose that $X_n(P) \approx \lambda$ for all $n \in \mathbb{N}$ and $p_k(X(P)) \approx \lambda^k$ for a fixed $k \in \mathbb{N}$. Then the following conditions are equivalent.

a) $p_k(X(P)) \otimes \Pi_k(X(P)) \approx X(P)$.

b) The set $A(X(P)) = \{y \in \mathbb{R}^{\infty} : X(P) * \varepsilon_y \approx X(P)\}$ contains $\mathbb{R}^k \times \{0\}^{\mathbb{N}}$.

c) $X(P) * i_k \circ p_k(X(P)) \approx X(P)$

where $i_k : \mathbb{R}^k \to \mathbb{R}^{\infty}$ is defined by $i_k(x_1,\ldots,x_k) = (x_1,\ldots,x_k,0,\ldots)$.

Proof:

c) <=> a) Choose another sequence of random variables Y_n such that $(X_n, Y_n, n \in \mathbb{N})$ are independent with distribution $X_n(P) = Y_n(P)$. Define $j_n : \mathbb{R}^{\infty} \to \mathbb{R}^{\infty}$, $j_n((x_k)_k) = (y_k)_k$ where $y_k = x_{k-n}$ for $k > n$ and $y_k = 0$ otherwise.

Put $X = \sum_{n=1}^{\infty} X_n a_n$, $Y = \sum_{n=1}^{\infty} Y_n a_n$. Then

$$Z = i_k \circ p_k(X) + j_k \circ \Pi_k(Y) \tag{3.11}$$

has the distribution $p_k(X(P)) \otimes \Pi_k(X(P))$.

Note that the following sets have P - probability 1.

$$M_1 = \{\omega : \sum_{n=1}^{m} a_n X_n(\omega) \text{ converges in } \mathbb{R}^{\infty}\},$$

$$M_2 = \{\omega : \sum_{n=1}^{m} a_n Y_n(\omega) \text{ converges in } \mathbb{R}^{\infty}\}.$$

In the sequel let us always choose $\omega \in M_1 \cap M_2$.

Define

$$V : \Omega \to \mathbb{R}^\infty, \quad V = (Y_n)_n \tag{3.12}$$

Let A be a Borel subset of $\mathbb{R}^\infty$. Then

$$0 = Y(P) * i_k \circ p_k(Y(P))(A)$$
$$= \int P(i_k \circ p_k(X) + \sum_{n=1}^{\infty} y_n a_n \in A) \; dV(P)(y_n)_n$$

iff

$$P(i_k \circ p_k(X) + \sum_{n=1}^{\infty} y_n a_n \in A) = 0 \quad V(P) \text{ a.e. .} \tag{3.13}$$

On the other hand (3.11) implies

$$Z = Y + i_k \circ p_k(X) - i_k \circ p_k(Y). \tag{3.14}$$

Hence

$$0 = P(Z \in A)$$
$$= \int P(\sum_{n=1}^{\infty} y_n a_n + i_k \circ p_k(X) - i_k \circ p_k(\sum_{n=1}^{\infty} y_n a_n) \in A) \, dV(P)(y_n)_n \tag{3.15}$$

Note that by our assumptions

$$0 = P(p_k(X) - p_k(\sum_{n=1}^{\infty} y_n a_n) \in i_k^{-1}(A - \sum_{n=1}^{\infty} y_n a_n))$$

iff $0 = P\,(p_k(X) \in i_k^{-1}(A - \sum_{n=1}^{\infty} y_n a_n))$.

Thus (3.15) is equivalent to

$$0 = P(i_k \circ p_k(X) \in A - \sum_{n=1}^{\infty} y_n a_n) \quad V(P) \text{ a.e. .} \tag{3.16}$$

If we now compare (3.13) and (3.16) the equivalence follows.

a) => b) and b) => c) are obvious since
$i_k \circ p_k(X(P))$ is concentrated on $\mathbb{R}^k \times \{0\}^{\mathbb{N}}$. □

Let V denote the subvector space of $\mathbb{R}^\infty$ which is generated by $\{a_n : n \in \mathbb{N}\}$. Note that always $V \subset A(X(P))$ holds. Thus we obtain a sufficient condition for the validity of Theorem (3.4) a) in terms of the coefficients a_n of X.

4. Exchangeability and absolute continuity

Let $(S^\infty, \mathfrak{F}^\infty)$ be the countable product space of a measurable space $(S, \mathfrak{F})$ and let ϕ^∞ be the countable product measure of a probability measure ϕ on $(S, \mathfrak{F})$. By definition $\mathcal{S}$ denotes the group of finite permutations Π of the coordinates of S. A probability measure P on $(S^\infty, \mathfrak{F}^\infty)$ is said to be exchangeable if P is invariant under all $\Pi \in \mathcal{S}$, i.e. $\Pi(P) = P$. The concept of exchangeability goes back to B. de Finetti, for more information cf. Dubins and Freedman [5], Freedman [10], Dubins [6]. Let $\mathcal{M}_1(S)$ denote the set of all probability measures on $(S, \mathfrak{F})$. Then $\mathfrak{F}^*$ denotes the σ-field on $\mathcal{M}_1(S)$ generated by the sets $\{\phi \in \mathcal{M}_1(S) : \phi(F) < t\}$, $F \in \mathfrak{F}$, $t \in \mathbb{R}$. Subsets of $\mathcal{M}_1(S)$ are always equipped with the trace σ-field of $\mathfrak{F}^*$. For more details concerning $\mathfrak{F}^*$ cf. Dubins and Freedman [4]. If $\mathfrak{F}$ is the Baire σ-fields on a completely regular space S then $\mathfrak{F}^*$, viewed on the set of all bounded Baire measures, is just the Baire σ-field with respect to the weak topology, cf. [11], Satz 2.3.

Let μ be a prior probability measure on $(\mathcal{M}_1(S), \mathfrak{F}^*)$. Then the mixture

$$P_\mu(A) = \int_{\mathcal{M}_1(S)} \phi^\infty(A)\, d\mu(\phi), \quad A \in \mathfrak{F}^\infty \tag{4.1}$$

is exchangeable. Conversely, an exchangeable P is called presentable if P admits a disintegration (4.1). The mixing measure μ is unique if P is presentable, cf. Dubins and Freedman [5]. It is known that each exchangeable $P \in \mathcal{M}_1(S^\infty, \mathfrak{F}^\infty)$ is presentable if S is a compact space equipped with the Baire σ-field. In general, examples of exchangeable measures are known being not presentable, cf. Dubins and Freedman [5].

Let $\mathcal{T}_\infty := \bigcap_{n=1}^\infty \Pi_n^{-1}(\mathfrak{F}^\infty)$ be the tail σ-field, Π_n is the projection introduced in part 3.

<u>(4.1) Lemma</u> (Breiman, LeCam, Lourraine Schwartz [1])

Define $\mathcal{T} := \{\{\phi \in \mathcal{M}_1(S) : \phi^\infty(D) = 1\} : D \in \mathcal{T}_\infty\}$.
Then $\mathcal{T}$ is a σ-field fulfilling $\mathfrak{F}^* = \mathcal{T}$.

<u>Proof:</u> cf. H. Strasser [18], p. 16. □

The Lemma can be used to compare presentable exchangeable measures.

(4.2) Theorem

Let P_{μ_i} be presentable exchangeable probability measures satisfying (4.1) for i = 1,2. Suppose that $\mu_1 = \mu_a + \mu_b$ denotes the Lebesgue decomposition of μ_1 with respect to μ_2 such that $\mu_a << \mu_2$, $\mu_b \perp \mu_2$. Then

$$\int \phi^\infty \, d\mu_a(\phi) << P_{\mu_2} \quad \text{and} \quad \int \phi^\infty \, d\mu_b(\phi) \perp P_{\mu_2}.$$

Especially: $P_{\mu_1} \perp P_{\mu_2}$ iff $\mu_1 \perp \mu_2$.

Proof: It is sufficient to show that $\mu_1 \perp \mu_2$ implies $P_{\mu_1} \perp P_{\mu_2}$. The rest of the proof is trivial. Suppose that $\mu_1(A) = 0$ and $\mu_2(A) = 1$ for some $A \in \mathfrak{F}^*$. Then by Lemma (4.1) there exists a tail event $D_A \in \mathcal{T}_\infty$ such that $A = \{\phi : \phi^\infty(D_A) = 1\}$. Hence $A^c = \{\phi : \phi^\infty(D_A) = 0\}$. Thus

$$P_{\mu_1}(D_A) = \int_{A^c} \phi^\infty(D_A) \, d\mu_1(\phi) = 0 \quad \text{and}$$

$$P_{\mu_2}(D_A) = \int_A \phi^\infty(D_A) \, d\mu_2(\phi) = 1 \quad \text{is proved. } \square$$

If $\mathfrak{F}$ is countably generated then the result is known, cf. Dubins [6]. The theorem applies to the following example.

(4.3) Example

Let $N(o,t)$ be the normal law with variance $t > 0$ on $\mathbb{R}$ and suppose that μ is a one-sided stable distribution on $\mathbb{R}_+$ with index $p \in (0,1)$ having the Laplace transform $L(t) = \exp(-t^p)$.

Then $P = \int_{\mathbb{R}_+} (N(o,t))^\infty \, d\mu(t)$ is a symmetric stable distribution on $\mathbb{R}^\infty$ with index 2p. Then for $s > 0$

$$P * (N(o,s))^\infty = \int_{\mathbb{R}_+} (N(o,t+s))^\infty \, d\mu(t) << P$$

but the measures are not equivalent. Note that all finite dimensional marginal distributions are equivalent. $P * (N(o,s))^\infty$ is selfdecomposable. This is an example showing that the convolution product of two symmetric stable distributions is not always equivalent or mutually singular with respect to a further symmetric stable distribution. The conjecture of Chatterji and Ramaswamy [3] concerning Hâjek's and Feldman's dichotomy for symmetric stable laws seems to be still open.

The next lemma shows that the assumptions of theorem (3.1) are fulfilled for certain exchangeable measures.

(4.4) Lemma

Let $\mathfrak{F}$ be countably generated. For $\phi_0 \in \mathcal{M}_1(S)$ define $\mathcal{M}_{\phi_0} = \{ \phi \in \mathcal{M}_1(S) : \phi \approx \phi_0\}$. If Q is defined by

$$Q = \int_{\mathcal{M}_{\phi_0}} \phi^\infty \, d\mu(\phi) \tag{4.2}$$

for some probability measure μ on $(\mathcal{M}_{\phi_0}, \mathfrak{F}^* \cap \mathcal{M}_{\phi_0})$ then

$$p_n(Q) \otimes \Pi_n(Q) \approx Q \quad \text{and} \tag{4.3}$$

$$p_n(Q) \approx \phi_0^n \quad \text{for each} \quad n \in \mathbb{N} \tag{4.4}$$

Proof: It is only necessary to prove (4.3). Note that

$$\mathcal{M}_{\phi_0} \times \mathfrak{F}^n \to [0,1] \tag{4.5}$$

$$(\phi, B) \to \phi^n(B)$$

is a stochastic kernel. It is well-known that there is a product measurable function f on $\mathcal{M}_{\phi_0} \times S^n$ such that

$$\frac{d\phi^n}{d\phi_0^n}(x) = f(\phi, x). \tag{4.6}$$

Use standard martingal arguments. Fubini's theorem implies

$$\mu \otimes \phi_0^n(\{f = 0\}) = 0 \tag{4.7}$$

since $\phi_0^n(\{ \frac{d\phi^n}{d\phi_0^n} = 0\}) = 0$ for all ϕ. Define

$$A_x := \{y \in S^\infty : (x,y) \in A\} \quad \text{for} \quad x \in S^n,\ A \subset S^\infty. \tag{4.8}$$

If $A \in \mathfrak{F}^{\infty}$ then

$$(\phi, x) \to \phi^{\infty}(A_x) \tag{4.9}$$

is product measurable on $\mathcal{M}_{\phi_0} \times S^n$. This is a property of the stochastic kernel $(\phi, B) \to \phi^{\infty}(B)$. Thus

$$Q(A) = \iint \phi^{\infty}(A_x)\, f(\phi, x)\, d\phi_0^n(x)\, d\mu(\phi) \tag{4.10}$$
$$= \iint \phi^{\infty}(A_x)\, f(\phi, x)\, d\mu(\phi)\, d\,\phi_0^n(x)\,.$$

On the other hand

$$\phi_0^n \otimes \Pi_n(Q)(A) = \int \Pi_n(Q)(A_x)\, d\phi_0^n(x) \tag{4.11}$$
$$= \iint \phi^{\infty}(A_x)\, d\mu(\phi)\, d\,\phi_0^n(x).$$

In order to prove (4.3) it is enough to show

$$Q(A) = 0 \quad \text{iff} \quad \phi_0^n \otimes \Pi_n(Q)(A) = 0.$$

In view of (4.7) the function $\phi^{\infty}(A_x)\, f(\phi, x)$ vanishes $\mu \otimes \phi_0^n$ almost everywhere (a.e.)

iff $\phi^{\infty}(A_x)$ vanishies $\mu \otimes \phi_0^n$ (a.e.). □

For more information concerning absolute continuity of mixed measures compare with Skorohod [17], § 18.

<u>(4.5) Example</u>

Let G be a second countable locally compact group with left Haar measure ω.

a) Suppose that $P = \bigotimes_{n=1}^{\infty} P_n$ is the product of probability measures $P_n \ll \omega$. Assume that $\phi_0 \in \mathcal{M}_1(G)$ has strictly positive ω - density and let μ be a prior distribution on G. Define $Q = \int_G (\varepsilon_x * \phi_0)^{\infty}\, d\mu(x)$ to be the mixture of the translation family $\{(\varepsilon_x * \phi_0)^{\infty} : x \in G\}$, $*$ denotes the convolution. Then either $P \ll Q$ or $P \perp Q$.

b) Suppose that $G = \mathbb{R}$. Then Q can be substituted by

$\int_{\mathbb{R}} (N_x)^\infty \, d\mu(x)$ where $x \to N_x$ is a kernel of normal distributions. The dichotomy result of a) follows.

5. Miscellaneous results

In this section it will be shown that lemma (4.1) can be used to prove further results for exchangeable measures and to obtain new proofs of known results. Our first attention is devoted to the representation of exchangeable measures which has been studied by Dubins and Freedman [5].

(5.1) Lemma

Suppose that $P \in \mathcal{M}_1(S^\infty)$ is exchangeable. Then the following assertions are equivalent.

a) P is presentable, (4.1).

b) Each tail event $D \in \mathfrak{T}_\infty$ such that $\{\Phi : \Phi^\infty(D) = 1\} = \emptyset$ is a P - null set.

Proof: a) => b) follows from (4.1).

b) => a). According to lemma (4.1) there exists for each $A \in \mathfrak{F}^*$ a tail event $D_A \in \mathfrak{T}_\infty$ such that $A = \{\phi : \phi^\infty(D_A) = 1\}$. Define

$$\mu(A) = P(D_A) \tag{5.1}$$

Let us prove that μ is well defined. Suppose that A admits a second representation $A = \{\phi : \phi^\infty(D) = 1\}$ for some $D \in \mathfrak{T}_\infty$.

Then $\{\phi : \phi^\infty(D_A \Delta D) = 1\} = \emptyset$ and $P(D_A \Delta D) = 0$ follows, (Δ is the symmetric difference). Next we prove that μ is additive. Therefore choose two disjoint sets A, B of $\mathfrak{F}^*$ and corresponding sets $D_A, D_B \in \mathfrak{T}_\infty$. Then

$$\mu(A) + \mu(B) = P(D_A) + P(D_B) = P(D_A \cup D_B) + P(D_A \cap D_B) \tag{5.2}$$

follows. Recall that $A \cup B = \{\phi : \phi^\infty(D_A \cup D_B) = 1\}$, $\{\phi : \phi^\infty(D_A \cap D_B) = 1\} = \emptyset$. Thus the result follows from

$$\mu(A \cup B) = P(D_A \cup D_B) \cdot \tag{5.3}$$

Finally it is proved that μ is σ - additive. Without restriction we may assume $D_{A_n} \downarrow$ where $A_n \downarrow \emptyset$ in $\mathfrak{F}^*$. Hence

$$\lim_{n\to\infty} \mu(A_n) = P\left(\bigcap_{n=1}^{\infty} D_{A_n} \right). \tag{5.4}$$

Since $\{\phi : \phi^{\infty} (\bigcap_{n=1}^{\infty} D_{A_n}) = 1\} = \emptyset$ the set $\bigcap_{n=1}^{\infty} D_{A_n}$ is a P - null set. Hence μ is a probability measure on $(\mathcal{M}_1(S), \mathfrak{F}^*)$. Define $Q = \int \phi^{\infty} d\mu$. Then $Q|_{\mathcal{T}_\infty} = P|_{\mathcal{T}_\infty}$ follows. It is well-known that two exchangeable probability measures are equal if they coincide on the tail σ - field, cf. Blum [0], p. 174. Thus the proof is finished. □

<u>(5.2) Theorem</u>

Each exchangeable $P \in \mathcal{M}_1(S^{\infty})$ can uniquely be decomposed in a convex combination $P = \alpha P_1 + (1-\alpha) P_2$, $0 \le \alpha \le 1$, such that the following conditions are satisfied.

(i) P_1, P_2 are exchangeable probability measures.

(ii) P_1 has a representation $P_1 = \int \phi^{\infty} d\mu$.

(iii) P_2 is mutually singular with respect to all exchangeable probability measures Q being presentable or $\alpha = 1$.

<u>Proof:</u> Let $\mathcal{R}$ denote the σ - ring $\mathcal{R} = \{D \in \mathcal{T}_\infty : \{\phi : \phi^{\infty}(D) = 1\} = \emptyset\}$. Choose $D_0 \in \mathcal{R}$ such that $P(D_0) = \sup\{P(D) : D \in \mathcal{R}\}$,

$$\tilde{P}_2 = P_{|D_0}, \quad \tilde{P}_1 = P - \tilde{P}_2 \tag{5.5}$$

Then $\tilde{P}_1, \tilde{P}_2$ are exchangeable and $\tilde{P}_1$ vanishes on $\mathcal{R}$. If $\tilde{P}_1 \neq 0$ then

$$P_1 = (P_1(S^{\infty}))^{-1} \tilde{P}_1 \tag{5.6}$$

satisfies the conditions of Lemma (5.1) and it is presentable. If Q is any presentable measure then $Q(D_0) = 0$ follows and $P - P_1$ is mutually singular with respect to Q. □

(5.3) Example

The exchangeable probability measures P being mutually singular with respect to all presentable Q have the following meaning in statistics. Consider the test problem {P} against $\{\phi^\infty : \phi \in \mathcal{M}_1(S)\}$. (P against independence of the one dimensional marginal distributions).
Then the following assertions are equivalent:

a) P is mutually singular with respect to all presentable probability measures Q.

b) For each prior distribution μ on $\mathcal{M}_1(S)$ there exists a sequence of tests φ_n based on p_n (the first n observations) such that

$$\lim_{n\to\infty} \int \varphi_n \, dP = 0$$

and the functions $\phi \to \int \varphi_n \, d\phi^\infty$ converge to 1 in $L_1(\mu)$.

(5.4) Remarks

a) If $\mathfrak{F}$ is countably generated then theorem (5.2) can be deduced from theorem (1.4) of Dubins [6].

b) Lemma (4.1) and (5.1) directly imply the following assertions of Dubins and Freedman [5], (3.4). If $P = \int \phi^\infty \, d\nu(\phi)$ for a content ν then

(i) P determines ν uniquely,
(ii) P is countably additive iff ν is.

c) In connection with finitely additive set functions the σ - field generated by $\mu \to \mu(A)$ on the set of contents has been studied in [11] by the author.
Finally let us make a didactic note. Dichotomy results are often connected with zero - one laws. The zero - one of Hewitt and Savage applies to the following example.

(5.5) Example

The symmetric random walk on $\mathbb{R}$ is recurrent.
Let $(X_n)_{n\in\mathbb{N}}$ be i.i.d. random variables with common distribution $\frac{1}{2}(\varepsilon_1 + \varepsilon_{-1})$. Then we shall prove that

$P(\{S_n = 0$ infinitely often (i.o.)$\}) = 1$ for $S_n = \sum_{i=1}^{n} X_i$.

Note that $\{S_n = 0 \ \text{i.o.}\}^c = A \cup B$,

$A = \{S_n < 0 \ \text{finally}\}$, $B = \{S_n > 0 \ \text{finally}\}$.

Since A, B are tail events with respect to S_n the zero - one law of Hewitt and Savage yields $P(A), P(B) \in \{0,1\}$. Since the random walk is symmetric and A, B are disjoint $P(A) = P(B) = 0$ follows.

Acknowledgement.
The author likes to thank U. Lüxmann for various hints concerning the literature .

References

[0] Blum, J.R. (1982). Exchangeability and Quasi - Exchangeability. In: Exchangeability in Prob. and Statist., G. Koch and F. Spizzichino (editors), North - Holland, 171 - 176.

[1] Breiman, L., LeCam, L. Schwartz, L. (1964). Consistent estimates and zero - one sets. Ann. Math. Statistics 35, 157 - 161

[2] Chatterji, S.D., Mandrekar, V. (1978). Equivalence and singularity of Gaussian measures and applications. In: Probabilistic analysis and related topics, 169 - 195. Edited by A.T. Bharucha - Reid.

[3] Chatterji, S.D., Ramaswamy, S. (1982). Mesures Gaussiennes et mesures produit. Lect. Notes Math. 920, 570 - 580. Springer.

[4] Dubins, L., Freedman, D. (1964). Measurable sets of measures. Pacific J. Math. 14. 1211 - 1222.

[5] Dubins, L. Freedman, D. (1979). Exchangeable processes need not be mixtures of independent identically distributed random variables. Zeitschr. Wahrscheinlichkeitstheorie verw. Geb. 48, 115 - 132.

[6] Dubins, L. (1983). Some exchangeable probabilities are singular with respect to all representable probabilities. Zeitschr. Wahrscheinlichkeitstheorie verw. Geb. 64, 1 - 5.

[7] Engelbert, H.J., Sirjaev, A.N. (1980). On absolute continuity and singularity of probability measure. Mathem. Statistics. Banach Center public., Vol. 6. PWN - Polish scientific publishers, Warsaw.

[8] Fernique, X. (1984). Comparaison de mesures Gaussiennes et de mesure produit. Ann. Inst. Henri Poincaré, Prob. Statist. 20, 165 - 175.

[9] Fernique, X. (1985). Comparison de mesures Gaussiennes et de mesures produit dans les espaces de Fréchet separables. Lecture Notes Math. 1153, 179 - 197, Springer Verlag.

[10] Freedman, D. (1980). A mixture of independent identically distributed random variables need not admit a regular conditional probability given the exchangeable σ-field. Zeitschr. Wahrscheinlichkeitstheorie 51, 239 - 248.

[11] Janssen, A. (1981). Meßbare Mengen von Maßen und Inhalten. Manuscripta Math. 34, 1 - 15.

[12] Janssen, A. (1982). Zero - one laws for infinitely divisible probability measures on groups. Zeitschr. Wahrscheinlichkeitstheorie 60, 119 - 138.

[13] Kabanov, J.M., Lipcer, R.S., Sirjaev, A.N. (1977). On the question of absolute continuity and singularity of probability measures. Math. USSR Sbornik 33, 203 - 221.

[14] Kakutani, S. (1948). On equivalence of infinite product measures. Ann. of Math. 49, 214 - 224.

[15] LeCam, L. (1969). Théorie asymptotique de la décision statistique. Les presses de l'université de Montréal, Canada.

[16] Ramaswamy, S. (1983). Gaussian measures and product measures, Preprint.

[17] Skorohod, A.V. (1974). Integration in Hilbert space. Ergebn. d. Mathem. 79, Springer.

[18] Strasser, H. (1985): Mathematical theory of Statistics, de Gruyter Studies in Mathematics 7.

ERGODIC AND MIXING PROPERTIES OF MEASURES ON LOCALLY COMPACT GROUPS

Eberhard Kaniuth

The ergodic and mixing properties of (probability) measures on locally compact abelian groups were characterized by Choquet and Deny [1] and by Foguel [3] in terms of the Fourier-Stieltjes transform. Let G be a locally compact abelian group, Γ its dual group and 1_G the identity of Γ. Let μ be a bounded complex Borel measure on G, and denote by $\hat{\mu}$ its Fourier-Stieltjes transform and by μ^n the n-times convolution product of μ.

<u>Theorem 1</u> (Foguel) Suppose that $\|\mu^n\| \leq c$ for all $n \in \mathbb{N}$ and some $c > 0$, and let

$$I_o = \{f \in L^1(G);\ \hat{f}(1) = \int_G f(x)dx = 0\}.$$

Then the following conditions are equivalent:

(i) $\|\mu^n * f\|_1 \to 0$ for every $f \in I_o$;

(ii) $|\hat{\mu}(\gamma)| < 1$ for all $\gamma \in \Gamma\setminus\{1_G\}$.

<u>Theorem 2</u> (Choquet, Deny) The following conditions are equivalent:

(i) for $\varphi \in L^\infty(G)$, $\mu * \varphi = \varphi$ implies that φ is constant;

(ii) $\hat{\mu}(\gamma) \neq 1$ for all $\gamma \in \Gamma \setminus \{1_G\}$.

In [9] new proofs of these results were given which are transparent from the point of view of harmonic analysis. The purpose of this note is to show that using concepts and recent results in non-commutative harmonic analysis immediately leads to certain generalizations to non-abelian groups. Of course, in the case of an arbitrary locally compact group G, the dual group Γ has to be replaced by the dual space $\hat{G}$ of G, i.e. the set of equivalence classes of irreducible unitary representations of G, and instead of $\hat{\mu}(\gamma)$ one has to consider the operator $\pi(\mu) = \int_G \pi(x)d\mu(x)$, $\pi \in \hat{G}$.

Let us recall some definitions. A locally compact group G has polynomial growth if for every compact subset K of G there is a polynomial P_K such that the Haar measure of K^n is bounded by $P_K(n)$ for all $n \in \mathbb{N}$. A Banach $*$-algebra A is called symmetric if every selfadjoint element in A has a real spectrum. The class of polynomially growing groups with symmetric L^1-algebra contains all compact extension of nilpotent locally compact groups [5; 7], and much progress has been made in studying the ideal structure in such algebras.

Theorem I. Suppose that μ is a central bounded complex Borel measure on the locally compact group G and that $\|\mu^n\| \leq c$ for all $n \in \mathbb{N}$ and some $c > 0$. Let $I_0 = \{f \in L^1(G); \int_G f(x)dx = 0\}$, and consider the following conditions:

(i) $\|\mu^n * f\|_1 \to 0$ for every $f \in I_0$;

(ii) $\|\pi(\mu)\| < 1$ for all $\pi \in \hat{G}$, $\pi \neq 1_G$ (the trivial representation).

Then (i) implies (ii), and the converse holds if G has polynomial growth and $L^1(G)$ is symmetric.

Theorem II. Consider the following two conditions for a locally compact group G.

(i) for $\varphi \in L^\infty(G)$, $\mu * \varphi = \varphi * \mu = \varphi$ implies that φ is constant;

(ii) $\pi(\mu) \neq I$ for all $\pi \in \hat{G}$, $\pi \neq 1_G$.

Then (i) implies (ii), and the converse holds if G has polynomial growth and $L^1(G)$ is symmetric.

We now introduce some concepts from ideal theory which will be useful in proving the non-trivial parts of Theorems I and II, namely the implications (ii) $\Rightarrow$ (i). The primitive ideal space of $L^1(G)$ consists of all L^1-kernels ker π, $\pi \in \hat{G}$ (if π is a unitary representation of G, then the corresponding $*$-representation of $L^1(G)$ is also denoted by π). For $M \subseteq L^1(G)$ and $E \subseteq \text{Prim } L^1(G)$ the hull of M and the kernel of E are defined by

$$h(M) = \{P \in \text{Prim } L^1(G);\ P \supseteq M\} \text{ and } k(E) = \cap\{P;\ P \in E\},$$

respectively. Prim $L^1(G)$ carries the hull-kernel-topology, i.e. the closure of $E \subseteq \text{Prim } L^1(G)$ is given by $\overline{E} = h(k(E))$. A closed subset E of Prim $L^1(G)$ is called a spectral set or set of synthesis if k(E) is the only closed ideal in $L^1(G)$ with hull equal to E. In general, $\emptyset$ and $\{I_0\}$ need not be spectral sets. The smallest dimensional connected solvable Lie group where $\emptyset$, and hence $\{I_0\}$, fails to be spectral, is the group of all matrices

$$\begin{matrix} 1 & x & z \\ 0 & a & y \\ 0 & 0 & 1 \end{matrix} \quad , \ x,y,z \in \mathbb{R},\ a > 0 \text{ [6, Corollary on p.131].}$$

On the other hand, if G is polynomially growing and has a symmetric L^1-algebra, then $\emptyset$ and $\{I_0\}$ are spectral sets [7, Theorem 2].

Proof of Theorem I. Notice first that, for $\pi \in \hat{G}$, $\pi(\mu) = c \cdot I$, a multiple of the identity operator in the Hilbert space H_π of π. Hence

$$|c|^n \|\pi(f)\| = \|\pi(\mu^n * f)\| \leq \|\mu^n * f\|_1.$$

If $\pi \neq 1_G$, then $\pi(f) \neq 0$ for some $f \in I_o$, so that (i) implies $\|\pi(\mu)\| = |c| < 1$. For the converse, consider the closed ideal

$$I = \{f \in I_o;\ \|\mu^n * f\|_1 \to 0\}$$

of $L^1(G)$. Using Dixmier's functional calculus [2] for the polynomially growing group G and the symmetry of $L^1(G)$, Ludwig [8] has shown the important fact that for any closed subset E of Prim $L^1(G)$, there exists a smallest ideal j(E) in $L^1(G)$ such that $h(j(E)) = E$. j(E) is generated by a set M of functions $f = f^* \in k(E)$ which can be factorized in the following way: there exists $g = g^* \in k(E)$ such that

$$f * g = g * f = f.$$

Moreover, a thorough look to the construction of $f \in M$ and a corresponding g shows that we can assume that $\|\rho(g)\| \leq 1$ for every representation ρ of G. Now, setting $h = \mu * g$, we have

$$\|\mu^n * f\|_1 = \|(\mu * g)^n * f\|_1 \leq \|f\|_1 \|h^n\|_1.$$

We claim that $\|h^n\|_1 \to 0$, i.e. $f \in I$. To this end, let A denote the closed subalgebra of $L^1(G)$ generated by h and h^*. Then A is commutative since h commutes with h^*. Denote by $\Delta(A)$ the Gelfand space of A, then by the spectral radius formula

$$\lim_{n\to\infty} \|h^n\|_1^{1/n} = \sup\{|\delta(h)|;\ \delta \in \Delta(A)\}.$$

On the other hand, since $L^1(G)$ is assumed to be symmetric, A is symmetric, and therefore every $\delta \in \Delta(A)$ is a $*$-homomorphismen and extends to an indecomposable positive linear functional on $L^1(G)$ [10, Corollary 4.7.7 and Theorem 4.7.11]. By the Gelfand-Naimark-Segal construction, this gives rise to an irreducible representation of G. Thus we obtain $|\delta(h)| \leq \|\rho(h)\|$ for some $\rho \in \hat{G}$. Now, $\|\pi(\mu)\| < 1$ for all $\pi \neq 1_G$, $h \in I_o$ and $\|\rho(g)\| \leq 1$ for every representation ρ of G. From this we conclude $|\delta(h)| < 1$ for all $\delta \in \Delta(A)$ and hence

$$\sup\{|\delta(h)|\ ;\ \delta \in \Delta(A)\} < 1.$$

This proves $\|h^n\|_1 \to 0$. We have shown so far that $M \subseteq I$. Finally, since $\{I_o\}$

is a spectral set, if follows that $I = I_o$.

Proof of Theorem II. Again, (i) ⇒ (ii) is easy. Indeed, if $\pi \in \hat{G}$ such that $\pi(\mu) = I$, then for all $\xi \in H_\pi$ and $\varphi(x) = \langle \pi(x)\xi, \xi \rangle$

$$\mu * \varphi(x) = \langle \pi(x)\xi, \pi(\mu)\xi \rangle = \varphi(x),$$

so that φ is constant. This yields $\pi = 1_G$.

Conversely, consider the translation-invariant weak-*-closed subspace $\phi = \{\varphi \in L^\infty(G); \mu * \varphi = \varphi = \varphi * \mu\}$ of $L^\infty(G)$. Then $I = \{f \in L^1(G); \int_G f(x)\varphi(x)dx = 0$ for all $\varphi \in \phi\}$ is a closed translation-invariant subspace, i.e. an ideal, in $L^1(G)$. To prove (i), we have to show that $I = I_o$ or $I = L^1(G)$. Since $\emptyset$ and $\{I_o\}$ are spectral sets, it is enough to see that $h(I) \subseteq \{I_o\}$. If $\pi \in \hat{G}$ such that $I \subseteq \ker \pi$, then every coordinate function $\varphi(x) = \langle \pi(x)\xi, \xi \rangle$, $\xi \in H_\pi$, belongs to ϕ. It follows that

$$\langle \pi(x)\xi, \pi(\mu)\xi \rangle = \langle \pi(x)\xi, \xi \rangle \text{ for all } x \in G,\ \xi \in H_\pi.$$

Since $\pi(\mu) \neq I$ for $\pi \neq 1_G$, this implies $\pi = 1_G$.

Remarks. a) In Theorem II, the assumptions that G has polynomial growth and $L^1(G)$ is symmetric, are only used to guarantee that $\emptyset$ and $\{I_o\}$ are spectral sets.

b) Unfortunately, we didn't succeed in weakening the assumption that μ is central in Theorem I. It should be mentioned that by a result of Greenleaf, Moskowitz, and Rothschild [4], central measures on connected groups are supported by the closed subgroup consisting of all elements having a relatively compact conjugacy class.

References

[1] Choquet, G., Deny, J.: Sur l'équation de convolution $\mu = \mu * \sigma$. C.R. Acad. Sci. Paris 250, 799-801 (1960)

[2] Dixmier, J.: Opérateurs de rang fini dans les représentations unitaires. Publ. math. Inst. Hautes Etudes Sci. 6, 305-317 (1960)

[3] Foguel, S.R.: On iterates of convolutions. Proc. Amer. Math. Soc. 47, 368-370 (1975)

[4] Greenleaf, F.P., Moskowitz, M., Rothschild, L.P.: Unbounded conjugacy classes in Lie groups and location of central measures.
Acta Math. 132, 225-243 (1974)

[5] Guivarch, Y.: Croissance polynomiale et périodes des fonctions harmoniques.
Bull. Soc. math. France 101, 333-379 (1973)

[6] Leptin, H., Poguntke, D.: Symmetry and nonsymmetry for locally compact groups.
J. Functional Analysis 33, 119-134 (1979)

[7] Ludwig, J.: A class of symmetric and a class of Wiener group algebras.
J. Functional Analysis 31, 187-194 (1979)

[8] Ludwig, J.: Polynomial growth and ideals in group algebras.
Manuscripta math. 30, 215-221 (1980)

[9] Ramsey, Th., Weit, Y.: Ergodic and mixing properties of measures on locally compact abelian groups.
Proc. Amer. Math. Soc. 92, 519-520 (1984)

[10] Rickart, C.E.: General theory of Banach algebras.
New York: van Nostrand 1960

Eberhard Kaniuth
Fachbereich Mathematik/Informatik
der Universität - Gesamthochschule Paderborn
Warburger Straße 100
D - 4790 Paderborn

ON JUMPS OF PATHS OF MARKOV PROCESSES

Jan Kisyński
Technical University of Lublin
ul. J. Dąbrowskiego 13
20-109 Lublin, Poland

Abstract. Let X be a cadlag Markov process with separable metric state space S, governed by semigroup of transition kernels $(N_t)_{t \geqslant 0}$. Let f be a bounded, non-negative, continuous function on S^2, vanishing in a uniform neighbourhood of the diagonal. Define $J_t f(x) = \frac{1}{t} \int_S N_t(x,dy) f(x,y)$ and suppose that $\sup\{|J_t f(x)| : t > 0,\ x \in S\} < \infty$ and that $Jf(x) = \lim_{t \downarrow 0} J_t f(x)$ exists for each $x \in S$. Then $E_x\left[\sum_{0<u\leqslant t} f(X_{u-}, X_u)\right] = \int_0^t N_u(x, Jf)\,du$ for each $t \geqslant 0$ and each $x \in S$.

1. The setting, some preliminaries, and the main result.

1.1. Assumptions about the Markov process and its semigroup. Let S be a separable metric space and $\mathcal{B}(S)$ its Borel σ-field. Let $X = (\Omega, \mathcal{F}, \mathcal{F}_t, X_t, \theta_t, P_x)$ be a time-homogeneous cadlag Markov process whose state space is $(S, \mathcal{B}(S))$, and whose life time is ∞, for each $\omega \in \Omega$. Concerning the transition kernels N_t, defined on $S \times \mathcal{B}(S)$ by $N_t(x,B) = P_x\{X_t \in B\}$, it is assumed that the map $S \ni x \longrightarrow N_t(x,B) \in [0,1]$ is Borel measurable whenever $t \in [0,\infty)$ and $B \in \mathcal{B}(S)$ are fixed.

The above are the only assumptions concerning the process X.

In particular, it is not assumed that the process is strongly Markov.

1.2. <u>The kernel K on $S\times[0,\infty)\times\mathcal{B}(S)$ and the generalized Fubini's theorem</u>. Under Assumptions 1.1 the equality $K(x,t,B) = N_t(x,B)$ defines a kernel on $S\times[0,\infty)\times\mathcal{B}(S)$. Indeed, it is obvious that if (x,t) is fixed then $K(t,x,\cdot)$ is a probability measure on $\mathcal{B}(S)$. It remains to show that if $B\in\mathcal{B}(S)$ is fixed, then $K(\cdot,\cdot,B)$ is a Borel function on $S\times[0,\infty)$. But if $\varphi\in C_b(S)$ then $K(x,t,\varphi) = N_t(x,\varphi) = E_x[\varphi(X_t)]$ is Borel measurable in x and cadlag in t, and consequently it is Borel measurable in (x,t). By a monotone class theorem [8; Chap. I, T 20], [2; Chap. I, No. 21-22], the same measurability in (x,t) remains true for each bounded Borel function φ on S, in particular for each indicator function of a Borel subset of S.

Denote by $M^1(S)$ the set of all probability measures on $\mathcal{B}(S)$. If $\mu\in M^1(S)$, $t\geqslant 0$, $h>0$, then the measures

$$\mu N_t \qquad \text{and} \qquad \mu_h = \frac{1}{h}\int_0^h \mu N_u\,du,$$

both belonging to $M^1(S)$, are defined by the conditions that

$$(\mu N_t)(B) = \int_S \mu(dx)N_t(x,B),$$

$$\mu_h(B) = \frac{1}{h}\int_0^h (\mu N_u)(B)\,du = \frac{1}{h}\iint_{S\times[0,h]} (\mu(dx)\times du)K(x,u,B).$$

Let φ be a non-negative, or a bounded, Borel function on S. By the generalized Fubini's theorem [9; Th. 35.11 and 35.14], GFT for short, applied to the kernel N_t,

(1.2.1) $$\langle \mu N_t, \varphi \rangle = \langle \mu , N_t \varphi \rangle$$

and, by GFT and FT applied to the kernel K and the product measure $\mu(dx) \times du$,

(1.2.2) $$\langle \mu_h, \varphi \rangle = \langle \mu, \frac{1}{h} \int_0^h K(\cdot, u, \varphi)\, du \rangle = \langle \mu, \frac{1}{h} \int_0^h N_u \varphi\, du \rangle.$$

Consequently, by the semigroup property,

(1.2.3) $$\langle \mu_h N_t, \varphi \rangle = \langle \mu_h, N_t \varphi \rangle = \langle \mu, \frac{1}{h} \int_t^{t+h} N_u \varphi\, du \rangle.$$

1.3. <u>The additive functionals</u> $P_t(f)$ <u>and their approximation</u>. For each $f \in C(S^2)$ and each $n = 1,2,\dots$ the formula

$$P_t^n(f) = \sum_{k=1}^{n} f(X_{\frac{k-1}{n}t}, X_{\frac{k}{n}t}), \qquad t \geqslant 0,$$

defines a cadlag $(\mathcal{F}_t)$-adapted process. Denote

$$d(f) = \inf\{\mathrm{dist}(x,y) : (x,y) \in S^2,\ f(x,y) \neq 0\}.$$

Then $d(f) > 0$ means that f vanishes in a uniform neighbourhood of the diagonal of S^2. Let now $f \in C(S^2)$ be such that $d(f) > 0$. Then, for each $\omega \in \Omega$ and each $t \in (0,\infty)$, the interval $(0,t]$ contains at the most finite number of points u at which dist $\mathrm{dist}(X_{u-}(\omega), X_u(\omega)) > d(f)$, and consequently the sum

$$P_t(f)(\omega) = \sum_{0<u\leq t} f(X_{u-}(\omega), X_u(\omega))$$

contains at the most finite number of terms different from zero. Define $P_0(f) = 0$.

<u>Lemma</u>. Under Assumptions 1.1 suppose that $f \in C(S^2)$ and $d(f) > 0$. Then, for each $t \geq 0$ and each $\omega \in \Omega$,

$$\lim_{n \to \infty} P_t^n(f)(\omega) = P_t(f)(\omega). \tag{1.3.1}$$

The proof of this Lemma is the same as that of 4.3 in [6]. If $f \in C(S^2)$ and $d(f) > 0$, then the process $P_t(f)$ is cadlag and, by the Lemma, it is $(\mathcal{F}_t)$-adapted. The obvious equalities $P_t(f) \circ \theta_s = \sum_{0<u\leq t} f(X_{(u+s)-}, X_{u+s}) = P_{t+s}(f) - P_s(f)$ show that $P_t(f)$ is an additive functional of the Markov process X.

1.4. <u>The operator</u> J <u>and the main result</u>. Denote by $C_b(S^2)$ the space of all bounded continuous real functions on S^2. Let $f \in C_b(S^2)$. For each $t > 0$ define the bounded Borel function $J_t f$ on S by

$$(J_t f)(x) = \frac{1}{t} \int_S f(x,y) N_t(x,dy) \tag{1.4.1}$$

and denote

$$M(f) = \sup\{|(J_t f)(x)| : t > 0, x \in S\}. \tag{1.4.2}$$

Denote by $\mathcal{D}$ the linear subset of $C_b(S^2)$ formed by all the functions $f \in C_b(S^2)$ for which $d(f) > 0$, $M(f) < \infty$ and $\lim_{t \downarrow 0} (J_t f)(x)$ exists for each $x \in S$. Let J be the linear operator defined on $\mathcal{D}$, which to any $f \in \mathcal{D}$ assigns the bounded Borel function Jf, defined on S by the formula

(1.4.3) $$(Jf)(x) = \lim_{t \downarrow 0} (J_t f)(x).$$

Denote by $\mathcal{D}^+$ the set of all non-negative functions belonging to $\mathcal{D}$.

Theorem. Under Assumptions 1.1 suppose that $f \in \mathcal{D}^+$. Then

(1.4.4) $$E_x P_t(f) = \int_0^t N_u(x, Jf)\,du$$

for each $x \in S$ and each $t > 0$.

2. Examples and remarks.

2.1. A corollary for Feller processes. Let S be a compact metric space and $(N_t)_{t \geq 0}$ a strongly continuous one-parameter semigroup of non-negative linear contractions of the space $C(S)$, such that $N_t 1 = 1$ for each $t \geq 0$, where 1 denotes the function equal identically to one. Let X be a cadlag Markov process with values in S, such that $(N_t \varphi)(x) = E_x[\varphi(X_t)]$ for every $t \geq 0$, $\varphi \in C(S)$ and $x \in S$. Denote by G the infinitesimal generator of the semigroup (N_t). If $\varphi \in C(S)$ and $\psi \in \mathcal{D}(G)$ are non-negative and have disjoint supports, then $\varphi \otimes \psi \in \mathcal{D}^+$, $J(\varphi \otimes \psi) = \varphi \cdot G\psi$ and the formula (1.4.4) takes the form

(2.1.1) $$E_x \sum_{0 < u \leq t} \varphi(X_{u-})\psi(X_u) = \int_0^t [N_u(\varphi \cdot G\psi)](x)\,du.$$

Let us remark that (2.1.1) is true also for φ and ψ which are not necessarily non-negative, under all remaining assumption unchanged. This can be proved by the same arguments as that used in [6; Sec. 5.6].

2.2. <u>Application to a criterion for continuity</u>. The equality (2.1.1) implies the following known result [7; p. 193, Th.5], [11; p.198, Th.2].

<u>Theorem</u>. Under assumptions of Section 2.1 suppose moreover that

(i) the operator G is local, and

(ii) for each $x \in S$ and each $\varepsilon > 0$ there is a non-negative function $\psi_{x,\varepsilon} \in \mathcal{D}(G)$ such that $x \notin \operatorname{supp} \psi_{x,\varepsilon}$ and $\psi_{x,\varepsilon}(y) > 0$ whenever $\operatorname{dist}(x,y) > \varepsilon$.

Then the Markov process X is almost surely continuous.

Indeed, for each $x \in S$ and each $\varepsilon > 0$ choose a non-negative $\varphi_{x,\varepsilon} \in C(S)$ such that $\operatorname{supp} \varphi_{x,\varepsilon} \cap \operatorname{supp} \psi_{x,\varepsilon} = \emptyset$ and $x \in V_{x,\varepsilon} = \{y \in S : \varphi_{x,\varepsilon}(y) > 0\}$. If $\varepsilon > 0$ is fixed and $V_{x_1,\varepsilon}, \ldots, V_{x_N,\varepsilon}$ is a finite covering of S, then

$$\{\operatorname{dist}(X_{u-}, X_u) > 2\varepsilon \text{ for some } u \in (0,t]\} \subset$$

$$\subset \{ \sum_{n=1}^{N} \sum_{0<u\leqslant t} \varphi_{x_n,\varepsilon}(X_{u-}) \psi_{x_n,\varepsilon}(X_u) > 0 \}$$

and the formula (2.1.1) shows that both these events are negligible.

2.3. <u>Continuity in the interior of the state space of a certain Markov process with non-local boundary condition</u>. Let $S = [0,\infty]$ be treated as one-point compactification of $[0,\infty)$. Denote by $C^n[0,\infty]$ the space of real functions continuous on $[0,\infty]$, whose derivatives up to order n are continuous on $[0,\infty)$ and have limit equal to zero at ∞. Let ν be a Borel probability measure on

$(0,\infty)$ and α a positive constant. Define the linear operator $G : \mathcal{D}(G) \longrightarrow C[0,\infty]$ by

$$\mathcal{D}(G) = \{\psi \in C^2[0,\infty] : \alpha\psi''(0) + \psi(0) = \langle \nu,\psi\rangle\},$$
$$G\psi = \tfrac{1}{2}\psi'' \quad \text{for} \quad \psi \in \mathcal{D}(G). \tag{2.3.1}$$

Using the Hille-Yosida generation theorem, it is not difficult to check that G is infinitesimal generator of a semigroup $(N_t)_{t\geq 0}$ with properties described in Section 2.1. Let X be a cadlag Markov process with values in $[0,\infty]$, whose transition semigroup is (N_t). Attention to this process is paid in the book of Lamperti [7; p.178]. Similar processes, but with compact state space which is countable and has a unique point of accumulation, occur in the book of Dynkin and Yuschkevitsch [3], in Chapter IV. An application of Dynkin s characteristic operator [7; p.225] shows that the cadlag Markov process X generated by the operator (2.3.1) cannot be a.s. continuous. Our equality (2.1.1) permitts to enlarge the knowledge about behaviour of this process by proving that it can jump only from 0 to the closure of $\operatorname{supp}\nu$. Indeed, if I_1 and I_2 are disjoint closed subintervals of $[0,\infty]$ such that $0 \notin I_1$ or $I_2 \cap \overline{\operatorname{supp}\nu} = \emptyset$, then it is possible to choose non-negative functions φ and ψ with disjoint supports, such that $\varphi \in C[0,\infty]$, $\psi \in \mathcal{D}(G)$, $\varphi > 0$ on I_1, and $\psi > 0$ on I_2. Since $\operatorname{supp} G\psi = \operatorname{supp}\psi'' \subset \operatorname{supp}\psi$, we have $\varphi\cdot G\psi \equiv 0$ and so, by (2.1.1), $E_x \sum_{0<u\leq t} \varphi(X_{u-})\psi(X_u) = 0$, whence

$$P_x\{(X_{u-}, X_u) \in I_1 \times I_2 \quad \text{for some} \quad u \in (0,t]\} = 0.$$

2.4. Relation to a theorem of N. Ikeda and S. Watanabe. If $f \in \mathcal{D}^+$ then the right side of (1.4.4) may be written as $E_x[Q_t(f)]$, where the continuous $(\mathcal{F}_t)$-adapted process

$$Q_t(f) = \int_0^t (Jf) \circ X_u \, du$$

is an additive functional of the process X. The equality (1.4.4) states that the cadlag additive functional

$$M_t(f) = P_t(f) - Q_t(f)$$

has the property that

$$E_x[M_t(f)] = 0$$

for each $x \in S$ and each $t \geqslant 0$. By the Markov property, see [1;p.23], the former implies that $M_{\cdot}(f)$ is a martingale with respect to the filtraction $(\mathcal{F}_t)$ and to the probability measure P_x, for each $x \in S$. It is the same with

$$M_t(f,\lambda) = \int_0^t e^{-\lambda u} dM_u(f) = P_t(f,\lambda) - Q_t(f,\lambda),$$

where $\lambda > 0$, and where

$$P_t(f,\lambda) = \int_0^t e^{-\lambda u} dP_u(f) = \sum_{0<u\leq t} e^{-\lambda u} f(X_{u-}, X_u),$$

$$Q_t(f,\lambda) = \int_0^t e^{-\lambda u} dQ_u(f) = \int_0^t e^{-\lambda u} (Jf) \circ X_u \, du.$$

Consequently $E_x[M_\tau(f,\lambda)] = 0$ or, which is the same,

$$E_x[P_\tau(f,\lambda)] = E_x[Q_\tau(f,\lambda)] \tag{2.4.1}$$

for each $x \in S$, each $\lambda > 0$, and for each stopping time τ of the filtration $(\mathcal{F}_{t+})$.

If, in addition to Assumptions 1.1, we assume that S is compact and that

(2.4.2) $\left\{ \begin{array}{l} \varphi \otimes \psi \in \mathcal{D}^+ \text{ whenever } \varphi \in C(S) \text{ and } \psi \in C(S) \text{ are} \\ \text{non-negative and have disjoint supports,} \end{array} \right.$

then there is a $[0,\infty]$-valued kernel n on $S \times \mathcal{B}(S)$, called the Levy kernel, such that

$$(Jf)(x) = \int_S f(x,y)n(x,dy)$$

for each $f \in \mathcal{D}^+$ and each $x \in S$. The integral operator on the right permits to extend the operator J onto functions f not belonging to $\mathcal{D}$. With such extended operator J, still under Assumptions 1.1 and additional assumptions that S is compact and that (2.4.2) holds, by a monotone class theorem (similarly as in [6; Sec. 5.5]), the equality (2.4.1) remains true for each non-negative Borel function f on S^2, vanishing on the diagonal. The former result was proved for cadlag Feller processes by N. Ikeda and S. Watanabe in [5] and [10], by a method other then ours.

The verification of the assumption (2.4.2) may be not easy. It seems for instance that in the case of the process from Section 2.3 the condition (2.4.2) may be proved only after one proves that the process really behaves so as described in [7; p. 178]. But it may be hard to prove rigorously such a behaviour when one starts with the only knowledge of the infinitesimal generator. On the contrary, an application of our Corollary 2.1 is easy in this case and gives a preliminary information, valuable for further study of the process.

3. Proof of the Theorem.

Following a suggestion of W. Feller [4; Chap. IX, §5], we approximate the additive functional $P_t(f)$ by the processes $P_t^n(f)$. The proof consists of five steps.

Step 1. For any bounded Borel function f on S^2 we have

$$\begin{aligned} E_x[P_t^n(f)] &= \sum_{k=1}^{n} \int_S N_{\frac{k-1}{n}t}(x,dy) \int_S N_{\frac{t}{n}}(y,dz) f(y,z) = \\ &= \sum_{k=1}^{n} \frac{t}{n} N_{\frac{k-1}{n}t}(x, J_{\frac{t}{n}} f), \end{aligned} \tag{3.1.1}$$

which approximates the desired result. In the sequel let $f \in \mathcal{D}^+$. Then, by (3.1.1),

$$E_x[P_t^n(f)] \leqslant tM(f). \tag{3.1.2}$$

Moreover,

$$(P_t^n(f))^2 = P_t^n(f^2) + 2 \sum_{1 \leqslant i < k \leqslant n} f(X_{\frac{i-1}{n}t}, X_{\frac{i}{n}t}) f(X_{\frac{k-1}{n}t}, X_{\frac{k}{n}t}),$$

where the sum contains $\frac{1}{2}(n^2 - n)$ terms. Since

$$E_x[P_t^n(f^2)] \leqslant (\sup f)\, tM(f)$$

and since, for $i < k$,

$$E_x\left[f\left(X_{\frac{i-1}{n}t}, X_{\frac{i}{n}t}\right) f\left(X_{\frac{k-1}{n}t}, X_{\frac{k}{n}t}\right)\right] =$$

$$= \int_S N_{\frac{i-1}{n}t}(x, dx_4) \int_S N_{\frac{t}{n}}(x_4, dx_3) f(x_4, x_3) \int_S N_{\frac{k-1-i}{n}t}(x_3, dx_2) \int_S N_{\frac{t}{n}}(x_2, dx_1) f(x_2, x_1)$$

$$= \frac{t}{n} \int_S N_{\frac{i-1}{n}t}(x, dx_4) \int_S N_{\frac{t}{n}}(x_4, dx_3) f(x_4, x_3) N_{\frac{k-1-i}{n}t}(x_3, J_{\frac{t}{n}} f)$$

$$\leq \frac{t}{n} M(f) \int_S N_{\frac{i-1}{n}t}(x, dx_4) \int_S N_{\frac{t}{n}}(x_4, dx_3) f(x_4, x_3)$$

$$= \frac{t}{n} M(f) \frac{t}{n} N_{\frac{i-1}{n}t}(x, J_{\frac{t}{n}} f) \leq \left(\frac{t}{n} M(f)\right)^2,$$

we conclude that

$$E_x\left[(P_t^n(f))^2\right] \leq t(\sup f) M(f) + (tM(f))^2$$

It follows from this estimation, from (1.3.1) and from (3.1.2) that

$$\lim_{n \to \infty} E_x\left[P_t^n(f)\right] = E_x\left[P_t(f)\right] \tag{3.1.3}$$

and

$$E_x\left[P_t(f)\right] \leq tM(f) \tag{3.1.4}$$

for each $f \in \mathcal{D}^+$, each $x \in S$ and each $t \geq 0$. Since, as evident from (3.1.1), the map $(x,t) \longrightarrow E_x\left[P_t^n(f)\right]$ is measurable, we see from (3.1.3) that also the map $(x,t) \longrightarrow E_x\left[P_t(f)\right]$ is measurable, for each $f \in \mathcal{D}^+$.

Step 2. By (3.1.1) and (1.2.1), for each $\mu \in M^1(S)$ and each $f \in \mathcal{D}^+$, we have

$$(3.2.1)\qquad \langle \mu, E_{\cdot}[P_t^n(f)]\rangle = \langle \frac{t}{n}\sum_{k=1}^{n} \mu N_{\frac{k-1}{n}t}, J_{\frac{t}{n}} f\rangle.$$

Denote by $M_0^1(S)$ the set of all those probability measures $\mu \in M^1(S)$ for which the map

$$[0,\infty) \ni t \longrightarrow \mu N_t \in M^1(S)$$

is continuous in the sense of total variation. By the former we mean that

$$\lim_{0\leq s \to t} \mathcal{V}(\mu N_s - \mu N_t) = 0$$

for each $t \geq 0$, where $\mathcal{V}(\cdot)$ denotes the total variation on S of a real-valued set function defined on $\mathcal{B}(S)$. If $\mu \in M_0^1(S)$, then

$$(3.2.2)\qquad \lim_{\varepsilon \downarrow 0} \omega_{\mu,t}(\varepsilon) = 0$$

for each $t > 0$, where

$$\omega_{\mu,t}(\varepsilon) = \sup\{\mathcal{V}(\mu N_u - \mu N_v) : 0\leq u\leq t,\ 0\leq v\leq t,\ |u-v|\leq\varepsilon\}.$$

If $\mu \in M_0^1(S)$ and $f \in \mathcal{D}^+$, then

$$|\langle \frac{t}{n}\sum_{k=1}^{n}\mu N_{\frac{k-1}{n}t} - \int_0^t \mu N_u du, J_{\frac{t}{n}} f\rangle| \leq t\,\omega_{\mu,t}(\tfrac{t}{n}) M(f),$$

and so

$$(3.2.3)\qquad \begin{aligned} &|\langle \frac{t}{n}\sum_{k=1}^{n}\mu N_{\frac{k-1}{n}t}, J_{\frac{t}{n}} f\rangle - \langle \int_0^t \mu N_u du, Jf\rangle| \leq \\ &\leq t\,\omega_{\mu,t}(\tfrac{t}{n})M(f) + |\langle \int_0^t \mu N_u du, J_{\frac{t}{n}} f - Jf\rangle| . \end{aligned}$$

By (3.2.2), the first term on the right side of (3.2.3) tends to zero as $n \to \infty$. The second term also tends to zero, since $J_{\frac{t}{n}}f$ converges to Jf pointwisely on S and since $0 \leq J_{\frac{t}{n}}f \leq M(f)$. Consequently we infer from (3.1.2), (3.1.3), (3.2.1) and (3.2.3) that

$$\langle \mu, E.[P_t(f)] \rangle = \langle \int_0^t \mu N_u du, Jf \rangle .$$

If $t>0$, then the right side of this may by written as $\langle t\mu_t, Jf \rangle$ which is equal to $\langle \mu, \int_0^t N_u(Jf)du \rangle$, by (1.2.2). Thus, finally

$$\langle \mu, E.[P_t(f)] \rangle = \langle \mu, \int_0^t N_u(Jf)du \rangle \tag{3.2.4}$$

for each $\mu \in M_0^1(S)$, each $f \in \mathcal{D}^+$ and each $t>0$. Our theorem will follow when we show that (3.2.4) remains true for each $f \in \mathcal{D}^+$, each $t>0$, and for each $\mu \in M^1(S)$.

Step 3. To that end, take any $\mu \in M^1(S)$. For any $h>0$, $t \geq 0$ and $s \geq 0$, by (1.2.3), we have

$$|\langle \mu_h N_t - \mu_h N_s, \varphi \rangle| = |\langle \mu, \frac{1}{h}(\int_{s+h}^{t+h} - \int_s^t)N_u \varphi \, du \rangle| \leq \frac{2}{h}|t-s| \sup|\varphi|$$

for each bounded Borel function φ on S. This shows that $\mu_h \in M_0^1(E)$. Thus, by (3.2.4), we have

$$\langle \mu_h, E.[P_t(f)] \rangle = \langle \mu_h, \int_0^t N_u(Jf)du \rangle \tag{3.3.1}$$

for every $f \in \mathcal{D}^+$, $\mu \in M^1(S)$, $h>0$ and $t>0$.

Step 4. By Fubini's theorem, by (1.2.2), and by the semigroup property,

$$\langle \mu_h, \int_0^t N_u(Jf)\,du \rangle = \int_0^t \langle \mu_h, N_u(Jf) \rangle\,du =$$

$$= \int_0^t \langle \mu, \frac{1}{h} \int_0^h N_v N_u(Jf)\,dv \rangle\,du = \langle \mu, \frac{1}{h} \int_0^h \int_0^t N_{u+v}(Jf)\,du\,dv \rangle,$$

whence

$$|\langle \mu_h - \mu, \int_0^t N_u(Jf)\,du \rangle| = |\langle \mu, \frac{1}{h} \int_0^h [(\int_u^{u+t} - \int_0^t) N_v(Jf)\,dv]\,du \rangle|$$

$$\leqslant 2hM(f),$$

and so

(3.4.1) $$\lim_{h \downarrow 0} \langle \mu_h, \int_0^t N_u(Jf)\,du \rangle = \langle \mu, \int_0^t N_u(Jf)\,du \rangle$$

for each $f \in \mathcal{D}^+$, each $\mu \in M^1(S)$, and each $t > 0$.

Step 5. It remains now only to show that a limit passage similar to (3.4.1) is possible also on the left side of (3.3.1). By (1.2.2)

$$\langle \mu_h, E.[P_t(f)] \rangle = \langle \mu, \frac{1}{h} \int_0^h N_u E.[P_t(f)]\,du \rangle$$

and, by additivity of $P_t(f)$,

$$N_u(x, E.[P_t(f)]) = E_x[P_t(f) \circ \theta_u] = E_x[P_{t+u}(f) - P_u(f)].$$

Consequently

$$\langle \mu_h, E.[P_t(f)] \rangle = \langle \mu, \frac{1}{h} \int_0^h E.[P_{t+u}(f) - P_u(f)]\,du \rangle$$

and so,

$$\langle \mu_h - \mu, E_{\cdot}[P_t(f)] \rangle = \langle \mu, \frac{1}{h}\int_0^h E_{\cdot}[P_{t+u}(f) - P_u(f) - P_t(f)]du \rangle$$

for every $\mu \in M^1(S)$, $f \in \mathcal{D}^+$, $h > 0$ and $t > 0$. To prove that the right side of this converges to zero as $h \downarrow 0$, for fixed $f \in \mathcal{D}^+$ and fixed $t > 0$, take into account the random variables

$$Y_h = \frac{1}{h}\int_0^h (P_{t+u}(f) - P_u(f))du - P_t(f), \qquad h > 0.$$

Measurability of these random variables follows from the fact that the additive functional $P_{\cdot}(f)$ is $(\mathcal{F}_t)$-adapted and cadlag, so that the integral in the definition of Y_h may be taken in the sense of Riemann. We have

$$(3.5.1) \qquad |Y_h| \leq P_{t+h}(f)$$

and so, by (3.1.2), (3.1.3), and by Fubini's theorem,

$$\langle \mu_h - \mu, E_{\cdot}[P_t(f)] \rangle = \langle \mu, E_{\cdot}[Y_h] \rangle.$$

By right continuity of the process $P_t(f)$,

$$\lim_{h \downarrow 0} Y_h(\omega) = 0$$

for each $\omega \in \Omega$, whence by (3.5.1), (3.1.2), (3.1.3), and by Lebesgue's dominated convergence theorem, $\lim_{h \downarrow 0} \langle \mu, E_{\cdot}[Y_h] \rangle = 0$. Consequently

$$(3.5.2) \qquad \lim_{h \downarrow 0} \langle \mu_h, E_{\cdot}[P_t(f)] \rangle = \langle \mu, E_{\cdot}[P_t(f)] \rangle$$

for each $\mu \in M^1(S)$, each $f \in \mathcal{D}^+$ and each $t \geqslant 0$.

The equalities (3.3.1), (3.4.1) and (3.5.2) show that in fact (3.2.4) is true for each $f \in \mathcal{D}^+$, each $t \geqslant 0$, and for each $\mu \in M^1(S)$. This completes the proof.

Acknowledgements. The author is grateful to Professor K. Helmes for his valuable scepticism concerning assumptions in Theorem 5.3 of [6] and to Professor E. Siebert for several stimulating discussions.

References

[1] R.M. Blumenthal and R.K. Getoor, Markov Processes and Potential Theory, Academic Press, 1968.

[2] C. Dellacherie and P.-A. Meyer, Probabilities and Potential, North Holland Mathematics Studies 29, 1978.

[3] E.B. Dynkin, A.A. Yuschkewitsch, Sätze und Aufgaben über Markoffsche Prozesse, Springer-Verlag, 1969.

[4] W. Feller, An Introduction to Probability Theory and its Applications, Wiley and Sons, Inc., 1966.

[5] N. Ikeda and S. Watanabe, On some relations between the harmonic measure and the Lévy measure for a certain class of Markov processes, J. Math. Kyoto Univ. 2 (1962), p. 79-95.

[6] J. Kisyński, On a formula of N. Ikeda and S. Watanabe concerning the Levy kernel, p. 260-279 in "Probability Measures on Groups VII", Lecture Notes in Mathematics, Vol. 1064, Springer-Verlag, 1984.

[7] J. Lamperti, Stochastic Processes, a Survey of the Mathematical Theory, Applied Mathematical Sciences, Vol. 23, Springer-Verlag, 1977.

[8] P.-A. Meyer, Probability and Potentials, Blaisdell Publishing Company, 1966.

[9] K.R. Parthasarathy, Introduction to Probability and Measure, 1980 (russian translation, "Mir", 1983).

[10] S. Watanabe, On discontinuous additive functionals and Lévy measures of a Markov process, Japanese J. Math. 34, 1964, p. 53-70.

[11] A.D. Ventcel, A course of the Theory of Stochastic Processes (in russian), "Nauka", Moscow, 1975.

RECURRENT RANDOM WALKS ON
HOMOGENEOUS SPACES

R. SCHOTT

ABSTRACT.

An homogeneous space M is recurrent if there exist a recurrent random walk on it. A partial classification of the recurrent homogeneous spaces was given in [4] using the notion of growth. In this paper we try to give a broad class of recurrent measures on homogeneous spaces of some groups of rigid type.

I. INTRODUCTION.

G is as usually a locally compact group with countable basis (L.C.B. group), (X_n) is a sequence of independent random variables on G with the same adapted law μ, $S_n = g \cdot X_1 \ldots X_n$ is called the right random walk on G starting on g at time zero. Now if H is a closed subgroup of G, we consider the induced random walk $Z_n = \pi(S_n) = \pi(g) \cdot X_1 \ldots X_n$ on the homogeneous space $M = {}_H\backslash^G$. We know that Loynes dichotomy theorem is not true in general on M and that additional assumptions have to be made on (G, H, μ), see [6] and [8] for more details.

A characterization of recurrence has been given in [4] for a broad class of homogeneous spaces using the growth function. It can also be proved that if M is not amenable then M is transient (see [3], [10]). Suppose that M is recurrent (i.e. that we can find a probability measure μ on G such that the induced random walk on M is recurrent), the question is now the following : which measures ν have the same property as μ ?

A necessary condition about ν were given in [2] for the motion group and in [7] for it's homogeneous spaces. The aim of this paper is to give a similar condition for the homogeneous spaces of some groups of rigid type. In the first part we remember some basic facts about the random walks on homogeneous spaces : notion of growth, classification theorem, moment of a probability measure on locally compact groups. The second part contains the main result. We prove that if $G = K \ltimes N$ (compact extension of a simply connected solvable Lie group) or $G = K \ltimes R$ (compact extension of a simply connected solvable Lie group of rigid type), if the homogeneous spaces M of G are recurrent, then all the probability measures ν which are adapted, spread out and have a moment of order $2+S$ $(S > 0)$ are recurrent (i.e. the random walk of law μ on M is recurrent).

II. Random walks on homogeneous spaces.

The aim of this section is just to remember some definitions and properties of the random walks on homogeneous spaces, for more details the reader can see [4], [6], [9].

Consider (G,H) were G is a locally compact group with countable basis, H a closed subgroup of G, such that there exist a G-invariant measure λ on $M = G/_H$ (or $_H\backslash G$).
Let V be a compact neighbourhood of e (the neutral element of G) which generates G and $x \in M$.

Définition II.1.
If $\limsup_{n \to +\infty} [\lambda(x \cdot V^n)]^{1/n} > 1$ *(resp. = 1) we say that M has exponential growth (resp. non exponential growth).*
If there exist $k \in \mathbb{N}$ *such that :*

$$0 < \liminf_{n \to +\infty} \frac{\lambda(x \cdot V^n)}{n^k} \leq \limsup_{n \to +\infty} \frac{\lambda(x \cdot V^n)}{n^k} < +\infty .$$

We say that M has polynomial growth of degree k.

Remark II.2.
We have proved in [4] that this notion of growth is independent of $x \in M$ and V, $G/_H$ and $_H\backslash G$ have the same growth. In the case of polynomial growth, k can be explicitely calculated and $k \in \mathbb{N}$ if G is a Lie group of rigid type.

Example II.3.
If $G = K \ltimes N$ then M has polynomial growth (see [4]).
If $G = K \ltimes N$ is a compact extension of a simply connected nilpotent Lie group N, and μ a probability measure which is adapted and spread out we know (see [6]) that all the states of the induced random walk on M are of the same type (i.e. all recurrent or all transient). More precisely we have proved that :

Theorem II.4.
Let $G = K \times N$ *and* μ *a probability measure on G which is adapted and spread out, then an homogeneous space M of G is recurrent and only if M has polynomial growth of degree less than or equal to 2.*

The complete proof of this result is in [4].

Remark II.5.
As we mentioned in the introduction, M recurrent means that we can find a probability measure μ such that the corresponding random walk Z_n on M is recurrent.

In order to characterize a broad class of measures which have the same property as μ, we remember the definition of the moment of order α $(\alpha > 0)$ for probability measures on locally compact groups (see [5] for more details).

Definition II.6.
Let S be a positive Borel function on the locally compact, compactly generated group G such that :

$\forall (x,y) \in G \times G$, $S(x \circ y) \leq S(x) + S(y)$ (*S is a subadditive function*).

A *probability measure* μ *has a moment of order* $\alpha > 0$ *if for each compact neighbourhood* V *of* e *(neutral element of* G*) we have :*

$$\int_G S_V^{\alpha}(g)\, d\mu(g) < +\infty \qquad (S_V \text{ is the restriction to } V) \quad .$$

Remark II.7.

i) This notion of moment is independent of the choice of V (see [5] for the proof).

ii) If G is a group which is compactly generated by V (i.e. $G = \bigcup_{n \geq 0} V^n$) let $S_V(g) = \inf\{n \geq 1,\ g \in V^n\}$
from the relation : $V^{m+n} \supset V^m \cdot V^n$
we obtain : $S_V(g \cdot h) \leq S_V(g) + S_V(h)$.

III. Recurrent random walks on homogeneous spaces.

In the section, we extend a result obtained in [7] for the homogeneous spaces of the motion group.

Theorem III.1.

Let $G = K \ltimes N$ *be a compact extension of a simply connected nilpotent Lie group* N , H *a closed subgroup of* G *such that the degree of polynomial growth of* $M = {}_H\backslash^G$ *(or* $G/_H$*) is less than or equal to 2,* μ *a probability measure on* G *which is adapted, spread out and admits a moment of order* $2+\alpha$ $(\alpha > 0)$ *. Then the random walk on* M *corresponding to* μ *is recurrent.*

Proof.

The proof is based on algebraic results contained in [4] about the structure of M and on results about the potential theory of random walks.
In the first step we show that we can suppose $H \subset N$.
In the second step we prove that the hypothesis concerning the growth of M implies that the dimension of M (in the topological sense) is equal to 1 or 2. We examine the two cases :

- if $\dim(M) = 1$, we can apply the method of the barrier function (method also used in [4], [7]). Here appears the necessity of the existence of a moment of order $2+\alpha$ $(\alpha > 0)$
- if $\dim(M) = 2$, then we use the central limit theorem for $SO(2) \times \mathbb{R}^2$ in order to prove that $U(C)$ the potential of all compact C of M is infinite.

Step 1.

We start with a general lemma.

Lemma III.2.

If G *is a group which is L.C.D.,* H *and* H_1 *closed subgroups of* G *such that* H *is uniform in* H_1 *(i.e.* $H\backslash H_1$ *is compact) then the random walks on* $H\backslash G$ *and* $H_1\backslash G$ *are of the same type (recurrent or transient).*

Proof.

Consider the mapping : $H\backslash G \xrightarrow{\Gamma} H_1\backslash G$ defined by $\Gamma(H\cdot g) = H_1\cdot g$, $\forall g \in G$,
Γ is continuous, surjective and equivariant since H is uniform in H_1.
If C is a compact subset of $H_1\backslash G$ then $\Gamma^{-1}(C)$ is a compact subset of $H\backslash G$.
Let $y \in H_1\backslash G$, $x \in \Gamma^{-1}(y)$ and W_n^y the random walk of law μ starting at y and Y_n^x the random walk of law μ on $H\backslash G$ starting at x . Then :

$$W_n^y \in C \iff Y_n^x \in \Gamma^{-1}(C) \qquad (1)$$

because Γ is equivariant.
The relation (1) proves that the random walks are of the same type (both recurrent or both transient). ∎

If $G = K \times N$ (compact exttension of the simply connected nilpotent Lie group N) , we know that : G is an algebraic group and therefore it's algebraic hull $\tilde{G}$ is equal to G and the algebraic hull $\tilde{H}$ of H is a subgroup of $\tilde{G}$.

- H is uniform in $\tilde{H}$
- $\tilde{H}$ is algebraic and distal (i.e. if $h \in \tilde{H}$,then the proper values of the matrix representation M_h of h , have a module equal to 1).

A result used in |4| concerning the structure of algebraic distal groups gives :
$H = K_0 \ltimes N_0$ where K_0 is a compact group and N_0 is a closed subgroup of N .

We apply now the lemma III.2 two times : first to $H\backslash K\ltimes N$ and $\tilde{H}\backslash K\ltimes N$ then to $\tilde{H}\backslash K\ltimes N$ and $N_0\backslash K\ltimes N$.
We obtain :

Proposition III.4.

The random walks on $H\backslash K\ltimes N$ *and* $N_0\backslash K\ltimes N$ *corresponding to the probability measure* μ *(see theorem III.1 for the assumptions about* μ*) are of the same type.*

Therefore it's sufficient to prove that the random walk on $N_0\backslash K\ltimes N$ is recurrent : this is the object of the next step.

Step 2.

If $G = K \ltimes N$, we know that all the homogeneous spaces M of G have polynomial growth (see [4]). The degree of growth M is the same as the degree k of growth of $N_0\backslash N$.

From the assumptions of the theorem III.1 and the previous remark, we deduce that k

is less than or equal to 2.

$k \leq 2$ implies obviously that $\dim({}_H\backslash^N) = 1$ or 2.

a) If $\dim({}_H\backslash^N) = 1$, then H is a normal subgroup of N. Therefore $M = {}_H\backslash^N$ is a commutative group of dimension 1. Now we can apply the same method as in [7] : the main idea of this approach is contained in the following lemma.

Lemma III.5.

Let μ be a probability measure on $\mathbb{R}$ and P the transition kernel of the associated right random walk. If there exist a function $f : \mathbb{R} \to \mathbb{R}^+$ such that $Pf(x) \geq f(x)$ for $x \notin K$ (K is a compact subset) and such that $f(x) \to +\infty$ as $|x| \to +\infty$. Then the random walk of law μ is recurrent.

Proof.

Suppose that the random walk is transient. Let K be a compact subset such that : $Pf(x) \leq f(x)$, $\forall x \notin K$.

If X is a random variable with law μ, we have :

$$E(f(x + X)) \leq f(x).$$

Let Y_n be the random walk of law μ and T_K it's entry time in K. Therefore :

$$E_x\{f(Y_{n \wedge T_K})\} \leq f(x) \qquad (1)$$

But :

$$Ex\,\{f(Y_{n \wedge T_K})\} = E_x\,\{f(Y_{T_K})\ ;\ n > T_K\} + E_x\,\{f(Y_n)\ ;\ n \leq T_K\}.$$

Y_n is supposed to be transient, this implies that there exist $x \notin K$ and $a > 0$ such that : $P_x(T_K = +\infty) = a$.

Therefore for all $M > 0$, there exist N such that :

$$n > N \implies E_x(f(Y_n),\ n \quad T_K) \geq \frac{Ma}{2}.$$

We can choose M such that : $\frac{Ma}{2} > f(x)$.

Finaly : $E_x\,\{f(Y_n),\ n < T_K\} > f(x) \qquad (2)$

(2) is in contradiction with the hypothesis and we conclude that the random walk is recurrent.

Remark III.6.

As in [7] we can see that for the function $f : \mathbb{R} \to \mathbb{R}^+$ defined by $f(x) = |x|^{\frac{1}{2}}$ the assumption of the lemma III.5 is fulfilled.

b) $\dim({}_H\backslash^N) = 2$.

We consider the action of N on ${}_H\backslash^N$, this means the mapping : $N \to {}_H\backslash^N$ defined by $x \to 0 \cdot x$.

If $\dim({}_H\backslash^N) = 2$ then ${}_H\backslash^N \quad \mathbb{R}^2$ (in the topological sense).

$0 \cdot x$ is a two dimensional vector whose components are both of degree 1.

Therefore H contains $N^2 = (N,N)$ (the first derivative of N) and ${}_H\backslash^N$ is a commutative group of rank 2.

We can again apply the method of [7] : the central limit theorem for random walks on $K \ltimes \mathbb{R}^2$ is the key of the proof.

If C is a compact subset of $M = {}_H\backslash^{K \ltimes N}$ we obtain :

$$U_x(C) = \sum_{n=0}^{\infty} \varepsilon_x * \mu^{*n}(C) \geq \frac{\beta}{n\varepsilon^2} \sum_{k=0}^{n} \mu^k(C_{2\varepsilon\sqrt{n}}) = \alpha(\varepsilon)$$

where $C_{2\varepsilon\sqrt{n}} = K \times C'_{2\varepsilon\sqrt{n}}$, β is a constant, $x \in M$

$C'_{2\varepsilon\sqrt{n}}$ is a disc of $\mathbb{R}^2$ whose radius is $2\varepsilon\sqrt{n}$.

With the help of the central limit theorem on $K \times \mathbb{R}^2$ we find :

$$\alpha(\varepsilon) \geq \frac{\beta'}{\varepsilon^2} \int_0^1 (1 - e^{-\beta'/x})\, dx \quad , \quad \beta' \, , \, \beta'' \text{ are positive constants and}$$

$$\lim \alpha(\varepsilon) = +\infty \quad \text{as} \quad \varepsilon \to 0 . \blacksquare$$

Remark III.7.

The semi-simple splitting technic (see [1]) asserts that if R is a simply connected Lie group which is solvable and of rigid type then there exist a compact group K and a unique simply connected nilpotent Lie group N such that :

$$K \ltimes R = K \ltimes N \quad .$$

This relation permits to say that the result of the theorem III.1 remains true for the homogeneous spaces of $G = K \times R$.

IV. Conclusion

In this paper we have given a broad class of recurrent measures for the homogeneous spaces of the groups which are compact extensions of simply connected solvable Lie groups of rigid type. At the moment we don't know if such kind of result can be obtained for the homogeneous spaces of connected Lie groups.

REFERENCES

[1] L. AUSLANDER, L.W. GREEN : G-induced flows. A.J.M., 88, p. 43-60 (1966)

[2] P. CREPEL : Marches aléatoires sur le groupe des déplacements de $\mathbb{R}^2$. Lecture Notes in mathematics n° 532, Springer Verlag.

[3] Y. DERRIENNIC, Y. GUIVARC'H : Théorème de renouvellement pour les groupes non moyennables. Note aux C.R.A.S. Paris, t. 277, p. 613-615.

[4] L. GALLARDO, R. SCHOTT : Marches aléatoires sur les espaces homogènes de certains groupes de type rigide. Astérisque n° 74, p. 149-170 (1980).

[5] Y. GUIVARC'H : Sur la loi des grands nombres et le rayon spectral d'une marche aléatoire. Astérisque n° 74, p. 47-98 (1980).

[6] H. HENNION, B. ROYNETTE : Un théorème de dichotomie pour une marche aléatoire sur un espace homogène. Astérisque n° 74, p. 99-122 (1980).

[7] A. HUARD : Récurrence des marches aléatoires des espaces homogènes récurrents du groupe des déplacements de $\mathbb{R}^d$. Astérisque n° 74, p. 139-148 (1980).

[8] D. REVUZ : Sur le théorème de dichotomie de Hennion-Roynette. Annales Institut Elie Cartan n° 7, p. 143-147 (1983).

[9] R. SCHOTT : Random walks on homogeneous spaces. Lecture Notes in mathematics n° 1064, p. 564-575.

[10] R. SCHOTT : Irrfahrten auf nicht mittelbaren homogenen Räumen. Arbeitsbericht 1-2, 1981, Seite 63-76, Mathematisches Institut Salzburg.

René SCHOTT
Université de Nancy I
UER Sciences Mathématiques
U.A. n° 750 du C.N.R.S.
B.P. 239
54506 Vandœuvre les Nancy Cedex

A Central Limit Theorem for Coalgebras

Michael Schürmann

Abstract. Using elementary properties of coalgebras, a limit theorem for linear functionals on a coalgebra is proved which generalizes several non-commutative central limit theorems [3, 5, 6, 9] .

A coalgebra is a triplet $(\mathcal{C}, \Delta, \delta)$ consisting of a complex vector space $\mathcal{C}$, a linear mapping

$$\Delta : \mathcal{C} \rightarrow \mathcal{C} \otimes \mathcal{C}$$

(here $\mathcal{C} \otimes \mathcal{C}$ denotes the algebraic vector space tensor product) and a linear functional δ on $\mathcal{C}$ such that

$$(\mathrm{Id} \otimes \Delta) \circ \Delta = (\Delta \otimes \mathrm{Id}) \circ \Delta$$

and

$$(\mathrm{Id} \otimes \delta) \circ \Delta = (\delta \otimes \mathrm{Id}) \circ \Delta = \mathrm{Id};$$

see [1, 8] . The "Fundamental Theorem on Coalgebras" says that given an element c of $\mathcal{C}$ the smallest subcoalgebra of $\mathcal{C}$ containing c is finite-dimensional as a vector space ([8] Theorem 2.2.1.). Let φ and ψ be two linear functionals on $\mathcal{C}$, that is $\varphi, \psi \in \mathcal{C}^*$ where $\mathcal{C}^*$ denotes the algebraic dual space of $\mathcal{C}$. Then the <u>convolution product</u> $\varphi * \psi \in \mathcal{C}^*$ of φ and ψ is defined by

$$\varphi * \psi = (\varphi \otimes \psi) \circ \Delta .$$

$\mathcal{C}^*$ becomes an associative, unital algebra with convolution as multiplication and with δ as its unit element. If we set

$$\mathcal{T}(\varphi) = (\mathrm{Id} \otimes \varphi) \circ \Delta,$$

then the mapping

$$\mathcal{T} : \mathcal{C}^* \rightarrow \mathcal{L}(\mathcal{C})$$

(where $\mathcal{L}(\mathcal{C})$ denotes the algebra of linear operators on $\mathcal{C}$) is an algebra homomorphism. Moreover we have

$$\delta \circ \mathcal{T} = \mathrm{Id}. \tag{1}$$

We denote by φ^{*n} the n-fold convolution of $\varphi \in \mathcal{C}^*$. As a consequence of the Fundamental Theorem on Coalgebras the series

$$\sum_{n=0}^{\infty} \frac{\varphi^{*n}}{n!}(c)$$

converges for all $\varphi \in \mathcal{C}^*$ and $c \in \mathcal{C}$. We denote its limit by $(\exp_* \varphi)(c)$, thus defining the convolution exponential $\exp_* \varphi \in \mathcal{C}^*$ of φ; see also [7].

Let $(\varphi_{nj})_{\substack{n \geq 1 \\ 1 \leq j \leq k(n)}}$, $k(n) \in \mathbb{N}$, be a double array of linear functionals on $\mathcal{C}$. We call (φ_{nj}) uniformly infinitesimal, if

$$\lim_{n \to \infty} \max_{1 \leq j \leq k(n)} |(\varphi_{nj} - \delta)(c)| = 0$$

for all $c \in \mathcal{C}$. Now we are ready to state our result which was proved in [3] for a special coalgebra using different methods.

Theorem 1. Let $(\mathcal{C}, \Delta, \delta)$ be a coalgebra and let (φ_{nj}) be a uniformly infinitesimal double array of linear functionals on $\mathcal{C}$ such that the functionals of each row commute (as elements of the convolution algebra $\mathcal{C}^*$). Let (φ_{nj}) also fulfill the condition

$$\sup_{n \geq 1} \sum_{1 \leq j \leq k(n)} |(\varphi_{nj} - \delta)(c)| < \infty$$

for all $c \in \mathcal{C}$. Then for a linear functional ψ on $\mathcal{C}$ the pointwise convergence of

$$\sum_{1 \leq j \leq k(n)} (\varphi_{nj} - \delta) \tag{2}$$

to ψ for $n \to \infty$ implies the pointwise convergence of

$$\prod^*_{1 \leq j \leq k(n)} \varphi_{nj} \tag{3}$$

to $\exp_* \psi$ for $n \to \infty$. (Here $\prod^*_{1 \leq j \leq k(n)} \varphi_{nj}$ stands for $\varphi_{n1} * \dots * \varphi_{nk(n)}$.)

Proof: We assume that (2) converges to ψ pointwise. Let c be a fixed element of $\mathcal{C}$. Denote by $\mathcal{D}_c$ the smallest subcoalgebra of $\mathcal{C}$ containing c (which is finite-dimensional by the Fundamental Theorem on Coalgebras.) The operators $\mathcal{T}(\varphi)$, $\varphi \in \mathcal{C}^*$, leave $\mathcal{D}_c$ invariant. We denote by T_{nj}, T the linear operators on $\mathcal{D}_c$ which are obtained by restricting the operators $\mathcal{T}(\varphi_{nj})$, $\mathcal{T}(\psi)$ respectively. If we take any norm on $\mathcal{D}_c$ we have

$$\lim_{n \to \infty} \max_{1 \leq j \leq k(n)} \| T_{nj} - \mathrm{Id} \| = 0 \tag{4}$$

$$\sup_{n \geq 1} \sum_{1 \leq j \leq k(n)} \| T_{nj} - \mathrm{Id} \| < \infty \tag{5}$$

and

$$\lim_{n \to \infty} \sum_{1 \leq j \leq k(n)} (T_{nj} - Id) = T \tag{6}$$

(norm convergence).

If T_{nj}, T are complex numbers, it is a well-known lemma (see for instance [4] p. 184) that (4), (5) and (6) imply

$$\lim_{n \to \infty} \prod_{1 \leq j \leq k(n)} T_{nj} = e^{T}. \tag{7}$$

But as the operators $T_{n1}, \ldots, T_{nk(n)}$ commute, this lemma in our case can be proved in exactly the same manner by using the logarithm. Applying the counit δ to equation (7) and using relation (1) and the fact that $\mathcal{T}$ is an algebra homomorphism, we arrive at the claimed pointwise convergence of (3). □

We call a coalgebra $(\mathcal{C}, \Delta, \delta)$ <u>graded</u>, if $\mathcal{C}$ is an $\mathbb{N}$-graded vector space, $\mathcal{C} = \bigoplus_{l=0}^{\infty} \mathcal{C}^{(l)}$, and Δ is homogeneous of degree 0, that is

$$\Delta\, \mathcal{C}^{(l)} \subset \bigoplus_{l_1 + l_2 = l} \mathcal{C}^{(l_1)} \otimes \mathcal{C}^{(l_2)};$$

see [2]. If $c \in \mathcal{C}$ is a homogeneous element, denote by deg(c) its degree. For $t \in \mathbb{C}$ define the linear operator $\alpha(t)$ on $\mathcal{C}$ by

$$\alpha(t)c = t^{\deg(c)}c$$

for c homogeneous. We have

$$(\varphi * \psi) \circ \alpha(t) = (\varphi \circ \alpha(t)) * (\psi \circ \alpha(t)) \tag{8}$$

for all $\varphi, \psi \in \mathcal{C}^*$.

<u>Theorem 2</u>. Let $(\mathcal{C}, \Delta, \delta)$ be a graded coalgebra and let $s \geq 1$ be an integer. If a linear functional φ on $\mathcal{C}$ vanishes on $\mathcal{C}^{(l)}$, $0 < l < s$, and agrees with δ on $\mathcal{C}^{(0)}$, then

$$\varphi^{*n} \circ \alpha(n^{-\frac{1}{s}})$$

converges pointwise to $\exp_* d_\varphi$ where d_φ is the linear functional on $\mathcal{C}$ vanishing on $\mathcal{C}^{(l)}$, $l \neq s$, and agreeing with φ on $\mathcal{C}^{(s)}$.

<u>Proof</u>: We set $k(n) = n$ and $\varphi_{nj} = \varphi \circ \alpha(n^{-\frac{1}{s}})$. Using the fact that δ vanishes on $\mathcal{C}^{(l)}$ for $l > 0$ (see [2] § 11.3), it is easy to check that the double array (φ_{nj}) fulfills the conditions of Theorem 1 with

$\psi = d_\varphi$. Now Theorem 1 together with (8) yields Theorem 2. □

Example

Let V be a complex vector space, $V \neq \{0\}$. Denote by $\mathcal{T}(V)$ the tensor algebra of V. We have the graduation

$$\mathcal{T}(V) = \bigoplus_{l=0}^{\infty} \mathcal{T}(V)^{(l)}$$

of $\mathcal{T}(V)$ where $\mathcal{T}(V)^{(l)}$, $l > 0$, is the linear span of all monomials $v_1 \otimes \ldots \otimes v_l$, $v_j \in V$, and $\mathcal{T}(V)^{(0)} = \mathbb{C}\,\mathbb{1}$. Define

$$\delta : \mathcal{T}(V) \rightarrow \mathbb{C}$$

to be the algebra homomorphism given by $\delta(v) = 0$, $v \in V$. We consider two different comultiplications on $\mathcal{T}(V)$. Denote by

$$\varepsilon : \mathbb{N} \times \mathbb{N} \rightarrow \{+1, -1\}$$

the mapping $\varepsilon \equiv 1$ or the mapping $\varepsilon(n,m) = (-1)^{nm}$. One can impose an algebra structure on the vector space $\mathcal{T}(V) \otimes \mathcal{T}(V)$ by setting

$$(a \otimes b)(a' \otimes b') = \varepsilon(\deg(b), \deg(a'))\, aa' \otimes bb'$$

for b, a' homogeneous; cf. [2], see also [7]. We denote this algebra by $\mathcal{T}(V) \otimes^{\varepsilon} \mathcal{T}(V)$. A comultiplication

$$\Delta^{\varepsilon} : \mathcal{T}(V) \rightarrow \mathcal{T}(V) \otimes \mathcal{T}(V)$$

is given by setting

$$\Delta^{\varepsilon}(v) = v \otimes \mathbb{1} + \mathbb{1} \otimes v,$$

$v \in V$, and by requiring Δ^{ε} to be an algebra homomorphism from $\mathcal{T}(V)$ to $\mathcal{T}(V) \otimes^{\varepsilon} \mathcal{T}(V)$. In the case $\varepsilon \equiv 1$ Theorem 1 gives a result which was proved in [3]. Theorem 2 yields the results of [5] ($\varepsilon \equiv 1$) and [9] ($\varepsilon \not\equiv 1$), the limit functional $\exp_* d_\varphi$ of Theorem 2 being in the case $s = 2$ a Boson ($\varepsilon \equiv 1$) or Fermion ($\varepsilon \not\equiv 1$) quasi-free state. The central limit theorem of [6] can easily be derived from the case $s = 2$, $\varepsilon \not\equiv 1$.

References

[1] Abe, E.: Hopf Algebras, Cambridge University Press (1980)

[2] Bourbaki, N.: Elements of Mathematics, Algebra, Chap. III, Hermann, Paris (1973)

[3] Canisius, J.: Algebraische Grenzwertsätze und unbegrenzt teilbare Funktionale, Diplomarbeit, Heidelberg (1979)

[4] Chung, K. L.: A Course in Probability Theory, Harcourt, Brace and World, New York (1968)

[5] Giri, N. and von Waldenfels, W.: An Algebraic Version of the Central Limit Theorem, Z. Wahrscheinlichkeitstheorie verw. Gebiete 42, 129-134 (1978)

[6] Hudson, R. L.: A Quantum-Mechanical Central Limit Theorem for Anti-Commuting Observables, J. Appl. Prob. 10, 502-509 (1973)

[7] Schürmann, M.: Positive and Conditionally Positive Linear Functionals on Coalgebras, in Lect. Notes in Math. 1136, Springer, New York, Heidelberg, Berlin, 475-492 (1985)

[8] Sweedler, M. E.: Hopf Algebras, Benjamin, New York (1969)

[9] von Waldenfels, W.: An Algebraic Central Limit Theorem in the Anticommuting Case, Z. Wahrscheinlichkeitstheorie verw. Gebiete 42, 135-140 (1978)

Michael Schürmann
Institut f. Ang. Mathematik
Universität Heidelberg
Im Neuenheimer Feld 294
D-6900 Heidelberg

HAAR MEASURES IN A REPRESENTATION AND A DECOMPOSITION PROBLEM

To the memory of Alfréd Haar (1885-1933) on the occasion of his 100th birthday

Gábor J. Székely
Department of Probability Theory
Loránd Eötvös University
Budapest, Hungary

"Alfred Haar ist einer der Mathematiker, deren Werk auf die neueste Entwicklung der Mathematik einen allgemein anerkannten großen Einfluß ausgeübt hat. Das gilt ... insbesondere für seine letzte Arbeit über kontinuierliche Gruppen, die übrigens seine Antrittsarbeit an der Ungarischen Akademie der Wissenschaften war."
(Alfred Haar:Gesammelte Arbeiten,1959,Vorwort)

1. A representation problem

Our first question has seemingly nothing to do with Haar measures but it will turn out that they are very closely related.

Problem 1. When is a commutative semigroup representable as a convolution semigroup of (probability) measures on a locally compact topological group?

The characteristic functions of these measures are complex valued functions. The convolution semigroup corresponds to the multiplicative structure of these characteristic functions thus if the desired representation exists then each element of the semigroup is either idempotent (if the corresponding characteristic function takes no other values than 0 and 1) or is of infinite order. Thus if S is a commutative semigroup such that

(i) every element of S is idempontent

or

(ii) S is torsionfree

then we can hope that the desired representation of S exists.

A semigroup S is called separative if the characters of S separate the elements of S. The classical paper Hewitt and Zuckermann (1956) (see also Clifford and Preston (1961), 4.3) proves that S is separative if and only if

(iii) $s_1^2 = s_2^2 = s_1 s_2$ implies $s_1 = s_2$.

Since our convolution semigroups obviously satisfy (iii) it is a necessary condition for the existence of the representation. A further necessary condition comes from Tortrat (1965):

(iv) if $s_1 s_2 = s_1$ then there exists an idempotent element $s \in S$ such that $ss_1 = s_1$ and $ss_2 = s$.

Our condition (i) obviously implies (iii) and (iv) and we shall prove below that (i) itself is a sufficient condition of the representation. However not even (ii), (iii) and (iv) together seem to be sufficient. I plan to return to this torsionfree case in another paper.

Theorem 1. A commutative semigroup S with condition (i) is always representable as a convolution semigroup of some probability measures on a compact group G. These probability measures are normed Haar measures on compact subgroups of G.

Proof. The convolution of two Haar measures on the compact subgroups K_1 and K_2 is also a Haar measure on the compact group generated by K_1 and K_2, thus it is enough to prove that S is representable as the "union" structure of certain compact subgroups of a compact group G ("union" means the generated subgroup).

By condition (i) we can define a partial ordering in S:

$$s_1 \leq s_2 \text{ if and only if } s_1 s_2 = s_1 .$$

The sets $A_s = \{x : x \in S, x \leq s\}$ form an algebraic lattice with respect to the union and intersection operations. By the classical representation theorem of Stone this lattice (as every lattice) can be represented as the union-intersection lattice of subsets of a set X, but we need more. Let $I_s \subset X$ denote the subset which corresponds to s in the Stone representation. Denote by C_2 the cyclic group having 2 elements. Then $G = C_2^{2^X}$ is obviously a compact group. Let $G_s \subset G$ be the compact subgroup of $X \to C_2$ functions taking the value 0 outside of I_s. The representation $s \to G_s$ obviously satisfies the requirements.

2. Decomposition of uniform distributions

Let U_H be the uniform distribution on a subset H of a commutative group G (i.e. if H has finitely many elements, say n, then let each element of H have probability $1/n$; in case H is an infinite set suppose that G is a locally compact topological group, H has positive and finite Haar measure and let U_H be the normed Haar measure on H $(U_H(H) = 1)$.

Problem 2. Under what conditions on H is U_H the convolution square of a probability distribution μ defined on certain subsets of G?

In other words when can we solve convolution equation

$$(*) \qquad U_H = \mu * \mu \quad .$$

If H is a compact subgroup of G then the Haar measure U_H is idempotent thus $\mu = U_H$ solves this equation. (For a description of idempotent measures on locally compact groups see Cohen (1960a,b).) What if H is not a subgroup (or a coset of a subgroup)? Problem 2 in its full generality seems to be rather hopeless. Not even the "simple" case when G is the mod m group of integers is settled. In the following we shall confine ourselves to the case of additive group R of real numbers.

Theorem 2. The convolution equation $(*)$ has a solution for finite $H \subset R$ if and only if H has only one element.

Proof. We show somewhat more. Let

$$(**) \qquad U_H = \mu * \mu * \ldots * \mu$$

where the number of convolution factors on the right hand side is $m > 1$. Then $(**)$ has a solution if and only if H has only one element. The "if" part is trivial (every real number can be divided by m).

For the proof of the "only if" part let $x_1 < \ldots < x_k$ be the ordered set of elements of H $(k > 1)$. Let $X_1, \ldots, X_n$ be independent random variables with distribution μ. The indirect reasoning shows that $\Pr(X_i = x_1/m) = \Pr(X_i = x_k/m) = k^{-1/m}$ $(i=1,2,\ldots,m)$, thus

$$\Pr\left(X_1+\ldots+X_m = \frac{x_1+(m-1)x_k}{m}\right) \geq \sum_{i=1}^{m} \Pr(X_i=x_1/m, X_1=\ldots=X_{i-1}=X_{i+1}=\ldots=X_m=x_k/m) = \frac{m}{k} > \frac{1}{k}$$

which is a contradiction.

A similar proof works for "most" non-finite sets $H \subset R$ but I could not prove the analogue of Theorem 2 for all Borel sets H with positive and finite Lebesgue measure. If H is an interval then our problem is trivial. E.g. in case $H = (-1,1)$ the characteristic function of U_H is $\sin t / t$ and all of its roots are of multiplicity 1 thus it cannot be the square (or any power greater than one) of another (analytic) characteristic function. The problem is more intricate if instead of the uniform distribution U_H we consider a "quasi-uniform" distribution V having a density function with the following property: it is bounded away from 0 and ∞ on the interval $(-1,1)$ and 0 outside of this interval.

Theorem 3. Let V be a quasi-uniform distribution on $(-1,1)$. If V is symmetrically distributed around 0 then the convolution equation $V = \mu * \mu$ has no solution.

Proof. By the symmetry of V the characteristic function $\hat{V}$ is real and in a neighbourhood of 0 it is positive thus in this neighbourhood the characteristic function $\hat{\mu}$ (whose existence is supposed indirectly) is also real. Since $\hat{\mu}$ is obviously analytic it is real everywhere, therefore $\hat{V} \geq 0$. By Corollary to XV.3. Theorem 3 of Feller (1966) if a characteristic function is nonnegative then it is integrable if and only if its density function is bounded. This boundedness was one of our conditions thus $\hat{V}$ is integrable therefore its density function is continuous which contradicts to our other condition.

Remark. I conjecture that there exist nonsymmetric quasi-uniform distributions having convolution square root.

REFERENCES

Clifford, A.H. and Preston, G.B. (1961), The Algebraic Theory of Semigroups, Rhode Island, Amer. Math. Soc.

Cohen, P.J. (1960a), On a conjecture of Littlewood and idempotent measures, Amer. J. Math. 82, 191-212.

Cohen, P.J. (1960b), On homomorphisms of group algebras, Amer. J. Math. 82 213-226.

Feller,W. (1966), An Introduction to Probability Theory and Its Applications II., New York, Wiley.

Hewitt,E. and Zuckermann,H.S. (1956), The ℓ_1-algebra of a commutative semigroup, Trans. Amer. Math. Soc. 83, 70-97.

Tortrat,A. (1965), Lois de probabilité sur un espace topologique complètement régulier et produits infinis à termes indépendants dans un groupe topologique, Ann. Inst. H.Poincaré 1, 217-237.

COMPACTNESS, MEDIANS AND MOMENTS

K. Urbanik
Institute of Mathematics, Wrocław University,
Plac Grunwaldzki 2/4, 50-384 Wrocław, Poland

Generalized convolutions were introduced in [2]. Let us recall some definitions. We denote by P the set of all probability measures defined on Borel subsets of the positive half-line R_+. The set P is endowed with the topology of weak convergence. For $\mu \in P$ and $a > 0$ we define the map T_a by setting $(T_a\mu)(E) = \mu(a^{-1}E)$ for all Borel subsets E of R_+. By δ_c we denote the probability measure concentrated at the point c.

A continuous in each variable separately commutative and assiociative P-valued binary operation $\circ$ on P is called *a generalized convolution* if it is distributive with respect to convex combinations and maps T_a $(a > 0)$ with δ_0 as the unit element. Moreover, the key axiom postulates the existence of norming positive constants c_n and a measure $\gamma \in P$ other than δ_0 such that $T_{c_n}\delta_1^{\circ n} \to \gamma$, where $\delta_1^{\circ n}$ is the n-th power of δ_1 under $\circ$. The measure γ is called *the characteristic measure* of $\circ$. It is defined uniquely up to a scale change T_a $(a > 0)$ and, by Proposition 4.5 in [3], fulfils the equation $T_a\gamma \circ T_b\gamma = T_{g_\kappa(a,b)}\gamma$, where $0 < \kappa \le \infty$, $g_\kappa(a,b) = (a^\kappa+b^\kappa)^{1/\kappa}$ if $0 < \kappa < \infty$ and $g_\infty(a,b) = \max(a,b)$. The constant κ is called *the characteristic exponent* of $\circ$. As examples of generalized convolutions we quote ordinary convolution, symmetric convolution, Kingman convolution, max-convolution and convolutions induced by Kendall operations on random sets.

It has been shown in [4], Chapter 2 that the generalized convolution $\circ$ can be extended to the space $\bar{P}$ of all Borel probability measures on the compactified half-line $\bar{R}_+ = [0,\infty]$. Since the space $\bar{P}$ is compact in the topology of weak convergence, this enables us to use compactness arguments and therefore is a useful tool in the study of generalized convolutions. We identify the space P with the subspace of $\bar{P}$ consisting of measures with zero mass at ∞. The algebraic properties of $\circ$ on P carry over to $\bar{P}$. Moreover, the extended convolution is continuous in each variable separately on $\bar{P}$. By Theorem 4.2 and Corollaries 3.2 and 3.5 in [4] for any $\mu \in P$ other than δ_0 we have

$$(1) \qquad \mu^{\circ n} \to \delta_c \quad \text{in } \bar{P}$$

where $0 < c \le \infty$.

Given $\mu \in P$ and a sequence $\{a_n\}$ of positive numbers by $G(\{a_n\},\mu)$ we shall denote the set of all cluster points in $\bar{P}$ of the sequence $T_{a_n}\mu^{\circ n}$. Of course, the set $G(\{a_n\},\mu)$ is compact in $\bar{P}$. We say that μ *belongs to the domain of attraction of a compact subset of* $P\setminus\{\delta_0\}$ if $G(\{a_n\},\mu) \subset P\setminus\{\delta_0\}$ for a norming sequence $\{a_n\}$. For the symmetric convolution this compactness property was introduced and studied by W. Feller in [1].

Given $\mu \in P$, by $A(\mu)$ we shall denote the set of all norming sequences $\{a_n\}$ for which the inclusion $G(\{a_n\},\mu) \subset P\setminus\{\delta_0\}$ is true. Furthermore two norming sequences $\{a_n\}$ and $\{b_n\}$ are said to be *equivalent*, in symbols $\{a_n\} \sim \{b_n\}$, if $c^{-1} \le a_n/b_n \le c$ $(n=1,2,\ldots)$ for a positive constant c. As a consequence of Lemma 1.1 in [3] we get the following statement.

Lemma 1. *Let* $\{a_n\} \in A(\mu)$. *Then* $\{b_n\} \in A(\mu)$ *if and only if* $\{b_n\} \sim \{a_n\}$.

Lemma 2. *If* $\{a_n\} \in A(\mu)$ *and* $b_n = \min\{a_j : 1 \le j \le n\}$ $(n = 1,2,\ldots)$, *then* $\{b_n\} \in A(\mu)$.

Proof. Setting $c_n = b_n/a_n$ we have $c_n \le 1$ and $T_{b_n}\mu^{\circ n} = T_{c_n}(T_{a_n}\mu^{\circ n})$ which yields the inclusion $G(\{b_n\},\mu) \subset P$. Further, observe that $b_n = a_{j_n}$ where $1 \le j_n \le n$ and, consequently, $T_{b_n}\mu^{\circ n} = T_{a_{j_n}}\mu^{\circ j_n} \circ \nu_n$ where $\nu_n = T_{a_{j_n}}\mu^{\circ(n-j_n)}$ if $j_n < n$ and $\nu_n = \delta_0$ if $j_n = n$. Applying Corollary 2.4 in [4] we conclude that $\delta_0 \notin G(\{b_n\},\mu)$ which completes the proof.

Lemma 3. *If* $\{a_n\} \in A(\mu)$ *and* $\overline{\lim\limits_{n\to\infty}} a_n > 0$, *then* $\{1\} \in A(\mu)$, $\mu^{\circ n} \to \delta_c$ $(0 < c < \infty)$ *and* $\circ$ *is the max-convolution.*

Proof. As an immediate consequence of Lemmas 1 and 2 we obtain the relation $\{1\} \in A(\mu)$. Thus, by (1), $G(\{1\},\mu) = \{\delta_c\}$ with $0 < c < \infty$. Hence it follows that δ_c is an idempotent under $\circ$. Now applying Theorems 4.1 and 4.2 in [4] we conclude that $\circ$ is the max--convolution.

Lemma 4. *Let* s *be a positive integer and* $\{a_n\} \in A(\mu)$. *Then for every* $\lambda \in G(\{a_n\},\mu)$ *there exist positive numbers* a,b *and* $\nu \in G(\{a_n\},\mu)$ *such that* $\lambda = T_a\nu^{\circ s}$ *and* $T_b\lambda^{\circ s} \in G(\{a_n\},\mu)$. *Moreover, denoting by* p_n *the integral part of* n/s *we have* $\{a_{p_n}\} \in A(\mu)$.

Proof. By Lemma 3 our statement is obvious if $\overline{\lim\limits_{n>\infty}} a_n > 0$. Assume now that $\lim\limits_{n\to\infty} a_n = 0$. Write $n = sp_n + r_n$ where $0 \le r_n < s$ and $c_n = a_n/ap_n$ $(n = 1,2,\dots)$. Then

$$(2) \qquad \rho_n = T_{a_n}\mu^{\circ r_n} \to \delta_0$$

and

$$(3) \qquad T_{a_n}\mu^{\circ n} = T_{c_n}(T_{a_{p_n}}\mu^{\circ p_n})^{\circ s} \circ \rho_n.$$

Observe that the set of all cluster points of the sequence $T_{a_{p_n}}\mu^{\circ p_n}$ is equal to $G(\{a_n\},\mu)$. Consequently, the set of all cluster points of the sequence $(T_{a_{p_n}}\mu^{\circ p_n})^{\circ s}$ is equal to $\{\nu^{\circ s}: \nu \in G(\{a_n\},\mu)\}$ and, according to Corollary 2.4 in [4], does not contain the measure δ_0. Hence, by (2) and (3), $c^{-1} \le c_n \le c$ for a positive constant c. In other words $\{a_n\} \sim \{a_{p_n}\}$ which, by Lemma 1, yields $\{a_{p_n}\} \in A(\mu)$. Moreover, we have the inclusions

$$G(\{a_n\},\mu) \subset \{T_a\nu^{\circ s}: \nu \in G(\{a_n\},\mu),\ c^{-1} \le a \le c\}$$

and

$$\{\lambda^{\circ s}: \lambda \in G(\{a_n\},\mu)\} \subset \{T_b\nu: \nu \in G(\{a_n\},\mu),\ c^{-1} \le b \le c\}$$

which complete the proof of the Lemma.

Observe that since $p_{sn} = n$, the above Lemma and Lemma 1 imply the following Corollary.

Corollary 1. *Let* s *be a positive integer and* $\{a_n\} \in A(\mu)$. *Then* $\{a_n\} \sim \{a_{sn}\}$.

Lemma 5. *Let* $\{a_n\} \in A(\mu)$. *Then all measures from* $G(\{a_n\},\mu)$ *vanish at the origin.*

Proof. Let $\lambda \in G(\{a_n\},\mu)$. It follows immediately from Lemma 4 and the compactness of $G(\{a_n\},\mu)$ that there exist a subsequence $s_1 < s_2 < \dots$ of positive integers, a sequence $\{b_k\}$ of positive numbers and probability measures ν, $\nu_k \in G(\{a_n\},\mu)$ $(k = 1,2,\dots)$ fulfilling the conditions

$$(4) \qquad \nu_k \to \nu$$

and

$$(5) \qquad \nu_k^{\circ s_k} = T_{b_k}\lambda \qquad (k=1,2,\dots).$$

Moreover, passing to a subsequence if necessary we may assume that

$$(6) \qquad T_{b_k}\lambda \to \rho \text{ in } \bar{P}.$$

Since $(T_{b_k}\lambda)(\{0\}) = \lambda(\{0\})$, it follows from (6) that $\lambda(\{0\}) \le$ $\le \rho(\{0\})$. Consequently, in order to prove the Lemma it is sufficient to show that $\rho(\{0\}) = 0$.

Given an arbitrary positive integer n, by μ_n we denote a cluster point in $\bar{P}$ of the sequence $\lambda_k^{\circ(s_k-n)}$ $(s_k > n)$. Taking into account (4), (5) and (6) we infer that

$$(7) \qquad \nu^{\circ n} \circ \mu_n = \rho \qquad (n=1,2,\dots).$$

The probability measures ρ and μ_n can be written in the form

$$(8) \qquad \rho = c\rho' + (1-c)\delta_\infty$$

and

$$\mu_n = c_n\mu_n' + (1-c_n)\delta_\infty$$

where $0 \leq c \leq 1$, $0 \leq c_n \leq 1$ and $\rho', \mu_n' \in P$. This implies, by (7),

(9) $$\rho = c_n\nu^{\circ n} \circ \mu_n' + (1-c_n)\delta_\infty .$$

Comparying this with (8) we have $c_n = c$ $(n = 1,2,\ldots)$. If $c = 0$, then, by (8), $\rho = \delta_\infty$ and, consequently, $\rho(\{0\}) = 0$. It remains the case $c > 0$. Then, by (8) and (9),

(10) $$\nu^{\circ n} \circ \mu_n' = \rho' \qquad (n = 1,2,\ldots)$$

which, by Corollary 2.3 in [4], shows that both sequences $\nu^{\circ n}$ and μ_n' are conditionally compact in P. Thus, by (1), $\nu^{\circ n} \to \delta_a$ where $0 < a < \infty$. In other words $G(\{1\},\nu) = \{\delta_a\} \subset P\setminus\{\delta_0\}$. Applying Lemma 3 we infer that $\circ$ is the max-convolution which, by (10), leads to the equality $\rho'([0,a)) = 0$. Thus $\rho(\{0\}) = 0$ which completes the proof.

Given $\lambda \in \bar{P}$, by $m(\lambda)$ and $M(\lambda)$ we shall denote the lowest and the greatest median of λ respectively. It is clear that the functions $\lambda \to m(\lambda)$ and $\lambda \to M(\lambda)$ are lower and upper semicontinuous respectively and

(11) $$m(T_a\lambda) = am(\lambda), \quad M(T_a\lambda) = aM(\lambda) \qquad (a > 0).$$

Given $\mu \in P$ other than δ_0, we put

$$l_*(\mu) = \underline{\lim}_{n\to\infty} \frac{M(\mu^{\circ 2n})}{M(\mu^{\circ n})}$$

and

$$l^*(\mu) = \overline{\lim}_{n\to\infty} \frac{M(\mu^{\circ 2n})}{M(\mu^{\circ n})} .$$

Since, by (1), $\underline{\lim}_{n\to\infty} m(\mu^{\circ n}) > 0$, the above definitions make sense. Further, we put $c_n(\mu) = M(\mu^{\circ n})^{-1}$ if $M(\mu^{\circ n}) > 0$ and $c_n(\mu) = 1$ otherwise. We shall also use the notation $l_*(\delta_0) = 1$ and $l^*(\delta_0) = \infty$. As an immediate consequence of Lemma 5, the compactness of

$G(\{a_n\},\mu)$ and the semicontinuity of medians we obtain the following statement.

Lemma 6. *Let* $\{a_n\} \in A(\mu)$. *Then*

$$m = \inf\{m(\lambda): \lambda \in G(\{a_n\},\mu)\} > 0$$

and

$$M = \sup\{M(\lambda): \lambda \in G(\{a_n\},\mu)\} < \infty.$$

Theorem 1. *If* μ *belongs to the domain of attraction of a compact subset of* $P\setminus\{\delta_0\}$, *then* $\{c_n(\mu)\} \in A(\mu)$.

Proof. Let $\{a_n\} \in A(\mu)$. Then using the notation of Lemma 6 and (11) we have the inequalities

$$\underline{\lim}_{n\to\infty} \frac{a_n}{c_n(\mu)} = \underline{\lim}_{n\to\infty} M(T_{a_n}\mu^{\circ n}) \geq m$$

$$\overline{\lim}_{n\to\infty} \frac{a_n}{c_n(\mu)} = \overline{\lim}_{n\to\infty} M(T_{a_n}\mu^{\circ n}) \leq M$$

which yield $\{a_n\} \sim \{c_n(\mu)\}$. Now our assertion follows from Lemma 1.

Theorem 2. *A measure* μ *from* P *belongs to the domain of attraction of a compact subset of* $P\setminus\{\delta_0\}$ *if and only if* $1^*(\mu) < \infty$.

Proof. The necessity. Suppose that μ belongs to the domain of attraction of a compact subset of $P\setminus\{\delta_0\}$. Then, by Theorem 1, $\{c_n(\mu)\} \in A(\mu)$ and, by Corollary 1, $\{c_n(\mu)\} \sim \{c_{2n}(\mu)\}$. Since

$$(12) \qquad \frac{M(\mu^{\circ 2n})}{M(\mu^{\circ n})} = \frac{c_n(\mu)}{c_{2n}(\mu)}$$

for n large enough, this yields $1^*(\mu) < \infty$.

The sufficiency. Suppose that $1^*(\mu) < \infty$. Then $\mu \neq \delta_0$ and, by (12), $\underline{\lim}_{n\to\infty} \frac{c_{2n}(\mu)}{c_n(\mu)} > 0$. Moreover $c_n(\mu) \leq m(\mu^{\circ n})^{-1}$ $(n = 1,2,\ldots)$. Applying Lemma 6 in [5] we conclude that $G(\{c_n(\mu)\},\mu) \subset P$. Further, for any $\lambda \in G(\{c_n(\mu)\},\mu)$ we have the inequality

$$M(\lambda) \geq \underline{\lim}_{n\to\infty} M(T_{c_n(\mu)}\mu^{\circ n}) = 1$$

which yields $\lambda \neq \delta_o$. Thus $\{c_n(\mu)\} \subset P\setminus\{\delta_o\}$ which completes the proof.

Lemma 7. *For every* $\mu \in P$ *the inequality* $1^*(\mu) \geq 1$ *is true.*

Proof. Suppose the contrary $1^*(\mu) < 1$. Then, by Theorem 2, μ belongs to the domain of attraction of a compact subset of $P\setminus\{\delta_o\}$. Applying Theorem 1 we conclude that $\{c_n(\mu)\} \in A(\mu)$. Moreover, by (12),

$$\underline{\lim}_{n\to\infty} \frac{c_{2n}(\mu)}{c_n(\mu)} > 1$$

which yields

(13) $$\overline{\lim}_{n\to\infty} c_n(\mu) = \infty.$$

Applying now Lemma 3 we infer that $\{1\} \in A(\mu)$ and, consequently, by Lemma 1, $\{1\} \sim \{c_n(\mu)\}$ which contradicts (13). The Lemma is thus proved.

Lemma 8. *If* $\{a_n\} \in A(\mu)$, $p > 0$, $n_k = 2^k n_o$ $(k = 1,2,\ldots)$ *and*

(14) $$\underline{\lim}_{k\to\infty} n_k a^p_{n_k} > 0,$$

then $\underline{\lim}_{n\to\infty} n\, a^p_n > 0$.

Proof. Put $b_n = \min\{a_j\colon 1 \leq j \leq n\}$. Then, by Lemmas 1 and 2, $\{a_n\} \sim \{b_n\}$ and, consequently,

(15) $$\underline{\lim}_{n\to\infty} \frac{b_n}{a_n} > 0.$$

Let $n_{k-1} \leq n \leq n_k$. Then we have the inequality

$$n\, a^p_n \geq n_{k-1}\, b^p_{n_k} = \frac{1}{2}\, n_k a^p_{n_k} \frac{b^p_{n_k}}{a^p_{n_k}}$$

which together with (14) and (15) yields the assertion of the Lemma.

For $\mu \in P$ we introduce the notation

$$p(\mu) = \sup\{s: \int_0^\infty x^s \mu(dx) < \infty, s > 0\}$$

and

$$q(\mu) = \sup\{s: \overline{\lim_{n\to\infty}}\, n^{-\frac{1}{s}} M(\mu^{\circ n}) < \infty, s > 0\}$$

where the supremum of the empty set is assumed to be 0. It has been shown in [5] that

$$q(\mu) = \min(p(\mu), \kappa)$$

where κ is the characteristic exponent of the generalized convolution in question.

Theorem 3. *For every* $\mu \in P$ *the inequalities*

$$l_*(\mu) \le 2^{\frac{1}{q(\mu)}} \le l^*(\mu)$$

are true.

Proof. First we shall prove the inequality

$$l_*(\mu) \le 2^{\frac{1}{q(\mu)}} .$$

It is obvious if $l_*(\mu) \le 1$. Write in the remaining case $l_*(\mu) = 2^{1/p}$ with $0 \le p < \infty$. Let $r > p$. Then there exists an index n_0 such that

$$\frac{M(\mu^{\circ 2n})}{M(\mu^{\circ n})} > 2^{\frac{1}{r}}$$

as $n \ge n_0$. Setting $n_k = 2^k n_0$ $(k = 1,2,\ldots)$ we get the inequality

$$M(\mu^{\circ n_k}) > c\, n_k^{\frac{1}{r}} \qquad (k = 1,2,\ldots)$$

with a positive constant c. Thus

$$\lim_{k\to\infty} n_k^{-\frac{1}{s}} M(\mu^{\circ n_k}) = \infty$$

for every $s > r$. This shows that $s \ge q(\mu)$. By the arbitrariness of s and $r > p$ we obtain the inequality $p \ge q(\mu)$ or, equivalently, $l_*(\mu) \le 2^{\frac{1}{q(\mu)}}$.

The inequality $2^{\frac{1}{q(\mu)}} \le l^*(\mu)$ is obvious if $l^*(\mu) = \infty$. Assume that $l^*(\mu) < \infty$. Then, by Theorems 1 and 2, $\{c_n(\mu)\} \in A(\mu)$. Moreover, by Lemma 7, $l^*(\mu) \ge 1$ and, consequently, $l^*(\mu) = 2^{1/p}$ with $0 < p \le \infty$. Given $0 < r < p$, there exists an index n_o such that, by (12),

$$\frac{c_{2n}(\mu)}{c_n(\mu)} > 2^{-\frac{1}{r}}$$

as $n \ge n_o$. Setting $n_k = 2^k n_o$ $(k = 1,2,\ldots)$ we get the inequality

$$c_{n_k}(\mu) > c\, n_k^{-\frac{1}{r}} \qquad (k = 1,2,\ldots)$$

for a positive constant c. Applying Lemma 8 we get the inequality

$$\underline{\lim}_{n\to\infty} n\, c_n^r(\mu) > 0$$

or, equivalently,

$$\overline{\lim}_{n\to\infty} n^{-\frac{1}{r}} M(\mu^{\circ n}) < \infty.$$

Thus $r \le q(\mu)$ which yields $p \le q(\mu)$. Consequently, $2^{1/q(\mu)} \le l^*(\mu)$ which completes the proof.

As an immediate consequence of Theorems 2 and 3 we get the following result.

Corollary 2. *If* μ *belongs to the domain of attraction of a compact subset of* $P\backslash\{\delta_o\}$, *then* $p(\mu) > 0$.

Lemma 9. *Let* $\{a_n\} \in A(\mu)$. *There exists then a positive constant* c *such that* $l^*(\lambda) \le c\, l^*(\mu)$ *for every* $\lambda \in G(\{a_n\},\mu)$.

Proof. Let s be a positive integer and $\lambda \in G(\{a_n\},\mu)$. By Lemma 4 there exists a positive number b such that $T_b\lambda^{\circ s} \in G(\{a_n\},\mu)$. Using the notation of Lemma 6 and (11) we have the inequality

$$\frac{M(\lambda^{\circ s})}{m(\lambda^{\circ s})} \le \frac{M}{m} = c < \infty \qquad (s = 1,2,\ldots). \tag{16}$$

Taking a subsequence $n_1 < n_2 < \dots$ for which $\rho_k = T_{a_{n_k}} \mu^{\circ n_k} \to \lambda$ we get the inequalities

$$m(\lambda^{\circ s}) \leq \underline{\lim}_{k\to\infty} M(\rho_k^{\circ s})$$

and

$$M(\lambda^{\circ s}) \geq \overline{\lim}_{k\to\infty} M(\rho_k^{\circ s})$$

Since, by (11),

$$\overline{\lim}_{k\to\infty} \frac{M(\rho_k^{\circ 2s})}{M(\rho_k^{\circ s})} \leq l^*(\mu),$$

we have

$$\frac{m(\lambda^{\circ 2s})}{M(\lambda^{\circ s})} \leq l^*(\mu) \qquad (s = 1,2,\dots)$$

which together with (16) implies

$$\frac{M(\lambda^{\circ 2s})}{M(\lambda^{\circ s})} \leq c\, l^*(\mu) \qquad (s = 1,2,\dots).$$

Consequently, $l^*(\lambda) \leq c\, l^*(\mu)$ which completes the proof.

As a consequence of the above Lemma and Theorems 2 and 3 we get the following Theorem.

Theorem 4. *Let* $\{a_n\} \in A(\mu)$. *Then*

$$\inf\{p(\lambda) : \lambda \in G(\{a_n\},\mu)\} > 0 .$$

REFERENCES

[1] W. Feller, On regular variation and local limit theorems, Proc. of the Fifth Berkeley Symposium on Math. Statist. and Prob. Vol. II, Part I, Probability Theory (1967), 373-388.

[2] K. Urbanik, Generalized convolutions, Studia Math. 23 (1964), 217-245.

[3] ———, Generalized convolutions IV, Studia Math., (in print).

[4] ———, Quasi-regular generalized convolutions, Coll. Math., (in print).

[5] ———, Limit behaviour of medians, Bull. of the Polish Acad. of Sciences, Mathematics (in print).

NON-COMMUTATIVE ALGEBRAIC CENTRAL LIMIT THEOREMS

Wilhelm von Waldenfels
Universität Heidelberg
Institut für Angewandte Mathematik
Im Neuenheimer Feld 294
6900 Heidelberg 1
Federal Republic of Germany

Abstract

We want to generalize some algebraic aspects of the weak law of large numbers and the central limit theorem to the non-commutative case. Let $\mathfrak{A}$ and $\mathfrak{B}$ two 2-graded algebras and $\omega: \mathfrak{A} \to \mathfrak{B}$ a linear even mapping preserving 1. Let $(a_i)_{i \in I}$ a family of homogeneous elements of $\mathfrak{A}$ and f a polynomial in the non-commutative indeterminates x_i, $i \in I$. Assume a fixed integral number $s \geq 1$. Assuming that $\omega(a_{i_1} \ldots a_{i_\ell}) = 0$ for $i_1, \ldots, i_\ell \in I$ and $1 \leq \ell \leq s-1$ we study

$$C_N^{(s)}(f) = M_N \circ \omega^{\otimes N} f(x_i \mapsto N^{-1/s}(a_i \otimes 1 \otimes \ldots \otimes 1 + \ldots + 1 \otimes \ldots \otimes 1 \otimes a_i))$$

for $N \to \infty$. At first it can be shown that it is sufficient to consider $\mathfrak{A}$ to be equal to the free algebra $\mathfrak{F}(I)$ generated by x_i, $i \in I$ and $\mathfrak{B}$ to be the free graded commutative algebra $\mathfrak{L}(W'(I))$ generated by the ξ_w, $w \in W'(I)$, where $W'(I)$ denotes the set of nonempty words of alphabet I. Then it is proved that $C_N^{(s)}(f) \to \mu \circ \gamma_s(f)$, where γ_s is higher Gaussian mapping $\mathfrak{F}(I) \to \mathfrak{L}(W'(I))$ and μ is a homomorphism containing all special informations on the a_i and ω. Finally a structure theorem for γ_s is proved. In the case s = 1 we obtain convergence to Grassmann numbers for the averages of odd a_i. The structure theorem in the case s = 2 induces some commutation relations, how they are known from quantum mechanics, the commutator between odd and even quantities being a Grassmann number.

Acknowledgement

The research reported here was supported by the Natural Sciences and Engineering Research Council of Canada, grant no. A2151, held by L.L. Campbell and by the Deutsche Forschungsgemeinschaft, Sonderforschungsbereich 123. The paper was written during my stay at the Department of Mathematics and Statistics, Queen's University, Kingston, Ontario, whom I would like to thank very much for their kind hospitality.

0. Introduction

The classical one-dimensional central limit theorem can be stated as follows. Assume a probability measure P on the real line such that $\int P(dx)x = 0$ and $\int P(dx)x^2 < \infty$ then for any continuous bounded $f:\mathbb{R} \to \mathbb{C}$ and for $N \to \infty$

$$\int P(dx_1)\ldots P(dx_N)\ f\ (N^{-1/2}(x_1+\ldots+x_N)) \to \int g_v(dx)\ f(x)$$

where g_v is a gaussian distribution, if $v = 0$ then $g_v = \delta_o$ and if $v > 0$ then $g_v(dx) = (2\pi v)^{-1/2} \exp(-x^2/2v)dx$.

This theorem can be formulated in an algebraic way. Assume a probability measure P whose all moments exist and associate to it a linear functional on the algebra $\mathcal{P} = \mathbb{C}[x]$ of the polynomials in one variable x by putting

$$\omega(f) = \int P(dx)\ f(x) \quad .$$

Let $\omega : \mathcal{P} \to \mathbb{C}$ be any functional with $\omega(1) = 1$ and $\omega(x) = 0$. Then for $f \in \mathcal{P}$ and $N \to \infty$

$$\omega^{\otimes N}(f(x \mapsto N^{-1/2}(x \otimes 1 \otimes \ldots \otimes 1 + \ldots + 1 \otimes \ldots \otimes 1 \otimes x)) \to \gamma_v(f)$$

Here

$$f \in \mathcal{P} \mapsto f(x \mapsto N^{-1/2}(x \otimes 1 \otimes \ldots \otimes 1 + \ldots + 1 \otimes \ldots \otimes 1 \otimes x))$$

is the algebra homomorphism replacing x in f by

$N^{-1/2}(x \otimes 1 \otimes \ldots \otimes 1 + \ldots + 1 \otimes \ldots \otimes 1 \otimes x)$

The functional $\gamma_v : \mathcal{P} \to \mathbb{C}$ is given by

$\gamma_v(1) = 1$,

$\gamma_v(x^{2n+1}) = 0$ for $n = 0,1,2,\ldots$

$\gamma_v(x^2) = v$

$\gamma_v(x^{2n}) = \dfrac{(2n)!}{2^n n!} v^n$ for $n = 0,1,2,\ldots$

If $v \geq 0$ then $\gamma_v(f) = \int g_v(dx) f(x)$.

This algebraic central limit theorem is weaker and stronger than the classical one. It is weaker, because it contains only assertions about the moments of probability distributions, it is stronger because it does not assume positivity but only that the functional maps 1 into 1 .

We want to generalize this algebraic central limit theorem as follows: Consider two 2-graded algebra $\mathcal{A}$ and $\mathcal{L}$, a family $(a_i)_{i \in I}$ of homogeneous elements in $\mathcal{A}$ and a linear even mapping $\omega : \mathcal{A} \to \mathcal{L}$ mapping 1 into 1 such that $\omega(a_{i_1} \ldots a_{i_\ell}) = 0$ for all $i_1, \ldots i_\ell \in I$ and $1 \leq \ell < s$, where s is a fixed number $s = 1,2,\ldots$ Denote by $\mathcal{F}(I) = \mathcal{F}(x_i : i \in I)$ the free algebra generated by x_i , $i \in I$ and look for the behavior of

(1) $C_N^{(s)}(f) = M_N \circ \omega^{\otimes N}(f(x_i \mapsto N^{-1/s}(a_i \otimes 1 \otimes \ldots \otimes 1 + \ldots + 1 \otimes \ldots \otimes 1 \otimes a_i))$

for $N \to \infty$. Here $\omega^{\otimes N} : \mathcal{A}^{\otimes N} \to \mathcal{L}^{\otimes N}$ is the tensorproduct and $M_N : \mathcal{L}^{\otimes N} \to \mathcal{L}$ is the multiplication mapping $b_1 \otimes \ldots \otimes b_N \mapsto b_1 \ldots b_N$.

For $s = 1$ we obtain the algebraic analogon of the weak law of large numbers, for $s = 2$ the analogue of the central limit theorem and for $s > 2$ higher central limit theorems which do not occur in probability theory because the vanishing of the second moments implies that the probability measure is trivial.

The case $\mathcal{L} = \mathbb{C}$ was studied in two previous papers [3] and [5]. In [3] we assumed $\mathfrak{A}$ to be non-graded (or, what is the same, consisting only of even elemets). In [5] was assumed that all a_i are odd. This paper is a generalization because we do not only assume that $\mathcal{L}$ may be different from $\mathbb{C}$, so e.g. non-commutative as well but we allow too that the a_i may be both, even and odd. In [1] the assumption of independence, i.e. to use $M_N \circ \omega^{\otimes N}$ as mapping from $\mathfrak{A}^{\otimes N} \to \mathcal{L}$ was weakened to a mixing condition. The paper contains a minor combinatorial error, which may easily be repaired by assuming $\mathcal{L}$ to be commutative.

In this paper essentially three theorems are stated. Introduce the algebra $\mathcal{L}(W'(I))$; the free graded commutative algebra generated by ξ_w, $w \in W'(I)$, where $W'(I)$ is set of non-empty words formed by the alphabet I. The first theorem reduces the situation to the case where $\mathfrak{A} = \mathcal{F}(I)$ and $\mathcal{L} = \mathcal{L}(W'(I))$ and $\omega = \rho$ with $\rho(x_{i_1} \dots x_{i_k}) = \xi_{i_1 \dots i_k}$. Remark that just by the general form of the problem we may work with a graded commutative algebra as codomain of ω. More explicitely we have the following statement. Consider $C_N(f) = M_N \circ \omega^{\otimes N}(f(x_i \mapsto a_i \otimes 1 \otimes \dots \otimes 1 + \dots + 1 \otimes 1 \otimes \dots 1 \otimes a_i))$ then

$$C_N(f) = \mu \circ M_N \circ \rho^{\otimes N}(f(x_i \mapsto x_i \otimes 1 \otimes \dots \otimes 1 + \dots + 1 \otimes \dots \otimes 1 \otimes x_i))$$

where M_N is the N-fold multiplication $\mathcal{L}(W'(I))^{\otimes N} \to \mathcal{L}(W'(I))$ and where the linear map $\mu : \mathcal{L}(W'(I)) \to \mathcal{L}$ is given by

$$\mu(\xi_{w_1} \dots \xi_{w_p}) = \sum_{\sigma \in \mathcal{S}_p} (-1)^{\zeta(\sigma,\alpha)} \omega(a_{w_{\sigma^{-1}(1)}}) \dots \omega(a_{w_{\sigma^{-1}(p)}})$$

where $\mathcal{S}_p$ denotes the permutation group of $\{1,\dots,p\}$ and where for $w = i_1 \dots i_k$ the expression a_w is shorthand for $a_{i_1} \dots a_{i_k}$ and ζ is the parity of σ with respect to $\alpha = (\text{grad } w_1, \dots, \text{grad} w_p)$ (cf. Chapter 1).

The second theorem is the central limit theorem. It states that under the conditions listed above (cf (1))

$$C_N^{(s)}(f) \to \mu \circ \gamma_s(f)$$

where $\gamma_s : \mathcal{F}(I) \to \mathcal{L}(W'(I))$ is given by

$$\gamma_s(x_{i_1}\dots x_{i_m}) = \begin{cases} 0 & \text{if } m \text{ is not a multiple of } s \\ \sum_{\{S_1,\dots,S_p\}} (-1)^{\zeta((S_1,\dots,S_p),\alpha)} \xi_{w|S_1}\cdots\xi_{w|S_p} \end{cases}$$

Here $\{S_1,\dots,S_p\}$ runs through all the partitions of $\{1,\dots,m\}$ with the proviso that for any partition $\{S_1,\dots,S_p\}$ one ordering of $S_1,\dots,S_p$ say $(S_1,\dots,S_p)$ has been chosen. $\xi((S_1,\dots,S_p),\alpha)$ is the parity of $(S_1,\dots,S_p)$ with respect to $\alpha = (\text{grad } x_{i_1},\dots, \text{grad } x_{i_m})$. We denote by w the word $i_1 \dots i_m$ and if $S \subset \{1,\dots,m\}$ we denote by $w|S$ the subword $\prod_{s\in S} i_s$.

The third theorem says something about the structure of γ_s. the mapping

$$M_2 \circ (\text{id} \otimes \gamma_s) : \mathcal{L}(W'(I)) \otimes \mathfrak{f}(I) \to \mathcal{L}(W'(I))$$

vanishes on the ideal generated by

$$1 \otimes f - \gamma_s(f) \otimes 1$$

where f runs through all elements of the form $[x_{i_1}, [x_{i_2},\dots, [x_{i_{s-1}}, x_{i_s}].]$ where for $f, g \in \mathfrak{f}(I)$ the graded commutator $[f,g]$ is defined by $[f,g] = fg - (-1)^{\text{grad } f \text{ grad } g} gf$.

In order to be able to employ probalilistic language we introduce the notion of law. Consider a situation as above: Two 2-graded algebras $\mathfrak{A}$ and $\mathcal{L}$ and a linear map $\omega : \mathfrak{A} \to \mathcal{L}$ preserving 1 and a family of homogeneous elements $(a_i)_{i\in I}$ of $\mathfrak{A}$ and the free algebra $\mathfrak{f}(x_i, i\in I)$ with grad x_i = grad a_i. The <u>law</u> of the family $(a_i)_{i\in I}$ is the linear mapping

$$f \in \mathfrak{f}(I) \to \omega(f(x_i \mapsto a_i))$$

where $f \mapsto f(x_i \mapsto a_i)$ is the homomorphism η replacing x_i by a_i in the polynomial f. So the law of the $(a_i)_{i\in I}$ is the functional $\omega \circ \eta : \mathfrak{f}(I) \to \mathcal{L}$.

Assume a sequence $\mathfrak{A}^{(N)}, \omega^{(N)}, (a_i^{(N)})_{i\in I}$, $N = 1,2,\ldots,$
We say that the sequence $(a_i^{(N)})_{i\in I}$ converges in law to $\mathfrak{A}, \omega, (a_i)_{i\in I}$ if

$$\omega^{(N)}(f(x_i \mapsto a_i)) \to \omega(f(x_i \mapsto a_i))$$

for $N \to \infty$ and all $f\in \mathfrak{F}(I)$. To make this statement meaningful we have to introduce some topology in $\mathfrak{L}$. We will not care about it because in our special case the convergence is trivial. For any f it takes place in a finite dimensional subspace of $\mathfrak{L}$.

Resuming the second and third theorem we may state the theorem in the following way. The quotient of $\mathcal{L}(W'(I)) \otimes \mathfrak{F}(I)$ by the ideal mentioned above is the generalized graded Weyl algebra $\mathfrak{w}_s(I)$ generated by Ξ_i, $i\in I$ such that

$$[\Xi_{i_1}, [\Xi_{i_2},\ldots,[\Xi_{i_{s-1}}, \Xi_{i_s}]\ldots] = \gamma_s([x_{i_1},[x_{i_2},\ldots,[x_{i_{s-1}}, x_{i_s}]\ldots])$$

for all $i_1,\ldots,i_s\in I$. Assuming $\omega(a_{i_1}\ldots a_{i_{\ell-1}}) = 0$ for $i_1,\ldots,i_\ell\in I$ and $1 \leq \ell \leq s-1$ for $N \to \infty$ the family

$$N^{-1/s}(a_i \otimes 1 \otimes \ldots \otimes 1 + \ldots + 1 \otimes\ldots\otimes 1 \otimes a_i)$$

on $\mathfrak{A}^{\otimes N}$ with law $M_N \circ \omega^{\otimes N}$ converges in law to Ξ_i on $\mathfrak{w}_s(I)$ with law $\mu \circ \gamma_s$.

Let us look closer to the cases $s = 1$ and $s = 2$. In the case $s = 1$ we do not need to introduce Ξ_i. We may state that

$$\frac{1}{N}(a_i \otimes 1 \otimes\ldots \otimes 1 + \ldots + 1 \otimes\ldots\otimes 1\otimes a_i) \to \xi_i$$

where the $\xi_i\in \mathcal{L}(W'(I))$ and the law is given by $\mu\circ\gamma_1$ and γ_1 is the algebra homomorphism

$$\gamma_1 : \mathfrak{F}(I) \to \mathcal{L}(W'(I))$$
$$x_i \mapsto \xi_i$$

In the case that a_i is odd, ξ_i is a Grassmann quantity.

In the case $s = 2$ the Ξ_i have the defining relations

$$[\Xi_i , \Xi_j] = \xi_{ij} - (-1)^{\operatorname{grad} i \ \operatorname{grad} j} \ \xi_{ji} \ .$$

and the law $\mu \circ \gamma_2$. If Ξ_i and Ξ_j are both even this equation reads

$$\Xi_i \Xi_j - \Xi_j \Xi_i = \xi_{ij} - \xi_{ji}$$

if both Ξ_i and Ξ_j are odd

$$\Xi_i \Xi_j + \Xi_j \Xi_i = \xi_{ij} + \xi_{ji}$$

and if Ξ_i is even and Ξ_j is odd

$$\Xi_i \Xi_j - \Xi_j \Xi_i = \xi_{ij} - \xi_{ji}$$

If both Ξ_i and Ξ_j are even or odd, the commutator resp. anticommutator is an even quantity commuting with all elements of $\mathcal{L}(W'(I))$. This is the usual form of commutation relations encountered in quantum physics. If Ξ_i is even and Ξ_j is odd then the commutator is a Grassmann number.

In Chapter 1 we bring some basic definitions, and treat the parity of mappings and the connections between graded symmetric tensors and the graded symmetric algebra. In Chapter 2 we prove the theorems and discuss at the end the behavior of $x_i \otimes 1 \otimes \ldots \otimes 1 + \ldots + 1 \otimes \ldots \otimes 1 \otimes x_i$ for large N .

1. Preliminaries

We assume that the vector spaces and algebras use complex numbers as scalars. The algebras are assumed to be associative and to have a unitelement 1 . All algebra homomorphisms map 1 into 1 .

Let Δ be a commutative monoid written additively with unitelement 1. A Δ-graded set is a set I together with a family $(I_\lambda)_{\lambda \in \Delta}$ of subsets, which are disjoint and whose union is I . If $i \in I_\lambda$ we call λ the degree of i and write $\deg i = \lambda$. A Δ-graded vector space is a vector space V together with a family $(V_\lambda)_{\lambda \in \Delta}$ of vector subspaces such that $V = \bigoplus_{\lambda \in \Lambda} V_\lambda$. If $x \in V_\lambda$ we call x

homogeneous and write $\deg x = \lambda$. A Δ-graded algebra is an algebra A together with a family of vector subspaces $(A_\lambda)_{\lambda \in \Delta}$ such that $A_\lambda A_\mu = A_{\lambda+\mu}$.

If $\Delta = \mathbb{Z}_2$ we say 2-graded instead of Δ-graded and write grad i and grad x instead of deg i and deg x.

If I is a set we denote by W(I) the free monoid generated by I , i.e. the set of all words formed out of the alphabet I . We denote by W'(I) the subset of non-empty words. If I is 2-graded, we consider on W(I) the Δ-gradation with $\Delta = \mathbb{N} \times \mathbb{Z}_2$, where $\mathbb{N} = \{0,1,2\ldots\}$. If $w = i_1 \ldots i_k$ one has $\deg w = (\#w, \operatorname{grad} w)$ with $\#w = k$ and $\operatorname{grad} w = \operatorname{grad} i_1 + \ldots + \operatorname{grad} i_k$, the sum taken in $\mathbb{Z}_2$.

By $\mathcal{W}(x_i, i \in I)$ or $\mathcal{W}_x(I)$ or $\mathcal{W}(I)$ we denote the vector space spanned by the indeterminates x_i, $i \in I$. If I is 2-graded, $I = I_0 \cup I_1$ so is $\mathcal{W}(I) = \mathcal{W}(I_o) \oplus \mathcal{W}(I_1)$, putting $\operatorname{grad} x_i = \operatorname{grad} i$.

By $\mathcal{F}(x_i, i \in I)$ or $\mathcal{F}_x(I)$ or $\mathcal{F}(I)$ we denote the free algebra generated by x_i, $i \in I$. If I is 2-graded then $\mathcal{F}(I)$ is Δ-graded with $\Delta = \mathbb{N} \times \mathbb{Z}_2$. If $w \in W(I)$, $w = i_1 \ldots i_n$ denote $x_w = x_{i_1} \ldots x_{i_h}$. The homogeneous elements of $\mathcal{F}(I)$ are linear combinations of x_w with fixed $\deg w = (\#w, \operatorname{grad} w)$.

A 2-graded algebra is called graded commutative if $[f,g] = fg - (-1)^{\operatorname{grad} f \operatorname{grad} g} gf = 0$ for homogeneous elements f,g . Let I be 2-graded index set. By $\mathcal{L}(\xi_i, i \in I)$ or $\mathcal{L}_\xi(I)$ or $\mathcal{L}(I)$ we denote the free graded commutative algebra generated by ξ_i . This is the algebra generated by ξ_i with the defining relations $[\xi_i, \xi_j] = 0$ for all $i, j \in I$, where $\operatorname{grad} \xi_i = \operatorname{grad} i$.

Let $\mathfrak{A}$ be a 2-graded algebra. Let S be a finite ordered set. The algebra $\mathfrak{A}^{\otimes S}$ has the vector space tensor product $\mathfrak{A}^{\otimes S}$ as underlying vector space. Let $(a_s)_{s \in S}$ and $(b_s)_{s \in S}$ be two families of homogeneous elements of $\mathfrak{A}$. Define $\alpha, \beta \in \mathbb{Z}_2^S$ by $\alpha(s) = \operatorname{grad} a_s$ and $\beta(s) = \operatorname{grad} b_s$. Then

$$\bigotimes_{s \in S} a_s \bigotimes_{s \in S} b_s = (-1)^{\vartheta(\alpha,\beta)} \bigotimes_{s \in S} a_s b_s$$

with

$$\vartheta(\alpha,\beta) = \sum_{s_1 > s_2} \alpha(s_1)\,\beta(s_2)$$

the sum being calculated in $\mathbb{Z}_2$.

If $s \in S$ define the homomorphisms $u_s : \mathfrak{A} \to \mathfrak{A}^{\otimes S}$

$$a \mapsto \bigotimes_{s' \in S} b_{s'} \text{, with } b_{s'} = \begin{cases} 1 & \text{for } s' \neq s \\ a & \text{for } s' = s . \end{cases}$$

Let S, S' two finite ordered sets and assume an application $f : S \to S'$. Let $\alpha \in \mathbb{Z}_2^{\,S}$ and define

$$\zeta(\alpha,f) = \sum_{\substack{s_1 < s_2 \\ \alpha(s_1) > \alpha(s_2)}} \alpha(s_1)\,\alpha(s_2)$$

The number $\zeta(f,\alpha)$ is called the <u>parity</u> of f with respect to α . If $S = S'$ and if f is a permutation and if $\alpha(s) = 1$ for all s , then $\zeta(\alpha,f)$ is the usual parity.

Let S,S' be two finite ordered sets and $f : S \to S'$ an application. Define $u(f) : \mathfrak{A}^{\otimes S} \to \mathfrak{A}^{\otimes S'}$ by

$$u(f) \bigotimes_{s \in S} a_s = \overrightarrow{\prod_{s \in S}} u_{f(s)}(a_s)$$

where $\overrightarrow{\Pi}$ signifies the product proceeding from left to right.

<u>Proposition 1.1</u> Assume a 2-graded algebra $\mathfrak{A}$, assume a family of homogeneous elements $(a_s)_{s \in S}$ and define $\alpha \in \mathbb{Z}_2^{\,S}$ by $\alpha(s) = \operatorname{grad} a(s)$. Then

$$u(f) \bigotimes_{s \in S} a_s = (-1)^{\zeta(f,\alpha)} \bigotimes_{s' \in S'} \overrightarrow{\prod_{s \in f^{-1}(s')}} a_s$$

We put $\prod_{s \in f^{-1}(s')} a_s = 1$ if $f^{-1}(s') = \emptyset$.

<u>Proof</u>: We proceed by induction on $\#S$. If $\#S = 1$, then $S = \{s\}$, then $u(f)\, a = u_{f(s)}\, a$ and nothing has to be proved. Assume the proof for $\#S = n$ and assume $\#S = n+1$, e.g. $S = \{1,\dots,n+1\}$. Then

$$u(f)(a_1 \otimes \dots \otimes a_{n+1}) = \prod_{s=1}^{n+1} u_{f(s)}(a_s) = u(g)(a_1 \otimes \dots \otimes a_n)\, u_{f(n+1)}(a_{n+1})$$

with $g = f|\{1,\dots,n\}$. We have

$$u(g)\ (a_1 \otimes \dots \otimes a_n) = (-1)^{\zeta(g,\beta)} \bigotimes_{s' \in S'} b_{s'}$$

with $\beta = (\alpha(1),\dots\alpha(n))$. Assume $f(n+1) = t$, then

$$u(f)(a_1 \otimes \dots \otimes a_{n+1}) = (-1)^{\nu} \bigotimes_{s' \in S'} c_{s'}$$

with

$$c_{s'} = \begin{Bmatrix} b_{s'} & \text{for } s' \neq t \\ b_t\ a_{n+1} & \text{for } s' = t \end{Bmatrix} = \prod_{s:\ f(s) = s'} a_s\ .$$

and

$$\nu = \zeta(g,\beta) + \sum_{s' > t} \operatorname{grad} b_{s'} \operatorname{grad} a_{n+1}$$

$$= \zeta(g,\beta) + \sum_{\substack{1 \leq s < n+1 \\ f(s) > t}} \alpha(s)\alpha(n+1) = \zeta(f,\alpha)$$

as

$$\operatorname{grad} b_{s'} = \sum_{s:f'(s)=s'} \alpha(s)\ .$$

We want now to derive some properties of $\vartheta(\alpha,\beta)$ and $\zeta(f,\alpha)$. We may interprete $\alpha \in \mathbb{Z}_2^S$ as a $\mathbb{Z}_2$-valued measure on S putting for $T \subset S$

$$\alpha(T) = \sum_{s \in T} \alpha(s)\ .$$

Using the notion of product measure we may write

$$\vartheta(\alpha,\beta) = \alpha \otimes \beta \{(s_1,s_2) \in S \times S : s_1 > s_2\}$$

$$\zeta(f,\alpha) = \alpha \otimes \alpha \{(s_1,s_2) \in S \times S : s_1 < s_2 , f(s_1) > f(s_2)\}$$

If $f:S \to S'$ is a mapping, we denote by $f(\alpha) \in \mathbb{Z}_2^{S'}$ the image of the measure α, i.e.

$$f(\alpha)(T) = \alpha(f^{-1}(T))$$

or

$$f(\alpha)(s') = \sum_{s \in f^{-1}(s')} \alpha(s) .$$

Lemma 1.1 Assume $f: S \to S'$ to be injective. Thus

$$\vartheta(f(\alpha), f(\beta)) = \vartheta(\alpha,\beta) + (\alpha \otimes \beta + \beta \otimes \alpha)\{s_1 < s_2 : f(s_1) > f(s_2)\}.$$

Proof: Define the function

$$H : S \times S \to \{-1, 0, 1\}$$

$$H(s_1,s_2) = \begin{cases} 1 & \text{if } s_1 < s_2 \\ 0 & \text{if } s_1 = s_2 \\ -1 & \text{if } s_1 > s_2 \end{cases}$$

and define

$$T_{\varepsilon_1,\varepsilon_2} = \{(s_1,s_2) \in S \times S : H(s_1,s_2) = \varepsilon_1, H(f(s_1),f(s_2)) = \varepsilon_2\}$$

with $\varepsilon_i \in \{-1,0,1\}$ for $i = 1,2$. Thus

$$\vartheta(\alpha,\beta) = (\alpha \otimes \beta)(T_{-1,1} \cup T_{-1,0} \cup T_{-1,-1})$$

and

$$\vartheta(f(\alpha),f(\beta)) = (\alpha \otimes \beta)(T_{1,-1} \cup T_{0,-1} \cup T_{-1,-1})$$

Now $T_{-1,0} = \emptyset$ because f is injective and $T_{0,-1} = \emptyset$ because f is a function. Then

$$\vartheta(f(\alpha), f(\beta)) - \vartheta(\alpha,\beta) = (\alpha \otimes \beta)(T_{1,-1}) - (\alpha \otimes \beta)(T_{-1,1})$$

Interchange the role of s_1 and s_2 and use the fact that $-1 = +1$ in $\mathbb{Z}_2$. This proves the lemma.

Lemma 1.2 Let S, S', S" be finite ordered sets. Let $f : S \to S'$ and $g : S' \to S''$ be applications and assume g to be injective. Let $\alpha \in \mathbb{Z}_2{}^S$. Then

$$\zeta(g \circ f, \alpha) = \zeta(f,\alpha) + \zeta(g,f(\alpha))$$

Proof: Using the same function H as in the proof of Lemma 1.1 we define

$$T_{\varepsilon_1,\varepsilon_2,\varepsilon_3} = \{(s_1,s_2) \in S \times S : H(s_1,s_2) = \varepsilon_1,\ H(f(s_1),f(s_2)) = \varepsilon_2$$
$$H(g(f(s_1)),g(f(s_2)) = \varepsilon_3\} .$$

Then

$$\zeta(f,\alpha) = (\alpha \otimes \alpha)\ \{H(s_1,s_2) = 1\ ,\ H(f(s_1),f(s_2)) = -1\}$$
$$= (\alpha \otimes \alpha)\ (T_{1,-1,1} \cup T_{1,-1,0} \cup T_{1,-1,-1})$$

and

$$\zeta(g,f(\alpha)) = (\alpha \otimes \alpha)\ \{H(f(s_1),f(s_2)) = 1\ ,\ H(g(f(s_1)),g(f(s_2)) = -1\}$$
$$= (\alpha \otimes \alpha)\ \{T_{1,1,-1} \cup T_{0,1,-1} \cup T_{-1,1,-1}\}$$

and similar

$$\zeta(g \circ f, \alpha) = (\alpha \otimes \alpha)(T_{1,1,-1} \cup T_{1,0,-1} \cup T_{1,-1,-1})$$

Observe that $T_{1,-1,0} = \emptyset$ because g is injective and that $T_{0,1,-1} = \emptyset$ because f is a function and $T_{1,0,-1} = \emptyset$ because g is a function and use

$$(\alpha \otimes \alpha)(T_{-1,1,-1}) = (\alpha \otimes \alpha)(T_{1,-1,1})$$

Proposition 1.2 If $f : S \to S'$ is injective then $u(f) : \mathfrak{A}^{\otimes S} \to \mathfrak{A}^{\otimes S'}$ is an algebra homomorphism.

Proof Let $(a_s)_{s\in S}$ and $(b_s)_{s\in S}$ two families of homogeneous elements of $\mathfrak{A}$. Then by proposition 1.1 and by the definition of the product

$$(u(f)(\otimes a_s))\ u(f)(\otimes b_s)) = (-1)^{\mu} \otimes c_{s'}$$

with

$$\mu = \zeta(f,\alpha) + \zeta(f,\beta) + \vartheta(f(\alpha)\ ,\ f(\beta))$$

and

$$c_{s'} = \begin{cases} 1 & \text{if } f^{-1}(s') = \emptyset \\ a_s b_s & \text{if } f^{-1}(s') = \{s\} \end{cases}$$

On the other hand

$$u(f)((\otimes\ a_s)(\otimes b_s)) = (-1)^{\nu}\ \otimes\ d_{s'}$$

one sees immediately that $d_{s'} = c_{s'}$ and that

$$\nu = \vartheta(\alpha,\beta) + \zeta(f,\alpha+\beta)$$

The equality $\mu = \nu$ follows from Lemma 1.1

Proposition 1.3 Let S, S', S'' be finite ordered sets and $f : S \to S'$ and $g : S' \to S''$ be applications and g injective. Then

$$u(g \circ f) = u(g) \circ u(f)$$

Proof: Let $(a_s)_{s\in S}$ be a family of homogeneous elements of . Then by proposition 1.1

$$u(f)(\bigotimes_{s\in S} a_s) = (-1)^{\zeta(f,\alpha)} \bigotimes_{s'\in S'} b_{s'}$$

with $\alpha = (\text{grad } a(s))_{s\in S}$ and $b_{s'} = \prod_{s\in f^{-1}(s')} a_s$. One has $\beta = (\text{grad } b_{s'})_{s'\in S} = f(\alpha)$. Hence by proposition 1.1 .

$$u(g) \circ u(f) = (-1)^{\zeta(f,\alpha) + \zeta(g,f(\alpha))} \bigotimes_{s'' \in S''} c_{s''} \quad .$$

with

$$c_{s''} = \prod_{s' \in g^{-1}(s'')} b_{s'} = \begin{cases} 1 & \text{if } g^{-1}(s'') = \emptyset \\ b_t & \text{if } g^{-1}(s'') = \{t\} \ . \end{cases}$$

Hence

$$c_{s}'' = \prod_{s \in f^{-1}(g^{-1}(s''))} a_s$$

The last equation together with lemma 2.1 finishes the proof.

We want now to treat graded symmetric tensors and their connection with the graded symmetric algebra. Let $\mathcal{V}$ be a 2-graded vector space $\mathcal{V} = \mathcal{V}_0 + \mathcal{V}_1$, and let σ be a permutation of $\{1,\dots,r\}$. Define

$$u(\sigma) : \mathcal{V}^{\otimes p} \to \mathcal{V}^{\otimes p}$$

$$a_1 \otimes \dots \otimes a_p \mapsto (-1)^{\zeta(\sigma,\alpha)} a_{\sigma^{-1}(1)} \otimes \dots \otimes a_{\sigma^{-1}(p)} \ ,$$

where the a_s are supposed to be homogeneous and $\alpha = (\text{grad } a_s)$. Lemma 1.2 yields as in the proof of proposition 1.3 for any two permutations

$$u(\sigma \circ \tau) = u(\sigma) \circ u(\tau) \ .$$

A tensor $t \in \mathcal{V}^{\otimes p}$ is called <u>graded symmetric</u> if $u(\sigma)t = t$ for all $\sigma \in \mathcal{Y}_p$, the group of all permutations of $\{1,\dots,p\}$. Call $\mathcal{V}_s^{\otimes p}$ the subspace of graded symmetric tensors.

Let

$$P_p = \frac{1}{p!} \sum_{\sigma \in \mathcal{Y}_p} u(\sigma)$$

Then P_p is a projector mapping $\mathcal{V}^{\otimes p}$ onto $\mathcal{V}_s^{\otimes p}$.

We consider in the tensor algebra $T(\mathfrak{W})$ the ideal $\mathfrak{I}$ generated by

$$a \otimes b \; - \; (-1)^{\text{grad } a \text{ grad } b} \, b \otimes a$$

where a and b run through all homogeneous elements of $\mathfrak{W}$. As $T(\mathfrak{W})$ is Δ-graded with $\Delta = \mathbb{N} \times \mathbb{Z}_2$, so is $\mathfrak{I}$ and the graded symmetric tensor algebra of $\mathfrak{W}$

$$S(\mathfrak{W}) = T(\mathfrak{W})/\mathfrak{I}$$

One has

$$S(\mathfrak{W}) = S_{c\ell}(\mathfrak{W}_0) \times \Lambda(\mathfrak{W}_1)$$

where $S_{c\ell}(\mathfrak{W}_0)$ is the classical symmetric tensor algebra of $\mathfrak{W}_0$ and $\Lambda(\mathfrak{W}_1)$ is the Grassmann algebra of $\mathfrak{W}_1$.

Lemma 1.4 Let $(p,\varepsilon) \in \Delta$ and $(T(\mathfrak{W}))_{p,\varepsilon}$ the space of tensors of degree (p,ε) and $(T_s(\mathfrak{W}))_{p,\varepsilon}$ the subspace of symmetric terms of degree (p,ε) and $\mathfrak{I}_{p,\varepsilon} = \mathfrak{I} \cap (T(\mathfrak{W}))_{p,\varepsilon}$. Then

$$\mathfrak{I}_{p,\varepsilon} = (1-P_p)(T(\mathfrak{W}))_{p,\varepsilon}$$

and hence

$$(T(\mathfrak{W}))_{p,\varepsilon} = \mathfrak{I}_{p,\varepsilon} \oplus (T_s(\mathfrak{W}))_{p,\varepsilon} \; .$$

Proof: Call $A = \mathfrak{I}_{p,\varepsilon}$ and $B = (1-P_p)(T(\mathfrak{W}))_{p,\varepsilon}$. We show at first $A \subset B$. An element of A is a linear combination of elements of the form

$$t = a_1 \otimes \dots \otimes a_{k-1} \otimes (a_k \otimes a_{k+1} - (-1)^{\text{grad } a_k \text{ grad } a_{k+1}} a_{k+1} \otimes a_k) \otimes a_{k+2} \otimes \dots \otimes a_p$$

$$= (1 - u(\tau))\,(a_1 \otimes \dots \otimes a_p) \; ,$$

where τ is the permutation interchanging k and k+1. As

$P_p\, u(\tau) = P_p$ we obtain $t = (1-P_p)\, t$, hence $t \in A$ and $A \subset B$.

We prove now $B \subset A$. Define

$$K = \{\sigma \in \mathcal{S}_p : (1-u(\sigma))t \in \mathcal{J}_{\varepsilon,p} \quad \text{for all} \quad t \in (T(\mathcal{V}))_{\varepsilon,p}\}$$

By the equation above we see that K contains all permutations of nearest neighbors. If $\sigma,\tau \in K$ then $\tau\sigma \in K$. For

$$(1 - u(\tau\sigma))t = (1 - u(\sigma)t + (1 - u(\tau))\, u(\sigma)t$$

so $K = \mathcal{S}_p$.

Hence

$$(1-P_p)t = \frac{1}{p!}\, \Sigma(1-u(\sigma))\; t \in A$$

and $B \subset A$.

Denote by $\varkappa$ the canonical map $T(\mathcal{V}) \to T(\mathcal{V})/\mathcal{J} = S(\mathcal{V})$. By lemma 1.4 the restriction of $\varkappa$ to $T_s(\mathcal{V})$ is bijective. Call $\iota : S(\mathcal{V}) \to T_s(\mathcal{V})$ its inverse, then

$$(1) \quad \iota\,(\varkappa(a_1)\ldots\varkappa(a_p)) = \frac{1}{p!} \sum_{\sigma \in \mathcal{S}_p} (-1)^{\zeta(\sigma,\alpha)} a_{\sigma^{-1}(1)} \otimes \ldots \otimes a_{\sigma^{-1}(p)}$$

for homogeneous elements $a_1,\ldots,a_p$ and $\alpha = (\text{grad } a_s)_{s = 1,\ldots,p}$.

As $u(\sigma)P_p = P_p$ we obtain

$$(2) \quad (-1)^{\zeta(\sigma,\alpha)}\, \varkappa(a_{\sigma^{-1}(1)}) \ldots \varkappa(a_{\sigma^{-1}(p)}) = \varkappa(a_1) \ldots \varkappa(a_p) .$$

2. Proof of the Theorems

Let $\mathcal{A}$ and $\mathcal{B}$ be 2-graded algebras and $\omega : \mathcal{A} \to \mathcal{B}$ an even linear mapping such that $\omega(1) = 1$. Even means that $\omega(\mathcal{A}_i) \subset \mathcal{B}_i$ for $i = 0,1$. We fix a set $(a_i)_{i \in I}$ of homogeneous elements of $\mathcal{A}$. We call I_0 (resp I_1) the subset corresponding to even (resp. odd) elements a_i . By this definition I get a 2-graded set. We consider the free algebra $\mathcal{F}(x_i : i \in I)$. This algebra is Δ-graded with $\Delta = \mathbb{N} \times \mathbb{Z}_2$. We consider for $f \in \mathcal{F}(I)$ the quantity

$$C_N(f) = M_N \circ \omega^{\otimes N}(f(x_i \mapsto a_i \otimes 1 \otimes \ldots \otimes 1 + \ldots + 1 \otimes 1 \otimes \ldots \otimes 1 \otimes a_i))$$

where

$$\omega^{\otimes N} : \mathcal{O}\!\mathcal{L}^{\otimes N} \to \mathcal{L}^{\otimes N}$$

$$a_1 \otimes \ldots \otimes a_N \to \omega(a_1) \otimes \ldots \otimes \omega(a_N) .$$

Here the fact comes in that ω is even, for if ω were odd $a(\pm 1)$-factor came in [5]. We denote by M the multiplication regardless what algebra we are considering, here

$$M_N : \mathcal{L}^{\otimes N} \to \mathcal{L}$$

$$b_1 \otimes \ldots \otimes b_N \to b_1 \ldots b_N .$$

We recall that $\mathcal{F}(I)$ is a bialgebra with regard to the comultiplication

$$\Delta : \mathcal{F}(I) \to \mathcal{F}(I) \otimes \mathcal{F}(I)$$

$$x_i \mapsto x_i \otimes 1 + 1 \otimes x_i$$

and the counit:

$$\delta : \mathcal{F}(I) \to \mathbb{C}$$

$$x_i \mapsto 0$$

As both Δ and δ are algebra homomorphisms they are determined by their values on the generators. The N-th iterate of Δ is

$$\Delta_N : \mathcal{F}(I) \to \mathcal{F}(I)^{\otimes N}$$

$$x_i \mapsto x_i \otimes 1 \otimes \ldots \otimes 1 + \ldots + 1 \otimes 1 \otimes \ldots \otimes 1 \otimes x_i$$

Denoting by η the algebra homomorphism

$$\eta : \mathcal{F}(I) \to \mathcal{O}\!\mathcal{L}, \quad x_i \mapsto a_i$$

we may write

$$C_N(f) = M_N \circ (\omega \circ \eta)^{\otimes N} \circ \Delta_N(f)$$

If $\mathcal{K}$ is an algebra define a multiplication called convolution in $\mathcal{L}(\mathcal{F}(I),\mathcal{K})$ the space of all linear mappings $\mathcal{F}(I) \to \check{\mathcal{K}}$, by $A,B \mapsto A * B = M_2 \circ (A \otimes B) \circ \Delta$. The unit element of convolution is $f \to 1\,\delta(f)$ called δ again. If $\mathcal{K}$ is graded commutative, so is $\mathcal{L}(\mathcal{F}(I), \check{\mathcal{K}})$.

Using convolution we may write

$$C_N(f) = (\omega \circ \eta)^{*N}(f) .$$

If $w \in W(I)$ denote $x_w = x_{i_1} \dots x_{i_n}$ if $w = i_1 \dots i_n$ and $x_\emptyset = 1$. Let $\mathcal{F}'(I)$ the subspace in $\mathcal{F}(I)$ spanned by x_w, $w \neq \emptyset$. Let j be the injection $\mathcal{F}'(I) \to \mathcal{F}(I)$ and $T(\omega \circ \eta \circ j)$ the canonical extension of the linear map $\omega \circ \eta \circ j : \mathcal{F}'(I) \to \mathcal{L}$ to an algebra homomorphism $T(\mathcal{F}'(I)) \to \mathcal{L}$. Recall that $W' = W'(I)$ was the set of non-empty words in $W = W(I)$. We consider $\mathcal{L}(W')$ which can be identified with the graded symmetric tensor algebra $S(\mathcal{F}'(I))$. The canonical mapping i from the graded symmetric tensors onto the symmetric tensor algebra becomes (cf. chapter 1) :

$$\iota : \mathcal{L}(W') \to T(\mathcal{F}'(I))$$

$$\iota(\xi_{w_1} \dots \xi_{w_p}) = \frac{1}{p!} \sum_{\sigma \in \mathcal{T}_p} (-1)^{\zeta(\sigma,\alpha)} x_{w_{\sigma^{-1}(1)}} \otimes \dots \otimes x_{w_{\sigma^{-1}(p)}}$$

with $\alpha = (\text{grad } w_1, \dots, \text{grad } w_p)$.

Define a linear mapping

$$\rho : \mathcal{F}(I) \to \mathcal{L}(W'(I))$$

$$1 \mapsto 1$$

$$x_w = x_{i_1} \dots x_{i_n} \mapsto \xi_w = \xi_{i_1 \dots i_n} \quad \text{for } w = i_1 \dots i_n \neq \emptyset$$

The following theorem states that the situation might be reduced to the case that $\mathcal{O} = \mathcal{F}(I)$, $\mathcal{L} = \mathcal{L}(W'(I))$ and $\omega = \rho$.

<u>Theorem 2.1</u> One has

$$C_N(f) = T(\omega\circ\eta\circ j) \circ \iota \circ \rho^{*N}(f) .$$

<u>Proof</u> Introduce the linear mapping $\phi : \mathcal{F}(I) \to \mathcal{F}'(I)$

$$x_w \mapsto x_w \ , w \neq \emptyset$$

$$1 \to 0$$

Then as $\omega(1) = 1$

$$\omega\circ\eta = \delta + \omega\circ\eta\circ j\circ\phi$$

and hence

$$(\omega\circ\eta)^{*N} = \delta + \sum_{p=1}^{N} \binom{N}{p} (\omega\circ\eta\circ j\circ\phi)^{*p}$$

$$= \delta + \sum_{p=1}^{N} \binom{N}{p} M_p \circ (\omega\circ\eta\circ j\circ\phi)^{\otimes p} \circ \Delta_p$$

$$= \delta + \sum_{p=1}^{N} \binom{N}{p} M_p \circ (\omega\circ\eta\circ j)^{\otimes p} \circ \phi^{\otimes p} \circ \Delta_p$$

Let $\sigma \in \mathcal{Y}_p$ be a permutation and let $u(\sigma)$ be defined as in chapter 1. Then

$$\phi^{\otimes p} \circ u(\sigma) = u(\sigma) \circ \phi^{\otimes p}$$

where $u(\sigma)$ on the left hand side is the mapping $\mathcal{F}(I)^{\otimes p} \to \mathcal{F}(I)^{\otimes p}$ defined in chapter 1 and $u(\sigma)$ on the right hand side is the corresponding mapping $\mathcal{F}'(I)^{\otimes p} \to \mathcal{F}'(I)^{\otimes p}$. Now $u(\sigma)$ is an algebra homomorphism by proposition 1.2. As $u(\sigma)\Delta x_i = \Delta x_i$ we obtain $u(\sigma) \circ \Delta_p = \Delta_p$ and hence

$$u(\sigma) \circ \phi^{\otimes p} \circ \Delta_p = \phi^{\otimes p} \circ \Delta_p .$$

So $\phi^{\otimes p} \circ \Delta_p$ maps $\mathcal{F}(I)$ into the space of graded symmetric tensors in $\mathcal{F}'(I)^{\otimes p}$. So we may write using equation (1) of chapter 1

$$\phi^{\otimes p} \circ \Delta_p = \iota \circ \mathcal{X} \circ \phi^{\otimes p} \circ \Delta_p$$

$$= \iota \circ M_p \circ (\mathcal{X} \circ \phi)^{\otimes p} \circ \Delta_p$$

$$= \iota \circ (\mathcal{X} \circ \phi)^{*p}$$

identifying $\mathcal{f}'(I)^{\otimes p}$ with the corresponding subspace in $T(\mathcal{f}'(I))$ and using the fact that $\mathcal{H}$ is an algebra homomorphism. To finally

$$(\omega\circ\eta)^{*N} = T(\omega\circ\eta\circ j)\circ\iota\circ(\delta + \sum_{p=1}^{N}\binom{N}{p}(\mathcal{H}\circ\phi)^{*p})$$

$$= T(\omega\circ\eta\circ j)\circ\iota\circ(\delta + \mathcal{H}\circ\phi)^{*N}$$

This finishes the proof as $\rho = \delta + \mathcal{H}\circ\phi$

Remark: Writing $\mu = T(\omega\circ\eta\circ j)\circ\iota$ and $\rho^{*N} = M_N\circ\rho^{\otimes N}\circ\Delta_N$ we obtain the form of the theorem stated in the introduction.

We may consider a word $w = i_1,\dots,i_n$ as a mapping $\{1,\dots,k\}\to I$. If $S\subset\{1,\dots,k\}$ we denote by $w|S$ the restriction of w to S, so if $S = \{j_1,\dots,j_\ell\}$ with $j_1,\dots,j_\ell$ then $w|S = i_{j_1},\dots,i_{j_\ell}$. If $S = \emptyset$ we put $w|\emptyset = \emptyset$.

If S is a set and F a function $S\to\{1,..,p\}$. Then $(F^{-1}(1),\dots,F^{-1}(p))$ is a sequence of subsets of S, which are pairwise disjoint and whose union is S. Vice versa, assume a sequence $(S_1,..,S_p)$ of subsets of S given, which are pairwise disjoint and whose union is S. Then there exists exactly one function $F:S\to\{1,..,p\}$ such that $F^{-1}(i) = S_i$, $i = 1,\dots,p$. We define for $\alpha\in\mathbb{Z}_2^S$ the parity of the sequence $(S_1,\dots,S_p)$ by $\zeta((S_1,\dots S_p),\alpha)=\zeta(F,\alpha)$.

Lemma 2.1 Let $w = i_1\dots i_k$ and $\alpha = (\text{grad } i_1,\dots,\text{grad } i_k)$. Then

$$\Delta_p x_w = \sum_{(S_1,\dots,S_p)} (-1)^{\zeta((S_1,\dots,S_p),\alpha)} x_{w|S_1}\otimes\dots\otimes x_{w|S_p}$$

where $S_1,\dots,S_p$ runs through all sequences of $\{1,\dots,k\}$, which are pairwise disjoint and whose union is $\{1,\dots,k\}$.

Proof: Using the notations of chapter 7 we obtain

$$\Delta_p x_w = (\Delta_p x_{i_1}) \dots (\Delta_p x_{i_k})$$

$$= \sum_{j_1,\dots,j_k = 1}^{p} u_{j_1}(x_{i_1}) \dots u_{j_k}(x_{i_k})$$

$$= \sum_{F : \{1,\dots,k\} \to \{1,\dots,p\}} u(F)(x_{i_1} \quad \dots \quad x_{i_k})$$

The lemma follows by proposition 1.1.

Lemma 2.2: Let $\phi : \mathcal{F}(I) \to \mathcal{L}(W')$ be a linear even map such that $\phi(1) = 0$. Assume $w = i_1 \dots i_k$ and $\alpha = (\text{grad } i_1,\dots,\text{grad } i_k)$. Then $\phi^{*p}(x_W) = p! \sum_{\{S_1,\dots,S_p\}} (-1)^{\zeta((S_1,\dots,S_p),\alpha)} \phi(x_{w|S_1})\dots\phi(x_{w|S_p})$

where $\{S_1,\dots,S_p\}$ runs through all partitions of $\{1,\dots,k\}$ choosing for any partition a sequence $(S_i,\dots,S_p)$ representing it. The expression behind the sign of summation Σ does not depend on the particular chosen sequence but only on the partition.

Proof: Apply lemma 1.2 and equation (2) of chapter 1.

Theorem 2.2: (Central Limit Theorems). Fix a natural number $s \geq 1$. Let $\omega(a_{i_1} \dots a_{i_\ell}) = 0$ for all $i_1,\dots,i_\ell \in I$ and $1 \leq \ell < s - 1$. Then for all $N = 1,2,3,\dots$ and $f \in \mathcal{F}(I)$

$$C_N^{(s)}(f) = M_N \circ \omega^{\otimes N} \cdot f(x_i \mapsto N^{-1/s}(a_i \otimes 1 \otimes \dots \otimes 1 + \dots + 1 \otimes \dots \otimes 1 \otimes a_i))$$

stay in a fixed finite dimensional subspace of $\mathcal{L}$ and converge for $N \to \infty$ to

$$T(\omega \circ \eta \circ j) \circ \iota \circ \gamma_s(f)$$

where $\gamma_s : \mathcal{F}(I) \to \mathcal{L}(W'(I))$ is the even linear mapping defined by

$$\gamma_s(X_w) = \begin{cases} 0 & \text{if } \# w \text{ is not multiple of } s \\ 1 & \text{if } w = \emptyset \\ \sum\limits_{\{S_1,\ldots,S_p\}} (-1)^{\zeta((S_1,\ldots,S_p),\alpha)} \xi_{w|S_1} \cdots \zeta_{w|S_p} & \\ \quad \text{if } x_w = x_{i_1} \ldots x_{i_{ps}} . & \end{cases}$$

The sum runs through all partitions $\{S_1,\ldots,S_p\}$ of $\{1,\ldots,p\ s\}$ such that $\# S_i = s$ for all $i = 1,\ldots,p$. For each partitions one choses one ordering of the subsets and defines ζ accordingly. We set $\alpha = (\text{grad } i_1,\ldots,\text{grad } i_{ps})$.

<u>Proof</u>: By theorem 2.1 and lemma 2.1 and 2.2, we obtain splitting $\rho = \delta + \rho'$ for $w = i_1 \ldots i_k$ and $\alpha = (\text{grad } i_1,\ldots,\text{grad } i_k)$

$$C_N^{(s)}(x_w) = N^{-k/s} C_N(x_w)$$
$$= N^{-k/s} T(\omega \circ \eta \circ j) \circ \iota \circ \rho^{*N}(x_w)$$

and

$$\rho^{*N}(x_w) = (\delta + \sum_{p=1}^{N} \binom{N}{p} \rho'^{*p})(x_w)$$

$$= \sum_{p=1}^{k} p! \binom{N}{p} \sum_{\{S_1,\ldots,S_p\}} (-1)^{\zeta((S_1,\ldots,S_p),\alpha)} \xi_{w|S_1} \cdots \xi_{w|S_p}$$

Due to the assumptions

$$T(\omega \circ \eta \circ j) \circ \iota\ (\xi_{w|S_1},\ldots,\xi_{w|S_p})$$

vanishes if one of the subsets $S_1,\ldots,S_p$ has strictly less than s elements. So in the sum only those partitions should be considered, where all S_i, $i = 1,\ldots,p$ have $\geq s$ elements. This implies $p \leq k/s$. On the other hand $N^{-k/s}\ p!\binom{N}{p}$ behaves like $N^{p-k/s}$ for $N \to \infty$. So for $N \to \infty$ only those terms survive, where $p \geq k/s$. This implies $p = k/s$ or $k = p.s$ and all S_i have s elements.

Let $\psi : \mathfrak{f}(I) \to \mathcal{L}(W'(I))$ a linear mapping such that $\psi(1) = 0$. Then by lemma 2.2 the expression $\psi^{*p}(x_w)$ vanishes for $p > \#w$. So power series in ψ are well defined.

Split

$$\rho = \delta + \rho_1 + \rho_2 + \ldots$$

with

$$\rho_n(x_w) = \begin{cases} \rho(x_w) = \xi_w & \text{if} \quad w = n \\ 0 & \text{if} \quad \#w \neq n . \end{cases}$$

One obtains immediately from lemma 2.2

(1) $\gamma_s = \exp_* \rho_s = \delta + \rho_s + \frac{1}{2!} \rho_s^{*2} + \ldots$

In a 2-graded algebra the commutator is defined by

$$[a,b] = ab - (-1)^{\text{grad } a.\text{grad } b}\, ba$$

for homogeneous elements a and b. This makes the algebra into a graded Lie algebra. The free graded Lie algebra $\mathcal{L}\mathfrak{f}(I) = \mathcal{L}\mathfrak{f}(x_i : i \in I)$ is defined as the smallest graded Lie algebra containing all x_i, $i \in I$. It is spanned by $x_i ; [x_i, x_j] ; [x_i [x_j, x_h]] ; \ldots$ It is Δ-graded with $\Delta = \mathbb{N} \times \mathbb{Z}_2$.

The following lemma is well-known and can be easily proved by induction.

<u>Lemma:</u> Assume $f \in \mathfrak{f}\mathcal{L}(I)$. Then

$$\Delta_p f = f \otimes 1 \otimes \ldots \otimes 1 + \ldots + 1 \otimes 1 \otimes \ldots \otimes 1 \otimes f .$$

We extend the ring of scalars of $\mathfrak{f}(I)$ from $\mathbb{C}$ to $\mathcal{L}(W')$ by introducing $\hat{\mathfrak{f}}(I) = \mathcal{L}(W'(I)) \otimes \mathfrak{f}(I)$ and identifying $\xi_w \otimes 1$ with ξ_w and $1 \otimes x_i$ with x_i. One has $[\xi_w, x_{w'}] = 0$ for all $w, w' \in W'(I)$. The tensor product $\hat{\mathfrak{f}}(I)$ is not an algebra in the usual sense as for $\lambda \in \mathcal{L}(W')$ and $f, g \in \hat{\mathfrak{f}}(I)$, all homogeneous,

$$\lambda(fg) = (\lambda f)g = (-1)^{\text{grad } f \text{ grad } \lambda} f(\lambda g)$$

but except this (±1) - factor $\hat{\mathcal{F}}(I)$ has all the properties of an algebra over $\mathcal{L}(W')$.

Let $\psi : \mathcal{F}(I) \to \mathcal{L}(W')$ be even and define $\hat{\psi} = M_2 \circ (id \otimes \psi) : \mathcal{L}(W') \otimes \mathcal{F}(I) \to \mathcal{L}(W')$. Then $\hat{\psi}$ is $\mathcal{L}(W')$ - linear, in the sense that

$$\hat{\psi}(\lambda f) = \lambda\hat{\psi}(f)$$

for $f \in \hat{\mathcal{F}}(I)$, $\lambda \in \mathcal{L}(W'(I))$. If no ambiguity occurs we identify ψ and $\hat{\psi}$.

<u>Theorem 2.3</u> The mapping

$$\hat{\gamma}_s : \hat{\mathcal{F}}(I) \to \mathcal{L}(W'(I))$$

vanishes on the ideal generated by

$$f - \hat{\gamma}_s(f)$$

where f runs through all elements of $\mathcal{F}\mathcal{L}(I)$ of degrees $(s,0)$ or $(s,1)$, where $\mathcal{F}\mathcal{L}(I)$ is considered as a subspace of $\hat{\mathcal{F}}(I)$ with respect to the imbedding $f \mapsto 1 \otimes f$.

<u>Proof</u> Let $f \in \mathcal{F}\mathcal{L}(I)$ of degree $(s,0)$ or $(s,1)$ we have to show that

$$\gamma_s(x_u f\, x_v) = \gamma_s(x_u \gamma_s(f) x_v)$$

for all words u and v . We may assume $\#u + \#v = ps$.

Using (1) the last equation becomes

$$(*) \qquad \frac{1}{(p+1)!}\, \rho_s^{*(p+1)}(x_u f\, x_v) = \frac{1}{p!}\, \rho_s^{*p}\, (x_u \gamma_s(f) x_v)$$

Assume $u = i_1 \dots i_k$, $v = j_1 \dots j_\ell$ with $k + \ell = ps$. Define

$$y_1 = x_{i_1}, \dots, y_k = x_{i_k},\ y_{k+1} = f\ ,\ y_{k+2} = x_{j_1}, \dots, y_{ps+1} = x_{j_\ell}\ .$$

and put $S = \{1,\dots,ps+1\}$, $S' = \{1,\dots,p+1\}$ and $t = k + 1$. The left hand side of (*) becomes

$$\frac{1}{(p+1)!} M_{p+1} \circ \rho_s^{\otimes p+1} \circ \Delta_{p+1} (x_u f x_v)$$

and by the last lemma and using the proof of lemma 2.1

$$\Delta_{p+1}(x_u f x_v) = \sum_{(S_1,\dots,S_{p+1})} (-1)^{\zeta((S_1,\dots,S_{p+1}),\alpha)} y_{S_1} \dots Y_{S_{p+1}}$$

where $(S_1,\dots,S_{p+1})$ runs through all sequences of subsets of S, which are pairwise disjoint and whose union is S. Using the reasoning of lemma 2.2 we obtain for the left hand side of (*)

$$\sum_{\{S_1,\dots,S_{p+1}\}} (-1)^{\zeta((S_1,\dots,S_{p+1}),\alpha)} \rho_s(y_{s_1}) \dots \rho_s(y_{s_{p+1}})$$

We chose S_1 so that $t = k + 1 \in S_1$. As deg f equals $(s,0)$ or $(s,1)$ we have $S_1 = \{t\}$ and $\{S_2,\dots,S_{p+1}\}$ is a partition of $S\setminus t$.

We have $\alpha = (\text{grad } i_1,\dots,\text{grad } i_k, \text{grad } f, \text{grad } j_1,\dots,\text{grad } j_\ell)$.

Call β the restriction of α to $S\setminus t$, then

$$\zeta((S_1,\dots,S_{p+1}),\alpha) = \zeta((S_2,\dots,S_{p+1}),\beta) + \mu$$

with $\mu = \text{grad } u \text{ grad } f$. For

$$\zeta((S_1,\dots,S_{p+1}),\alpha) = \sum_{1\leq \ell<k \leq p+1} (\alpha \otimes \alpha)\{(s_1,s_2) \in S_k \times S_\ell , s_1 < s_2\}$$

$$= \sum_{2\leq \ell<k\leq p+1} (\alpha \otimes \alpha)\{(s_1,s_2) \in S_k \times S_\ell , s_1 < s_2\}$$

$$+ \sum_{1<k\leq p+1} (\alpha \otimes \alpha)\{(s_1,s_2) \in S_k \times \{t\} , s_1 < s_2\}$$

$$= \sum_{2\leq \ell<k\leq p+1} (\beta \otimes \beta)\{(s_1,s_2) \in S_k \times S_\ell , s_1 < s_2\} + (\alpha \otimes \alpha) \{(s,t):s<t\}$$

So, finally using again lemma 2.2 the left hand side of (*) becomes

$$(-1)^{\mu} \rho_s (y_t) \sum_{\{S_2,\dots,S_{p+1}\}} (-1)^{\zeta((S_2,\dots,S_{p+1}),\beta)} \rho_s(y_{S_2}) \dots \rho_s(y_{S_{p+1}})$$

$$= (-1)^{\mu} \rho_s(f) \frac{1}{p!} \rho_s^{*p}(x_u x_v)$$

Use the fact that ρ_s^{*p} is $\mathcal{L}(W')$-linear and the commutation relations of $\rho_s(f)$ with x_u and obtain the right hand side of (*) .

We want now to look in the convergence problem from a different point of view. Split $\rho = \delta + \rho'$ and define

$$\sigma = \log \rho = \rho' - \frac{1}{2}\rho'^{*2} + \frac{1}{3}\rho'^{*3} - t .$$

Then

$$\rho = \exp_*\sigma$$

and by lemma 2.2

$$\rho(x_w) = \sum_p \sum_{\{S_1,\ldots,S_p\}} (-1)^{\zeta((S_1,\ldots,S_p),\alpha)} \sigma(x_{w|S_1}) \ldots \sigma(x_{w\|S_p}) ,$$

So σ is nothing else than the cummulant functional or the cluster expansion of ρ (cf[4]) . Split

$$\sigma = \sigma_1 + \sigma_2 + \ldots$$

with

$$\sigma_n(x_w) = \begin{cases} \sigma(x_w) & \text{if } \#w = n \\ 0 & \text{if } \#w \neq n . \end{cases}$$

For $t \in \mathbb{C}$ denote by $\alpha(t)$ the homomorphism

$$\alpha(t) : \mathcal{F}(I) \to \mathcal{F}(I)$$
$$x_i \mapsto t\, x_i$$

Then

$$\rho \circ \alpha(t) = \delta + t\rho_1 + t^2\rho_2 + \ldots$$

$$\sigma \circ \alpha(t) = t\sigma_1 + t^2\sigma_2 + \ldots$$

Recall the definition of convergence in law from the introduction. Consider the functionals $\rho^{\otimes N}$ on $\mathcal{F}(I)^{\otimes N}$ and the elements

$$x_i^{(N)} = \frac{1}{N}(x_i \otimes 1 \otimes \ldots \otimes 1 + \ldots + 1 \otimes 1 \otimes \ldots 1 \otimes x_i)$$

The law of $x_i^{(N)}$ is

$$\rho^{*N} \circ \alpha(N^{-1}) = (\rho \circ \alpha(N^{-1}))^{*N} = \exp_* N(\sigma \circ \alpha(N^{-1}))$$

$$= \exp_* (\sigma_1 + N^{-1} \sigma_2 + N^{-2}\sigma_3 + \ldots)$$

and converges to

$$\exp_*\sigma_1 = \exp_*\rho_1 = \gamma_1$$

as $\rho_1 = \sigma_1$. As γ_1 is the homomorphism $\mathcal{F}(I) \to \mathcal{L}(W')$ $x_i \mapsto \xi_i$ we may assume the limit ξ_i to be in $\mathcal{L}(W')$. The law is γ_1. So

$$x_i^{(N)} \to \xi_i = \rho(x_i)$$

in law. This is the weak law of large numbers. We might have obtained this result directly by theorem 2.2. In fact the method used here could be applied to an alternative proof of that theorem.

We consider the sequence $\mathcal{F}(I)^{\otimes N}$, $\hat{\rho}^{\otimes N}$ and $\tilde{x}_i(N) = \sqrt{N}(x_i^{(N)} - \xi_i) = \frac{1}{\sqrt{N}}((x_i - \xi_i) \otimes 1 \otimes \ldots \otimes 1 + \ldots + 1 \otimes \ldots \otimes 1 \otimes (x_i - \xi_i))$ so the $\tilde{x}_i^{(N)}$ are the fluctuations of $x_i^{(N)}$ around ξ_i. The law of $\tilde{x}_i^{(N)}$ is

$$M_N \circ \rho^{\otimes N} \circ \beta^{\otimes N} \circ \Delta_N \circ \alpha(N^{-1/2}) = (\rho \circ \beta \circ \alpha(N^{-1/2}))^{*N}$$

where β is the algebra homomorphism

$$\beta : \hat{\mathcal{F}}(I) \to \hat{\mathcal{F}}(I)$$

$$x_i \mapsto x_i - \xi_i$$

Now

$$\rho \circ \beta = \exp_*(-\rho_1) * \rho$$

because

$$\exp_*(-\rho_1) * \rho = M_2 \circ (\mathrm{id} \otimes \rho) \circ (\exp_*(-\rho_1) \otimes \mathrm{id}) \circ \Delta$$

and

$$(\exp_*(-\rho_1) \otimes \mathrm{id}) \circ \Delta = \beta$$

as it is an algebra homomorphism sending $x_i \mapsto x_i \otimes 1 + 1 \otimes x_i \mapsto - \xi_i \otimes 1 + 1 \otimes x_i = x_i - \xi_i = \beta(x_i)$ and because $M_2 \circ (\mathrm{id} \otimes \rho)$ can be identified with ρ . Using the commutativity of convolution for even mapping one gets

$$\rho \circ \beta = \exp_* (\sigma_2 + \sigma_3 + \ldots)$$

So finally the law of $(\tilde{x}_i^{(N)})_{i \in I}$ is

$$\exp_* N(N^{-1} \sigma_2 + N^{-3/2}\sigma_3 + \ldots)$$

converging to

$$\exp_*\sigma_2 = \psi \circ \gamma_2 ,$$

where ψ is the algebra homomorphism

$$\psi : \mathcal{L}(W') \to \mathcal{L}(W')$$

$$\xi_w \to 0 \quad \text{if} \quad \#w \neq 2$$

$$\xi_{ij} \to \xi_{ij} - \xi_i\xi_j = \rho(x_i x_j) - \rho(x_i)\rho(x_j)$$

Recalling theorem 2.3 we may say that

$$\tilde{x}_i^{(N)} \to \Xi_i$$

in law, where the Ξ_i are the generators of the algebra $\mathcal{W}(I)$ with defining relations

$$[\Xi_i, \Xi_j] = \xi_{ij} - (-1)^{\mathrm{grad}\, i\, \mathrm{grad}\, j}\, \xi_{ji}$$

and law $\psi \circ \gamma_2$.

The same result could have been obtained by using the central limit theorem for $s = 2$, $\mathfrak{A} = \mathcal{F}(I)$, $\mathfrak{L} = \mathfrak{L}(W')$ and $\omega = \rho \circ \beta$.

LITERATURE

[1] Accardi, L. and Bach, A. Quantum central limit theorems for strongly mixing random variables. Z. Wahrscheinlichkeitstheorie verw. Gebiete 68, 393-402 (1985).

[2] Bourbaki, N. Algebra I, chapters 1-3 Addison-Wesley, Reading, Massachusetts 1974.

[3] Giri, N. and von Waldenfels, W. An algebraic version of the central limit theorem. Z. Wahrscheinlichkeitstheorie verw. Gebiete. 42, 129-134 (1978).

[4] Ruelle, D. Statistical mechanics, rigorous results, New York, W.A. Benjamin 1969.

[5] Scheunert, M. Theory of Lie superalgebras: an introduction. Lecture Notes in Mathematics 716. Springer-Verlag. Berlin-Heidelberg-New York 1979.

[6] von Waldenfels, W. An algebraic central limit theorem in the anticommuting case. Z. Wahrscheinlichkeitstheorie verw. Gebiete 42, 135-140 (1978).

A DESCRIPTION OF THE MARTIN BOUNDARY FOR NEAREST NEIGHBOUR RANDOM WALKS ON FREE PRODUCTS

Wolfgang WOESS

Institut für Mathematik und Angewandte Geometrie,
Montanuniversität, A-8700 Leoben, Austria

1. *Introduction*

Let Γ_i , $i=1,2$, be two finitely generated discrete groups with common intersection $\{e\}$, the unit element. We assume that at least one of the groups has more than two elements. Denote by

$$\Gamma = \Gamma_1 * \Gamma_2$$

the free product of Γ_1 and Γ_2 . Each element $x \neq e$ of Γ has a unique representation

(1.1) $x = x_{i_1} x_{i_2} \cdots x_{i_s}$, where $s \geq 1$, $i_k \in \{1,2\}$, $i_{k+1} \neq i_k$ and $x_{i_k} \in \Gamma_{i_k} - \{e\}$.

In particular, the Γ_i are subgroups of Γ . On Γ , we consider a probability measure μ which can be written as

(1.2) $\mu = a_1\mu_1 + a_2\mu_2$, $a_i > 0$, $a_1 + a_2 = 1$,

where each μ_i is a finitely supported probability measure on Γ_i whose support generates Γ_i as a semigroup. Then μ generates a Markov chain X_n , $n=0,1,2,\ldots$, on Γ , whose n-step transition probabilities are governed by the convolution powers of μ :

(1.3) $p^{(n)}(x,y) = \Pr[X_{k+n}=y \mid X_k=x] = \mu^{(n)}(x^{-1}y)$.

In view of (1.2), (X_n) is called a *nearest neighbour random walk* on Γ . As a consequence of our assumptions, the power series

(1.4) $G(x,y|z) = \sum_{n=0}^{\infty} p^{(n)}(x,y)z^n$, $x, y \in \Gamma$, $z \in \mathbb{C}$,

have a common radius of convergence $r = 1/\limsup \mu^{(n)}(e)^{1/n}$ (called "convergence norm" in [17]). As Γ is nonamenable [5], we have $1 < r < \infty$ [8], and $G(x,y|r) < \infty$ for all $x, y \in \Gamma$ [9]. For $z > 0$, a function $h : \Gamma \to \mathbb{R}$ is called *z-harmonic* , if

(1.5) $h(x) = z \cdot \sum_{y\in\Gamma} p(x,y)h(y)$ for all $x \in \Gamma$.

We are interested in a description of the cone H_z^+ of all positive z-harmonic functions on Γ . By [15], H_z^+ is nonvoid if and only if $0 < z \leq r$, and the abstract tool to describe H_z^+ is the *Martin boundary* :

see [11] for all details. For $0 < z \le r$, define the *Martin kernel*

(1.6) $K(x,y|z) = G(x,y|z)/G(e,y|z)$, $x, y \in \Gamma$.

The *Martin compactification* $\hat{\Gamma}_z$ of (Γ,μ,z) is characterized by the following properties:

(I) $\hat{\Gamma}_z$ is a compact metrizable space containing Γ as a dense subset.

(II) For each $x \in \Gamma$, $K(x,.|z)$ extends continuosly to $\hat{\Gamma}_z$.

(III) If $K(.,\alpha|z) \equiv K(.,\beta|z)$ for $\alpha, \beta \in \hat{\Gamma}_z$, then $\alpha = \beta$.

The *Martin boundary* is then the set $M_z = \hat{\Gamma}_z - \Gamma$. By the *Poisson - Martin* representation theorem, for every h in H_z^+ there is a positive Borel measure ν_h on M_z , such that

(1.7) $h(x) = \int_{M_z} K(x,\alpha|z)\, \nu_h(d\alpha)$ for all $x \in \Gamma$.

A point $\alpha \in M_z$ is called *extreme* , if $K(.,\alpha|z)$ is extreme in the convex set $\{ h \in H_z^+ \mid h(e)=1 \}$. The measure ν_h is unique when a priori restricted to the extreme points. Furthermore, there is a random variable X_∞ taking values in M_1 , such that $\lim X_n = X_\infty$ almost surely in the topology of $\hat{\Gamma}_1$. The harmonic function $h \equiv 1$ in H_1^+ is represented by the *hitting distribution* ν_1 which is given by

(1.8) $\nu_1(B) = \Pr[\, X_\infty \in B \mid X_0 = e \,]$, B a Borel set in M_1 .

It is desirable to obtain a less abstract, geometric (or algebraic) description of the Martin boundary: for random walks on discrete groups, this goal has been achieved in several cases [6,12,4,16,13]. In the present paper (§4) we give a description of the Martin boundary of (Γ,μ,z) in terms of the Martin boundaries of (Γ_i,μ_i,ζ_i) , where ζ_i $(i=1,2)$ are assigned to z in a canonical way [17] (see §3 below). In combination with the results quoted above, this gives a geometric description of M_z in a large class of cases. Furthermore, the extreme points of M_z are determined (§5) and the hitting distribution is calculated (§6).
To "visualize" the Martin boundary, we shall use two graph-theoretical concepts: the *Cayley graph* and the *space of ends* of a group.-

2. *The ends of a free product*

If Γ is an arbitrary finitely generated group and A is a finite set of generators, then the *Cayley graph* $C(\Gamma,A)$ is the graph with vertex set Γ and unoriented edges $[x, xa] \equiv [xa, x]$, where $x \in \Gamma$ and $a \in A$. Note that $C(\Gamma,A)$ is connected. By an *infinite path* we mean a sequence $\pi = [x_0, x_1, x_2, \ldots]$ of successively contiguous vertices (elements of Γ) without repetitions. Two infinite paths are *equivalent*, if for any finite subset U of Γ, they can be connected by some finite path lying entirely in $C(\Gamma,a) - U$. An *end* of Γ is an equivalence class of infinite paths. The set of all ends of Γ is denoted by Ω. The group acts on Ω in a natural way: if $\omega \in \Omega$ is represented by $\pi = [x_0, x_1, x_2, \ldots]$ and $x \in \Gamma$, then $x\omega$ is the equivalence class of $x\pi = [xx_0, xx_1, xx_2, \ldots]$. If we remove a finite subset U of Γ, then $C(\Gamma,A)$ splits into finitely many connected components. If X is the set of vertices of such a component, then we add to X all ends which can be represented by an infinite path lying entirely within this component. Thus, we obtain a set $\overline{X} \subset \Gamma \cup \Omega$. The family of all sets $\overline{X}$ which can be obtained in this way from some finite $U \subset \Gamma$ constitutes a basis of a topology which makes $\overline{\Gamma} = \Gamma \cup \Omega$ a totally disconnected, compact Hausdorff space with Ω as a compact subspace and Γ as a discrete, dense subspace.

The space of ends of a group does not depend on the choice of the finite generating set. For a finite group, $\Omega = \emptyset$. An infinite group has one, two or infinitely many ends. These concepts and results are due to [7].

For the free product $\Gamma = \Gamma_1 * \Gamma_2$, define the subsets

$$(2.1) \quad \Gamma^{(i)} = \{e\} \cup \{\, x \neq e \mid i_s \neq i \text{ in representation (1.1)}\} \qquad (i = 1,2)$$

The group Γ consists of infinitely many *levels* $x\Gamma_i$, where $x \in \Gamma^{(i)}$ $(i = 1,2)$. To describe the Cayley graph of Γ, choose finite generating sets A_i of Γ_i (e.g. the supports of the μ_i), and let $A = A_1 \cup A_2$. Then on each level, $C(\Gamma,A)$ is a copy of $C(\Gamma_1,A_1)$ or $C(\Gamma_2,A_2)$, respectively, and each $x \in \Gamma$ is the point of intersection of precisely two such copies (cosets of Γ_i): $\{x\} = x\Gamma_1 \cap x\Gamma_2$. From this, we infer the following representation of Ω, the space of ends of Γ, in terms of the spaces of ends Ω_i of the Γ_i.-

<u>Lemma 1.</u> $\Omega = \Omega^{(0)} \cup \Omega^{(1)} \cup \Omega^{(2)}$, where

$$\Omega^{(0)} = \{\, x_{j_1} x_{j_2} x_{j_3} \cdots \mid j_k \in \{1,2\},\ j_{k+1} \neq j_k \,\} \qquad \textit{(infinite reduced words)}$$

and, for $i = 1,2$

$$\Omega^{(i)} = \bigcup \{\, x\Omega_i \mid x \in \Gamma^{(i)} \} .$$

To be precise, by $\omega = x_{j_1} x_{j_2} x_{j_3} \cdots$ we mean the equivalence class consisting of all those paths in $C(\Gamma,A)$, which meet all but finitely many of the points $x_{j_1} x_{j_2} \cdots x_{j_k}$ $(k=1,2,\ldots)$; if $\omega_i \in \Omega_i$ $(i \in \{1,2\})$ is represented by a path π in $C(\Gamma_i,A_i)$ and $x\in\Gamma^{(i)}$, then $x\omega_i$ denotes the equivalence class of $x\pi$ in $C(\Gamma,A)$.

Definiton. If $x\in\Gamma^{(i)}$, then the *projection* $\varphi_{x,i}:\Gamma\cup\Omega\to\Gamma_i\cup\Omega_i$ is defined as follows: if the presentation of $\xi\in\Gamma\cup\Omega$ in the sense of (1.1) or Lemma 1, respectively, starts with $x\xi_i$, where $\xi_i\in(\Gamma_i-\{e\})\cup\Omega_i$, then $\varphi_{x,i}(\xi)=\xi_i$; otherwise $\varphi_{x,i}(\xi)=e$.

We can describe the topology of $\Gamma\cup\Omega$ in terms of the topologies of $\Gamma_i\cup\Omega_i$ via convergence (denoted $\overset{e}{\to}$):

Lemma 2. Let (ξ_n) be a sequence in $\Gamma\cup\Omega$ and $\omega\in\Omega$.

(i) If $\omega=x_{j_1}x_{j_2}\cdots\in\Omega^{(0)}$, then $\xi_n\overset{e}{\to}\omega$ if and only if for each k, $\varphi_{u_k,i_{k+1}}(\xi_n)=x_{i_{k+1}}$ for all but finitely many n, where $u_k=x_{i_1}\cdots x_{i_k}$.

(ii) If $\omega=x\omega_i$, where $x\in\Gamma^{(i)}$, $\omega_i\in\Omega_i$, then $\xi_n\overset{e}{\to}\omega$ if and only if $\varphi_{x,i}(\xi_n)\overset{e}{\to}\omega_i$.

Note that $\Omega^{(i)}$ is void if and only if Γ_i is finite, and that $\Omega^{(0)}$ is dense in Ω. The importance of the space of ends for the description of the Martin boundary relies on the following result.-

Theorem 1. [14] The identity on Γ extends to a continuous surjection $\tau_z:\hat{\Gamma}_z\to\Gamma\cup\Omega$ $(0<z\le r)$, which maps M_z onto Ω.

Therefore we shall write $\alpha\in\omega$ if $\tau_z(\alpha)=\omega$ for $\alpha\in M_z$; we can imagine α as a subclass of equivalent paths of ω. We shall see that for $\omega\in\Omega^{(0)}$ there is exactly one $\alpha\in\omega$, and write $\alpha=\omega$ in this case.

As μ has the "nearest neighbour" property, the random walk respects the structure of levels of $C(\Gamma,A)$: in order to pass from a level to an adjacent one, the walk must visit their unique common point. An outline of the consequences of this fact is given in the next section.-

3. *Generating functions of transition probabilities*

Besides the n-step transition probabilities, we introduce the following quantities and their generating functions:

$$f^{(n)}(x,y)=\Pr[X_n=y,\ X_m\neq y \text{ for } m=0,\dots,n-1\mid X_0=x]\qquad (n\ge 0)$$

(3.1)

$$F(x,y|z)=\sum_{n=0}^{\infty}f^{(n)}(x,y)z^n\quad,\ x,y\in\Gamma\ ,\ z\in\mathbb{C}\ .$$

Observe that $F(x,y|z)$ is the $Q(x,y|z)$ of [17]. For $|z|\le r$, $F(x,y|z)$

is convergent, $F(x,x|z) = 1$, and $G(x,y|z) = F(x,y|z)G(y,y|z)$. Thus,

(3.2) $K(x,y|z) = F(x,y|z)/F(e,y|z)$, $0 < z \le r$.

If $x \neq e$ has reduced representation $x = x_{i_1} x_{i_2} \cdots x_{i_s}$, then [17, Lemma 4]

(3.3) $F(e,x|z) = F(e,x_{i_1}|z)F(e,x_{i_2}|z)\cdots F(e,x_{i_s}|z)$

Now, each μ_i defines a random walk on Γ_i with corresponding generating functions $G_i(x_i,y_i|z)$, $F_i(x_i,y_i|z)$, convergence norm r_i and Martin kernel $K_i(x_i,y_i|z)$. A detailed treatment of the relations between these quantities and generating functions and those corresponding to (Γ,μ) is given in [17], see also [2]. We shall need the following result.-

<u>Proposition 1.</u> [17, §3] For $0 \le z \le r$, $i = 1,2$, there exists a number $\zeta_i = \zeta_i(z)$, $0 \le \zeta_i \le r_i$, such that for each $x_i \in \Gamma_i \subset \Gamma$,

$$F(e,x_i|z) = F_i(e,x_i|\zeta_i) .$$

Furthermore, $z \mapsto \zeta_i$ is continuous and strictly increasing for $0 < z \le r$, $\zeta_i(0) = 0$; $\zeta_i(r) = r_i$ implies $G_i(e,e|r_i) < \infty$.

In fact, $\zeta_i(r) = r_i$ may happen in some particular cases: see [1] and §7 below. In the sequel, the Martin boundary for (Γ_i,μ_i,ζ_i) will be denoted by M_{i,ζ_i} .

4. *The Martin boundary*

We have now accomplished the collection of the tools and notions that are necessary for our goals.

<u>Proposition 2.</u> The Martin Kernel is locally constant near each $\omega \in \Omega^{(0)}$: let $y_n \overset{e}{\to} \omega \in \Omega^{(0)}$ $(y_n \in \Gamma)$, and $v \in \Gamma$. Then there is an index n_v , such that for all $n \ge n_v$, $K(v,y_n|z)$ has the constant value given below in (4.1) and (4.2), respectively.

<u>Proof.</u> Let $\omega = x_{j_1} x_{j_2} x_{j_3} \cdots$. If $v = e$ then $K(v,.|z) \equiv 1$. Suppose $v \neq e$, $v = v_{i_1} v_{i_2} \cdots v_{i_s}$ reduced as in (1.1).

<u>Case 1.</u> If $\Gamma_{i_1} \neq \Gamma_{j_1}$, then for all $n \ge n_0$, the reduced representation of y_n starts with x_{j_1} by Lemma 2, and by (3.3),

$F(v,y_n|z) = F(e,v^{-1}y_n|z) = F(e,v^{-1}|z)F(e,y_n|z)$. Hence

(4.1) $K(v,y_n|z) = F(e,v^{-1}|z)$ $(n \ge n_0)$.

<u>Case 2.</u> If $\Gamma_{i_1} = \Gamma_{j_1}$ then choose k maximal such that $x_{j_1} \cdots x_{j_k} =$

$v_{i_1}\cdots v_{i_k}$, $0 \le k \le s$, and set $v_{i_{k+1}} = e$ if $k = s$. Then by Lemma 2, for all $n \ge n_k$, y_n starts with $u_{k+1} = x_{j_1}\cdots x_{j_{k+1}}$, and by (3.3),

$$F(v,y_n|z) = F(e,v^{-1}y_n|z) = F(e,v_{i_s}^{-1}\cdots v_{i_{k+2}}^{-1}(v_{i_{k+1}}^{-1}x_{j_{k+1}})|z)F(e,u_{k+1}^{-1}y_n|z) \quad ,$$

$F(e,y_n|z) = F(e,u_{k+1}|z)F(e,u_{k+1}^{-1}y_n|z)$, and thus, for $n \ge n_k$

$$(4.2) \quad K(v,y_n|z) = F(e,v_{i_s}^{-1}\cdots v_{i_{k+2}}^{-1}(v_{i_{k+1}}^{-1}x_{j_{k+1}})|z)/F(e,x_{j_1}\cdots x_{j_{k+1}}|z) \qquad \square$$

Thus, in view of Theorem 1, to each end in $\Omega^{(0)}$ there corresponds exactly one point of M_z , and we can identify $\tau_z^{-1}(\Omega^{(0)})$ with $\Omega^{(0)}$.

Proposition 3. If (y_n) is a sequence in Γ converging to $\alpha \in M_z$ then either

(a) $\alpha \in \Omega^{(0)}$ or

(b) $\tau_z(\alpha) = x\omega_i$, where $x \in \Gamma^{(i)}$ and $\omega_i \in \Omega_i$ $(i \in \{1,2\})$, and $\varphi_{x,i}(y_n)$ converges to some $\alpha_i \in M_{i,\zeta_i}$ in the Martin topology of (Γ_i,μ_i,ζ_i) ; in particular, $\alpha_i \in \omega_i$ with respect to Γ_i .

Proof. By Theorem 1, $\tau_z(y_n) \overset{e}{\to} \omega$ for some $\omega \in \Omega$. If $\omega \in \Omega^{(0)}$, case (a), then $\alpha = \omega$ according to Prop.2 and the above remark. Otherwise, $\omega = x\omega_i$ as stated in (b), and by Lemma 2, $\varphi_{x,i}(y_n) = y_n^{(i)} \overset{e}{\to}$ $\overset{e}{\to} \omega_i$. In particular, for all n larger than some n_0 , the reduced representation of y_n starts with $xy_n^{(i)}$. Choose $x_i \in \Gamma_i$. Then, by assumption, $K(xx_i,y_n|z) \to K(xx_i,\alpha|z)$ as $n \to \infty$. As in the proof of Prop.2, one calculates, using (3.3) and Prop.1,

$$K(xx_i,y_n|z) = F(e,x_i^{-1}y_n^{(i)}|z)/F(e,xy_n^{(i)}|z) =$$

$$= F_i(e,x_i^{-1}y_n^{(i)}|\zeta_i)/F_i(e,y_n^{(i)}|\zeta_i)F(e,x|z) = K_i(x_i,y_n^{(i)}|\zeta_i)/F(e,x|z)$$

for $n \ge n_0$. Thus, as $n \to \infty$, $K_i(x_i,y_n^{(i)}|\zeta_i)$ converges for every x_i in Γ_i ; in other words, $y_n^{(i)}$ converges to some $\alpha_i \in M_{i,\zeta_i}$ in the Martin topology. $\square$

If $x \in \Gamma^{(i)}$ then we can equip $x\Gamma_i$ with the Martin topology of $(\Gamma_i, \mu_i,\zeta_i)$ via the mapping $x_i \mapsto xx_i$ $(x_i \in \Gamma_i)$. We can extend this as a homoeomorphism to the boundary M_{i,ζ_i} , and if $\alpha_i \in M_{i,\zeta_i}$, then $x\alpha_i$ denotes the corresponding point of the image xM_{i,ζ_i} . In the situation of case (b) above, we shall write $\alpha = x\alpha_i$, thus embedding xM_{i,ζ_i} into M_z .

Definition. As in §2, we now define a projection $\psi_{x,i}$ from $\Gamma \cup M_z$

onto $\Gamma_i \cup M_{i,\zeta_i}$ $(x \in \Gamma^{(i)})$: on Γ, the definition remains the same as that of $\varphi_{x,i}$; $\psi_{x,i}(x\alpha_i) = \alpha_i$ for $\alpha_i \in M_{i,\zeta_i}$, and $\psi_{x,i}(\beta) = \varphi_{x,i}(\tau_z(\beta))$ if $\beta \in M_z - xM_{i,\zeta_i}$.

<u>Proposition 4.</u> If $x \in \Gamma^{(i)}$ and (y_n) is a sequence in Γ such that $\psi_{x,i}(y_n)$ tends to $\alpha_i \in M_{i,\zeta_i}$, then $K(v,y_n|z)$ converges for every $v \in \Gamma$. The limit is given below in (4.3) and (4.4), respectively.

<u>Proof.</u> As in the proof of Prop.2, assume $v \neq e$, $v = v_{i_1} \cdots v_{i_s}$. Furthermore, let $y_n^{(i)} = \psi_{x,i}(y_n)$ $\left(= \varphi_{x,i}(y_n)\right)$ as above. Then, for all $n \geq n_0$ (n_0 suitable), the reduced representation of y_n starts with $xy_n^{(i)}$ and this is not a prefix of v.

<u>Case 1.</u> If x is not a prefix of v, or if $x = e$ and $\Gamma_i \neq \Gamma_{i_1}$, then as in (4.1) and (4.2),

$$(4.3) \quad K(v,y_n|z) = K(v,x|z) \qquad (n \geq n_0).$$

<u>Case 2.</u> If $x = v_{i_1} \cdots v_{i_t}$, $0 < t \leq s$ ($v_{i_{t+1}} = e$ if $t = s$), or if $x = e$ $(t = 0)$ and $\Gamma_i = \Gamma_{i_1}$, then by (3.3) and Prop.1, we have for $n \geq n_0$

$$K(v,y_n|z) = F(v,xy_n^{(i)}|z)/F(e,xy_n^{(i)}|z) =$$

$$= F(e,v_{i_s}^{-1} \cdots v_{i_{t+2}}^{-1}|z)F(e,v_{i_{t+1}}^{-1} y_n^{(i)}|z)/F(e,v_{i_1} \cdots v_{i_t}|z)F(e,y_n^{(i)}|z) =$$

$$= K_i(v_{i_{t+1}}, y_n^{(i)}|\zeta_i) \cdot \left(F(e,v_{i_s}^{-1} \cdots v_{i_{t+2}}^{-1}|z)/F(e,v_{i_1} \cdots v_{i_t}|z)\right).$$

Hence,

$$(4.4) \quad K(v,y_n|z) \to K_i(v_{i_{t+1}},\alpha_i|\zeta_i) \cdot K(v,x|z)/F(v_{i_{t+1}},e|z) \qquad \square$$

Combining Propositions 2,3 and 4 yields our main result.-

<u>Theorem 2.</u> Up to homoeomorphism, the Martin boundary can be identified with the disjoint union

$$M_z = \Omega^{(0)} \cup M_z^{(1)} \cup M_z^{(2)} ,$$

where $\Omega^{(0)}$ is as in Lemma 1 and, for $i = 1,2$,

$$M_z^{(i)} = \cup \{ xM_{i,\zeta_i} \mid x \in \Gamma^{(i)} \} .$$

The Martin topology on $\Gamma \cup M_z$ is the weakest topology such that τ_z and all the projections $\psi_{x,i}$ $(x \in \Gamma^{(i)}, i = 1,2)$ are continuous.

Indeed, a sequence (η_n) in $\hat{\Gamma}_z$ converges to $\alpha \in M_z$ if and only if either

(i) $\alpha = \omega \in \Omega^{(0)}$ and $\tau_z(\eta_n) \overset{e}{\to} \omega$ or

(ii) there are $x \in \Gamma^{(i)}$ and $\alpha_i \in M_{i,\zeta_i}$ such that $\alpha = x\alpha_i$ and $\psi_{x,i}(\eta_n) \to \alpha_i$ in $\Gamma_i \cup M_{i,\zeta_i}$.

5. *The extreme points of the boundary*

In this section we describe the extreme points of M_z via those of M_{i,ζ_i} , $i = 1,2$. For $U \subset \Gamma$ and $x, y \in \Gamma$ denote

$$(5.1)\quad \begin{aligned} \alpha_n^U(x,y) &= \Pr[\,X_n = y\,,\,X_m \notin U \text{ for } m=0,\dots,n-1 \mid X_0 = x\,] \quad (n \geq 0)\,, \\ \alpha^U(x,y|z) &= \sum_{n=0}^{\infty} \alpha_n^U(x,y) z^n \,. \end{aligned}$$

In particular, $\alpha^{\{y\}}(x,y|z) = F(x,y|z)$. If $y \in \Gamma_i$, one calculates using the formulas of [17,§3]:

$$(5.2)\quad \alpha^{\Gamma_i - \{e\}} = \mu_i(y)\zeta_i/(1-\mu_i(e)\zeta_i) \quad , \quad \zeta_i = \zeta_i(z)\,.$$

If $h \in H_z^+$ $(0 < z \leq r)$ and $U \subset \Gamma$, then

$$(5.3)\quad h(x) \geq \sum_{y \in U} \alpha^U(x,y|z)h(y)\,.$$

Theorem 3. (a) If $\alpha = \omega \in \Omega^{(0)}$, then α is extreme in M_z .

(b) If $\alpha \in M_z^{(i)}$, $\alpha = x\alpha_i$ $(x \in \Gamma^{(i)}$, $i \in \{1,2\})$ then α is extreme in M_z if and only if α_i is extreme in M_{i,ζ_i} .

Proof. (We use [11,Lemma 10-30], compare [4].)
If $\mu = c\overline{\mu} + (1-c)\delta_e$ (δ_e the Dirac measure at e), then $H_z(\mu) = H_{\overline{z}}(\overline{\mu})$ for $\overline{z} = cz/(1-(1-c)z)$. Therefore we may assume for the proof, that $\mu(e) = 0$. First, suppose there are $g, h \in H_z^+$, $g(e) = h(e) = 1$, and $0 < \lambda < 1$ such that $K(.,\alpha|z) = \lambda \cdot g + (1-\lambda)\cdot h$.

(a) $\alpha = x_{j_1}x_{j_2}x_{j_3}\dots \in \Omega^{(0)}$: let $v \in \Gamma$, set $u_k = x_{j_1}\cdots x_{j_k}$ and choose k large enough such that $K(v,\alpha|z) = K(v,u_k|z)$ (Prop.2). Observe that $K(u_k,\alpha|z) = 1/F(e,u_k|z)$. Then by (5.3),

$$\begin{aligned} K(v,\alpha|z) &= \lambda\cdot g(v) + (1-\lambda)\cdot h(v) \geq F(v,u_k|z)\big(\lambda\cdot g(u_k) + (1-\lambda)h(u_k)\big) = \\ &= F(v,u_k|z)K(u_k,\alpha|z) = K(v,\alpha|z)\,. \end{aligned}$$

Hence, $g(v) = F(v,u_k|z)g(u_k)$, and, in the same way, $g(e) = F(e,u_k|z)\, g(u_k)$. This yields $g(v) = g(v)/g(e) = K(v,u_k|z) = K(v,\alpha|z)$. The same holds for h , and we obtain $g = h = K(.,\alpha|z)$.

(b) $\alpha = x\alpha_i$. If x is not a prefix of $v \in \Gamma$ (Prop.4, case 1), then one proves exactly as above $g(v) = h(v) = K(v,\alpha|z)$, in particular

$g(x) = h(x) = 1/F(e,x|z)$, as $1 = g(e) = F(e,x|z)g(x)$ (the same holds for h). Now set $g_i(x_i) = g(xx_i)F(e,x|z)$, $h_i(x_i) = h(xx_i)F(e,x|z)$ for x_i in Γ_i . Then, as $\mu(e) = 0$ by assumption, we obtain from (4.4), (5.2) and (5.3)

$$K_i(x_i,\alpha_i|\zeta_i) = F(e,x|z)K(xx_i,\alpha|z) = F(e,x|z)\big(\lambda\cdot g(xx_i) + (1-\lambda)\cdot h(xx_i)\big) \geq$$

$$\geq F(e,x|z)\cdot\zeta_i\cdot \sum_{y_i\in\Gamma_i} \mu_i(y_i)\big(\lambda\cdot g(xx_iy_i) + (1-\lambda)h(xx_iy_i)\big) =$$

$$= F(e,x|z)\cdot\zeta_i\cdot \sum_{y_i\in\Gamma_i} \mu_i(y_i)\, K(xx_iy_i,\alpha|z) = \zeta_i\cdot \sum_{y_i\in\Gamma_i} \mu_i(y_i)K_i(x_iy_i,\alpha_i|\zeta_i)$$

$$= K_i(x_i,\alpha_i|\zeta_i) \quad .$$

In other words, $g_i , h_i \in H^+_{i,\zeta_i}$, $g_i(e) = h_i(e) = 1$ and $K_i(.,\alpha_i|\zeta_i) = \lambda\cdot g_i + (1-\lambda)\cdot h_i$.

If α_i is extreme , then $g_i = h_i = K_i(.,\alpha_i|\zeta_i)$.
From this and (4.4) we infer that $g(v) = h(v) = K(v,\alpha|z)$ if $v = v_{i_1}\cdots v_{i_s}$ (reduced) has x as a prefix, $x = v_{i_1}\cdots v_{i_t}$, and $s = t+1$.
We proceed by induction on s : let $\overline{v} = vv_{i_{s+1}}$, x a prefix of v , $t < s$ and $g(v) = h(v) = K(v,\alpha|z)$. Then $K(\overline{v},\alpha|z) = F(\overline{v},v|z)K(v,\alpha|z)$ by (4.4). This and the inequality argument used above yields $g(\overline{v}) = F(\overline{v},v|z)g(v) = K(\overline{v},\alpha|z)$ by the induction hypothesis, the same holds for h.
Thus $\alpha = x\alpha_i$ is extremal if α_i is.

To prove the converse, suppose that α_i is not extremal, $K_i(.,\alpha_i|\zeta_i) = \lambda\cdot g_i + (1-\lambda)\cdot h_i$ is a nontrivial convex combination in H^+_{i,ζ_i} .
Then define for $v \in \Gamma$

$$g(v) = K(v,x|z) \qquad \text{in the situation of (4.3) and}$$

$$g(v) = g_i(v_{i_{t+1}})\cdot K(v,x|z)/F(v_{i_{t+1}},e|z) \qquad \text{in the situation of (4.4).}$$

Define h in the same way. By lengthy, but easy calculations involving the formulas of [17,§3] one obtains $g , h \in H^+_z$, $g(e) = h(e) = 1$. Now, by (4.3) and (4.4), $K(.,\alpha|z) = \lambda\cdot g + (1-\lambda)\cdot h$, and α is not extremal.□

6. *The hitting distribution*

We now show how one can calculate the hitting distribution ν_1 on M_1 as defined in (1.8). In this section, we always assume $\zeta_i = \zeta_i(1)$, $i = 1,2$.

Lemma 3. For $i = 1,2$, $\zeta_i < 1$ if $z = 1$.

Proof. From [17,Propositions 4a and 3] it follows that

$\zeta_i(1) = a_i/(1-H_j(1))$, where $j = 3-i$ and

(6.1) $H_j(1) = a_j \sum_{x \in \Gamma_j} \mu_j(x)F(x,e|1) \leq a_j$

In particular, $\zeta_i(1) \leq 1$. Suppose $\zeta_i(1) = 1$. Then $H_j(1) = 1 - a_i = a_j$, and by (6.1) and Prop.1, $1 = F(x,e|1) = F_j(x,e|\zeta_j)$ for each x in the support of μ_j . Hence,

$$G_j(e,e|\zeta_j) = 1/\Big(1 - \sum_{x\in\Gamma_j} \mu_j(x)F_j(x,e|\zeta_j)\Big) = \infty .$$

However, by the nonamenability of Γ , $1 < r$, hence $\zeta_j < r_j$ by Prop.1, and $G_j(e,e|\zeta_j) < \infty$, a contradiction. □

For $i = 1,2$, define

(6.2) $E_i = \{ \eta \in \Gamma \cup M_1 \mid \psi_{e,i}(\eta) \neq e \}$.

By Th.2, xE_i is an open set in the Martin topology for each $x \in \Gamma^{(i)}$.
If $\alpha = x_{j_1}x_{j_2}x_{j_3}\cdots \in \Omega^{(0)} \subset M_1$, then the sets $x_{j_1}\cdots x_{j_k}E_{j_{k+1}}$ $(k = 0,1,2,\dots)$ constitute a neighbourhood basis of α .
If $\alpha = x\alpha_i \in M_1^{(i)}$ and $\{ U_k \mid k=0,1,2,\dots\}$ is a neighbourhood basis of α_i in $\Gamma_i \cup M_{i,\zeta_i}$, then the sets $xU_k \cup \bigcup \{ xx_iE_j \mid x_i \in U_k \}$ $(k = 0,1,2,\dots ;\ j = 3-i)$ constitute a neighbourhood basis of α .
In particular, xM_{i,ζ_i} is a Borel set of M_1 .

<u>Theorem 4.</u> (a) The support of ν_1 is M_1 .

(b) $\nu_1(xM_{i,\zeta_i}) = 0$ for all $x \in \Gamma^{(i)}$, $i = 1,2$.

(c) ν_1 is determined by the values

$$\nu_1(xE_i) = F(e,x|1)/\big(1 + (1-\zeta_i)G_i(e,e|\zeta_i)\big) \quad , x \in \Gamma^{(i)} , \ i = 1,2.$$

<u>Proof.</u> As (c) yields $\Omega^{(0)} \subset \mathrm{supp}(\nu_1)$ and $\Omega^{(0)}$ is dense in M_1 , (a) will follow from (c).

<u>(b)</u> By the nearest neighbour property and the group invariance of the random walk we have

$$\nu_1(xM_{i,\zeta_i}) = F(e,x|1) \Pr [X_\infty \in xM_{i,\zeta_i} \mid X_0=x] = F(e,x|z)\, \nu_1(M_{i,\zeta_i})$$

If $\lim X_n \in M_{i,\zeta_i}$, then there must be infinitely many different points $\psi_{e,i}(X_n)$, $n=1,2,\dots$, and - again by the nearest neighbour property - the walk must pass each one before reaching the next one, that is, $\psi_{e,i}(X_n) = X_n$ for infinitely many n . Hence,

$$\nu_1(M_{i,\zeta_i}) \leq \Pr [X_n \in \Gamma_1 \text{ infinitely often} \mid X_0=e] .$$

From Prop. 1 and Lemma 3 we obtain

$$\sum_{n=0}^{\infty} \Pr[X_n \in \Gamma_i \mid X_0 = e] = \sum_{x \in \Gamma_i} G(e,x|1) = \sum_{x\in\Gamma_i} F(e,x|1)G(x,x|1) =$$

$$= G(e,e|1) \sum_{x\in\Gamma_i} G_i(e,x|\zeta_i)/G_i(x,x|\zeta_i) =$$

$$= \big(G(e,e|1)/G_i(e,e|\zeta_i)\big) \sum_{n=0}^{\infty} \sum_{x\in\Gamma_i} \mu_i^{(n)}(x)\, \zeta_i^n =$$

$$= G(e,e|1)/G_i(e,e|\zeta_i)\,(1-\zeta_i) < \infty \quad .$$

Now the Borel-Cantelli Lemma implies (b).

(c) According to the above description of a neighbourhood basis at each point of M_1, and in view of (b), ν_1 is in fact determined by the values $\nu_1(xE_i)$, $x \in \Gamma^{(i)}$, $i = 1,2$. By the nearest neighbour property,

$$\nu_1(xE_i) = F(e,x|1)\, \nu_1(E_i)$$

Now, if we set $j = 3-i$, then

$$\nu_1(E_i) = \nu_1(M_{i,\zeta_i}) + \sum_{x\in\Gamma_i} \nu_1(xE_j) = \sum_{x\in\Gamma_i} F(e,x|1)\, \nu_1(E_j) =$$

$$= \sum_{x\in\Gamma_i} F(e,x|1)\, \big(1 - \nu_1(E_i)\big) \quad ,$$

as $E_1 \cup E_2 = \Gamma \cup M_1 - \{e\}$. As above, we obtain

$$\nu_1(E_i)/\big(1-\nu_1(E_i)\big) = 1/(1-\zeta_i)G_i(e,e|\zeta_i) \quad ,$$

and the proposed formula follows. □

By (b), $X_\infty \in \Omega^{(0)}$ almost surely, and the hitting probability ν_1 can be restricted to a probability ν on $\Omega^{(0)}$ by defining $\nu(xE_i \cap \Omega^{(0)}) = \nu_1(xE_i)$. In other words, the *Poisson boundary* of the random walk can be almost surely identified with $(\Omega^{(0)}, \nu)$, in accordance with [10,Th.7].

7. *Remarks and examples*

A) The associativity of the free product allows us to extend all the above results to free products $\Gamma_1 * \Gamma_2 * \ldots * \Gamma_k$, $k > 2$. The parts $\Omega^{(0)}$, $M_z^{(1)}$ and $M_z^{(2)}$ of M_z differ for $(\Gamma_1*\Gamma_2)*\Gamma_3$ and $\Gamma_1*(\Gamma_2*\Gamma_3)$, but their respective unions can be identified in a natural way.

B) If the Γ_i $(i = 1,2)$ admit "uniformly spanning trees" in the sense of [13], then $M_{i,\zeta_i} = \Omega_i$ and $M_z = \Omega$ for each z, $0 < z \le r$, in accordance with [13]. The Poisson boundary can be identified with (Ω, ν_1). This is not in contrast with its description as $(\Omega^{(0)}, \nu)$ given above, as $\nu_1(\Omega - \Omega^{(0)}) = 0$.

C) If $\Gamma_i = \mathbb{Z}^{d_i}$ $(d_i \geq 2)$ and μ_i has finite support generating Γ_i as a semigroup, then it is known[3,12] that for $0 < \zeta < r_i$, $M_{i,\zeta}$ is homoeomorphic to the unit sphere S_{d_i-1}, and the functions $K_i(.,\alpha|\zeta)$ $(\alpha \in M_{i,\zeta})$ are precisely the exponentials

(7.1) $h_c(x) = \exp(<c\,,x>)$,

where c is a vector in $\mathbb{Z}^{d_i}$ satisfying

(7.2) $\sum_x \mu_i(x)\, h_c(x) = 1/\zeta$.

The precise correspondence between the "directions of convergence" $\alpha \in M_{i,\zeta}$ of the Martin kernel, the vectors c satisfying (7.2), and S_{d_i-1}, can be derived from [12] by applying the results given there to the probability measure $\overline{\mu}_i$ defined by $\overline{\mu}_i(x) = \zeta\,\mu_i(x)\,h_c(x)$, where c is arbitrarily chosen to satisfy (7.2). Furthermore, if $\zeta = r_i$, then (7.2) has a unique solution $c_i = c(r_i)$, and the exponential h_{c_i} is the only positive r_i-harmonic function up to multiplication with a constant.
Now let $\Gamma = \Gamma_1 * \Gamma_2$ and $\mu = a_1\mu_1 + a_2\mu_2$. Then for $0 < z < r$, $\zeta_i = \zeta_i(z) < r_i$ and the parts $M_z^{(i)}$ of the Martin boundary consist of countably many copies of S_{d_i-1}, each one representing one end of Γ $(i = 1,2)$. If for $i \in \{1,2\}$ $d_i \leq 4$, then $\zeta_i(r) < r_i$ [17,Remark D], so that $M_r^{(i)}$ can be described as above. However, for $d_i \geq 5$ it may occur that $\zeta_i(r) = r_i$ [1], and in this case $M_r^{(i)}$ "collapses" to $\Omega^{(i)}$ as defined in Lemma 1.

D) In all known cases of groups, M_z does not depend on z, $0 < z \leq r$ (with a possible exception at $z = r$), nor on the particular choice of μ (with finite support generating Γ as a semigroup), but only on the group structure. Can this be confirmed by a general theorem ?

E) One would also like to describe the Martin boundary on a free product $\Gamma = \Gamma_1 * \Gamma_2$, when μ has finite support generating Γ as a semigroup, but is not necessarily of "nearest neighbour" type. This has been achieved for the Poisson boundary, which can always be identified with $\Omega^{(0)}$ [10], even for infinitely supported μ with finite first moment. However, before describing the whole Martin boundary, it seems that one needs a "canonical" representation theorem in the sense of *D)*.

References

[1] CARTWRIGHT,D.I.: Some remarks about random walks on free products of discrete groups. Preprint, Univ.Sidney (1986).

[2] CARTWRIGHT,D.I. - SOARDI,P.M.: Random walks on free products, quotients and amalgams. To appear, Nagoya Math.Journal

[3] CHOQUET,G. - DENY,J.: Sur l'équation de convolution $\mu=\mu*\sigma$. CRAS Paris 250 (1960) 799-801.

[4] DERRIENNIC,Y.: Marche aléatoire sur le groupe libre et frontière de Martin. Z.Wahrscheinlichkeitstheorie 32 (1975) 261-276.

[5] DIXMIER,J.: Les moyennes invariantes dans les sémigroupes et leurs applications. Acta Sci.Math.Szeged 12 A (1950) 213-227.

[6] DYNKIN,E.B. - MALYUTOV,M.B.: Random walks on groups with a finite number of generators. Soviet Math.Doklady 2 (1961) 399-402.

[7] FREUDENTHAL,H.: Über die Enden diskreter Räume und Gruppen. Comment.Math.Helvet. 17 (1944) 1-38.

[8] GERL,P.: Probability measures on semigroups. Proc.AMS 40 (1973) 527-532.

[9] GUIVARC'H,Y.: Sur la loi des grands nombres et le rayon spectral d'une marche aléatoire. Astérisque 74 (1980) 47-98.

[10] KAIMANOVICH,V.A.: An entropy criterion for maximality of the boundary of random walks on discrete groups. Soviet Math.Doklady 31 (1985) 193-197.

[11] KEMENY,J.G. - SNELL,J.L - KNAPP,A.W.: Denumerable Markov Chains, 2nd edition. Springer, New York-Heidleberg-Berlin 1976.

[12] NEY,P. - SPITZER,F.: The Martin boundary for random walk. Trans. AMS 121 (1966) 116-132.

[13] PICARDELLO,M.A. - WOESS,W.: Martin boundaries of random walks: ends of trees and groups. Preprint, Univ.Roma (1985).

[14] PICARDELLO,M.A. - WOESS,W.: Random walks, harmonic functions, and ends of graphs. Preprint, Montanuniv.Leoben (1986).

[15] PRUITT,W.E.: Eigenvalues of nonnegative matrices. Ann Math.Statist. 35 (1964) 1797-1800.

[16] SERIES,C.: Martin boundaries of random walks on Fuchsian groups. Israel J.Math. 44 (1983) 221-242.

[17] WOESS,W.: Nearest neighbour random walks on free products of discrete groups. To appear, Boll.Un.Mat.Ital.

ON HYPERBOLIC HYPERGROUPS

Hansmartin Zeuner
Mathematisches Institut
der Universität Tübingen
Auf der Morgenstelle 10

Introduction

Let X,Y be d-dimensional random variables with rotation invariant distributions and $\mu := P_{||X||}$, $\nu := P_{||Y||}$. Then the distribution of $||X+Y||$ is the measure

$$c \iiint_0^\pi \varepsilon_{\sqrt{x^2+y^2+2xy\cos\theta}} (\sin\theta)^{d-2}\, d\theta\, \mu(dx)\, \nu(dy)$$

where $c = \Gamma(\frac{d}{2})/\sqrt{\pi}\,\Gamma(\frac{d-1}{2})$ and $d \geq 2$. It is clear that the square root arises from the cosine law $z^2 = x^2 + y^2 + 2xy\cos\theta$ and that this convolution of measures on $\mathbb{R}_+$ is associative. Kingman |10| showed that a convolution operation, defined by

$$\mu * \nu := \iiint \varepsilon_{\sqrt{x^2+y^2+2xy\cos\theta}} P(d\theta)\, \mu(dx)\, \nu(dy) \quad (\mu,\nu \in M^b(\mathbb{R}_+))$$

for some probability measure P on $[0,\pi]$, is associative if and only if $P = \varepsilon_0$ or $P = \varepsilon_{\pi/2}$ or $P = \frac{1}{2}(\varepsilon_0 + \varepsilon_\pi)$ or P has Lebesgue-density $k_a \sin^a$ with $a>-1$, and used this to derive interesting probabilistic results. The same was done by Bingham [1] using the spherical cosine law $\cos z = \cos x \cos y + \sin x \sin y \cos\theta$ (see also Gasper [7]).

In this article the hyperbolic case will be treated. It will be shown that the convolution

$$(\mu,\nu) \to \mu \star \nu := \iiint \varepsilon_{\operatorname{arch}(\cosh x \cosh y + \sinh x \sinh y \cos\theta)} P(d\theta)\, \mu(dx)\, \nu(dy)$$

$(\mu,\nu \in M^b(\mathbb{R}_+))$ for some fixed $P \in M^1([0,\pi])$ satisfies

$$(\mu \star \nu) \star \eta = \mu \star (\nu \star \eta) \qquad \text{for } \mu,\nu,\eta \in M^b(\mathbb{R}_+)$$

if and only if P is one of the measures ε_0, $\varepsilon_{\pi/2}$, $\frac{1}{2}(\varepsilon_0 + \varepsilon_\pi)$ or $k_a \sin^a \lambda_{[0,\pi]}$ with $a>-1$ (where $k_a = \Gamma(\frac{a}{2}+1)/\sqrt{\pi}\,\Gamma(\frac{a+1}{2})$).

The proof of the "if"-part of this theorem is contained in a more general result of Flensted-Jensen and Koornwinder [5]. The main tool is the product formula for Jacobi functions which in [5] is derived from a generalization of Bateman's formula. In the present article a different proof of the product formula is given using the convolution structure introduced above.

Unless $P = \varepsilon_o$ or $P = \varepsilon_{\pi/2}$ these convolutions are hypergroups in the sense of [2] and [8]. They occur as a special case of the convolution structures described by Chébli [3]. If a is an integer, the hypergroup corresponding to $P=k_a \sin^a \lambda_{[0,\pi]}$ agrees with the double coset convolutions $SO(a+2,1)_1//SO(a+2)$ (see [4], chapitre III and [9], 9.5, 14.2 and 15.5).

1. The Main Result

1.1 Let K be a locally compact topological space, $M^b(K)$ the set of all bounded, signed Borel measures and $M^1(K)$ the subset of probability measures with the weak topology. In the sequel we will consider bilinear applications $(\mu,\nu) \to \mu*\nu$ from $M^b(K)xM^b(K)$ into $M^b(K)$ such that $(x,y) \to \varepsilon_x*\varepsilon_y$ is a continuous mapping from KxK into $M^1(K)$. It is clear that this operation is uniquely determined by the convolutions $\varepsilon_x*\varepsilon_y$ for $x,y \in K$. It is called commutative if $\mu*\nu = \nu*\mu$ for $\mu,\nu \in M^b(K)$ and it is called associative if $(\mu*\nu)*\eta = \mu*(\nu*\eta)$ for every $\mu,\nu,\eta \in M^b(K)$. This condition is clearly equivalent to $(\varepsilon_x*\varepsilon_y)*\varepsilon_z = \varepsilon_x*(\varepsilon_y*\varepsilon_z)$ for $x,y,z \in K$.

1.2 Consider a probability measure P on $[-1,1]$. Let us define a convolution operation $\underset{P}{*}$ on $M^b(\mathbb{R}_+)$ by

$$\varepsilon_x \underset{P}{*} \varepsilon_y := \int \varepsilon_{\mathrm{arch}(\mathrm{ch}\,x\,\mathrm{ch}\,y\,+\,\lambda\,\mathrm{sh}\,x\,\mathrm{sh}\,y)}\,P(d\lambda) \quad \text{for } x,y \in \mathbb{R}_+$$

where sh and ch are the hyperbolic sine- and cosine functions and $\mathrm{arch} = \mathrm{ch}^{-1}$. Clearly this operation is commutative.

For every $\alpha>0$ let $P_\alpha \in M^1([-1,1])$ be the probability measure with the density $c_\alpha(1-\lambda^2)^{\alpha-1}$ with respect to the Lebesgue measure $\lambda_{[-1,1]}$ where $c_\alpha := \Gamma(\alpha+\frac{1}{2})/\sqrt{\pi}\,\Gamma(\alpha)$. Furthermore define $P_o := \frac{1}{2}(\varepsilon_{-1}+\varepsilon_1) = \lim_{\alpha\to 0} P_\alpha$.

1.3 <u>Theorem</u>:

The operation $\underset{P}{*}$ *is associative if and only if* P *is one of the measures* ε_1, ε_o, *or* P_α $(\alpha\geq 0)$.

1.4 <u>Corollary</u>:

$(\mathbb{R}_+, \underset{P}{*})$ *is a hypergroup if and only if* $P \in \{P_\alpha : \alpha \geq 0\}$.

2. Proof of the "If"-Part of the Theorem

2.1 If $P=\varepsilon_1$ then $\varepsilon_x \underset{P}{*} \varepsilon_y = \varepsilon_{x+y}$ and the operation is clearly associative. For $P=\varepsilon_0$ we obtain $\varepsilon_x \underset{P}{*} \varepsilon_y = \varepsilon_{\mathrm{arch}(\mathrm{ch}\, x\, \mathrm{ch}\, y)}$. This convolution is isomorphic to the preceding example via the application $\ln\circ\mathrm{ch}$ and hence it is associative. These are no hypergroups since there is no x such that $0 \in \mathrm{supp}(\varepsilon_x \underset{P}{*} \varepsilon_1)$.

2.2 In the case $P = P_0 = \frac{1}{2}(\varepsilon_1 + \varepsilon_{-1})$ we obtain $\varepsilon_x \underset{P}{*} \varepsilon_y = \frac{1}{2}(\varepsilon_{|x-y|} + \varepsilon_{x+y})$. With this convolution $\mathbb{R}_+$ is isomorphic to the double coset hypergroup of the semidirect product of $\mathbb{R}$ and $\mathbb{Z}_2$ modulo $\mathbb{Z}_2$.

2.3 Now consider $\alpha>0$ and $P := P_\alpha$. Since the associativity of $* := \underset{P}{*}$ has already been proved in [5] we will only sketch a proof of this fact using different methods.

Let $\lambda \in \mathbb{C}$ and define an application $\phi_\lambda : \mathbb{R}_+ \to \mathbb{C}$ by the formula

$$\phi_\lambda(x) = \int (\mathrm{ch}\, x + \mu\, \mathrm{sh}\, x)^{\sqrt{\alpha^2-\lambda}-\alpha} P_\alpha(d\mu)$$
$$= 2^\alpha c_\alpha (\mathrm{sh}\, x)^{-2\alpha-1} \int_{-x}^{x} \exp(u\sqrt{\alpha^2-\lambda})(\mathrm{ch}\, x - \mathrm{ch}\, u)^{\alpha-1}\, du.$$

It can be shown that $\phi_\lambda(0) = \|\phi_\lambda\|_\infty = 1$ if $4\alpha^2 \operatorname{Re}\lambda \geq (\operatorname{Im}\lambda)^2$, that ϕ_λ is real-valued if $\lambda \geq 0$ and non negative if $\lambda \in [0,\alpha^2]$. By an explicit calculation one obtains that ϕ_λ is the unique solution of the differential equation

$$L\phi_\lambda = \lambda\,\phi_\lambda \quad , \quad \phi_\lambda(0) = 1$$

where L is the differential operator

$$f \to Lf := -f'' - 2\alpha \frac{\mathrm{ch}}{\mathrm{sh}} f'$$

(compare [4], appendice C). ϕ_λ is a Jacobi function.

2.4 Let ω be the measure with Lebesgue-density $\mathrm{sh}^{2\alpha}$ on $\mathbb{R}_+$. Then it is easily seen that for every $f,g \in L^1(\omega)$ the function $f*g : \mathbb{R}_+ \to \mathbb{C}$ defined by

$$f*g(u) := \int_0^\infty f(x)\, \mathrm{sh}^{2\alpha} x \int g\, d\varepsilon_u * \varepsilon_x\, dx$$
$$= \int_0^\infty g(y)\, \mathrm{sh}^{2\alpha} y \int f\, d\varepsilon_u * \varepsilon_y\, dy \quad \text{for almost every } u \geq 0$$

is a density of $(f\omega)*(g\omega)$ with respect to ω.

By the same formula $f*g$ can be defined if f and g are continuous functions and one of them is compactly supported. By differentiation under the integral sign and partial integration one calculates the following identity:

2.5 Lemma:

Let f be a two times continuously differentiable function on $\mathbb{R}_+$ *with* $f'(0) = 0$, $g \in C(\mathbb{R}_+)$ *and* f *or* g *compactly supported. Then* $f*g$ *is two times continuously differentiable and* $L(f*g) = (Lf)*g$.

Now we are able to prove the following proposition:

2.6 Proposition

Let ϕ *be a continuous function on* $\mathbb{R}_+$. *Then the following assertions are equivalent:*

(i) $\int \phi \, d\varepsilon_x * \varepsilon_y = \phi(x)\,\phi(y)$ *for* $x,y \in \mathbb{R}_+$ *and* ϕ *is not identically* 0,

(ii) $\phi(0) = 1$ *and for every continuous function* f *with compact support there exists* $c_f \in \mathbb{C}$ *such that* $f*\phi = c_f \phi$,

(iii) ϕ *is two times continuously differentiable,* $\phi(0) = 1$ *and there exists a complex number* λ *such that* $L\phi = \lambda\phi$,

(iv) $\phi = \phi_\lambda$ *for some* $\lambda \in \mathbb{C}$.

Proof: "(i)⇒(ii)": Since

$$\phi*f(u) = \int f(y)\, sh^{2\alpha}y \int \phi \, d\varepsilon_u*\varepsilon_y \, dy$$
$$= \int f(y)\, sh^{2\alpha}y \,\phi(u)\,\phi(y)\, dy$$
$$= \phi(u) \int f(y)\,\phi(y)\, sh^{2\alpha}y\, dy,$$

we have to choose $c_f := \int f(y)\,\phi(y)\, sh^{2\alpha}y\, dy$.

"(ii)⇒(i)": This is easily proved by approximation of ε_y by probability measures $f_n\omega$ where f_n is continuous and compactly supported.

"(ii)⇒(iii)": Let f be a two times continuously differentiable function with compact support, $f'(0)=0$ and $c_f \neq 0$. Then by 2.5 $\phi = c_f^{-1}\cdot(f*\phi)$ is in $C^2(\mathbb{R}_+)$ and

$$L\phi = L(c_f^{-1}\cdot f*\phi) = c_f^{-1}\cdot(Lf)*\phi = c_{Lf}c_f^{-1}\phi .$$

Therefore we have to choose $\lambda := c_{Lf}c_f^{-1}$.

"(iii)⇒(ii)": Let f be a continuous function on $\mathbb{R}_+$ with compact support. Then by 2.5 we have

$$L(f*\phi) = f*(L\phi) = \lambda\, f*\phi .$$

The unique solution of this differential equation having the value

$f\ast\phi(0)$ at 0 is the function $f\ast\phi(0)\cdot\phi$. Hence we have to choose $c_f := f\ast\phi(0)$.

"(iii)$\Leftrightarrow$(iv)": This follows from 2.3.

2.7 For every $\lambda,x,y,z \in \mathbb{R}_+$ we obtain from 2.6 the identity

$$\int \phi_\lambda \, d(\varepsilon_x \ast \varepsilon_y)\ast\varepsilon_z = \int \phi_\lambda \, d(\varepsilon_x\ast\varepsilon_y)\cdot \phi(z) = \phi(x)\cdot\phi(y)\cdot\phi(z)$$
$$= \int \phi_\lambda \, d\,\varepsilon_x\ast(\varepsilon_y\ast\varepsilon_z).$$

The uniqueness theorem for the Mehler transform implies that

$$(\varepsilon_x\ast\varepsilon_y)\ast\varepsilon_z = \varepsilon_x\ast(\varepsilon_y\ast\varepsilon_z).$$

3. Proof of the "Only If"-Part of the Theorem

We consider a probability measure $P \in M^1([-1,1])$ with $P \neq \varepsilon_0, \varepsilon_1, P_0$ such that $\underset{P}{\ast}$ is associative. In the sequel it will be shown that $P = P_\alpha$, where

$$\alpha := \frac{1}{2}\left(\int \lambda^2 \, P(d\lambda)\right)^{-1} - \frac{1}{2}. \tag{1}$$

First we define for every integer $n \geq 0$ a function $\psi_n : [1,\infty[\to \mathbb{R}_+$ by

$$\psi_n(x) := \int (x + \lambda\sqrt{x^2-1})^n \, P(d\lambda) \qquad (x \in [1,\infty[\,).$$

Now we use the associativity of $\underset{P}{\ast}$ to prove the following lemma:

3.1 Lemma:

For every integer $n \geq 0$ *and every* $x \in [1,\infty[$ *we have*

$$\psi_n(x)^2 = \int \psi_n(x^2 + \lambda(x^2-1)) \, P(d\lambda).$$

Proof: For every $x,z \in [1,\infty[$ and every $f \in C([1,\infty[)$ we have

$$\iint f(x(xz+\lambda\sqrt{x^2-1}\sqrt{y^2-1}) + \mu\sqrt{x^2-1}\sqrt{(xz+\lambda\sqrt{x^2-1}\sqrt{z^2-1})^2-1}) \, P(d\lambda)P(d\mu)$$

$$= \int f\circ ch \, d\,\varepsilon_{\operatorname{arch} x} \underset{P}{\ast} (\varepsilon_{\operatorname{arch} x} \underset{P}{\ast} \varepsilon_{\operatorname{arch} z})$$

$$= \int f\circ ch \, d\,(\varepsilon_{\operatorname{arch} x} \underset{P}{\ast} \varepsilon_{\operatorname{arch} x}) \underset{P}{\ast} \varepsilon_{\operatorname{arch} z}$$

$$= \iint f((x^2+\lambda(x^2-1))z + \mu\sqrt{(x^2+\lambda(x^2-1))^2-1}\sqrt{z^2-1}) \, P(d\lambda)\,P(d\mu)$$

Choosing $f(u) := (u/z)^n$ and letting $z \to \infty$ we obtain

$$\iint [x(x+\lambda\sqrt{x^2-1}) + \mu\sqrt{x^2-1}\,(x+\lambda\sqrt{x^2-1})]^n \, P(d\lambda)\,P(d\mu)$$

$$= \iint [(x^2+\lambda(x^2-1)) + \mu\sqrt{(x^2+\lambda(x^2-1))^2-1}\,]^n \, P(d\mu)\,P(d\lambda).$$

The first of the double integrals can be separated and has the value $\psi_n(x)^2$. The second is equal to $\int \psi_n(x^2+\lambda(x^2-1)) \, P(d\lambda)$. □

3.2 Now we continue the proof of the theorem. In order to show $P = P_\alpha$ it is sufficient to prove

$$\int_{-1}^{1} \lambda^n \, P(d\lambda) = \int_{-1}^{1} \lambda^n \, P_\alpha(d\lambda) \qquad \text{for } n \geq 0. \qquad (2).$$

This will be done by induction. The case $n = 0$ is trivial. Now assume that $n \geq 1$ and that $\int \lambda^j \, P(d\lambda) = \int \lambda^j \, P_\alpha(d\lambda)$ for every integer $j < m$. We have to show that $\beta = 0$ where $\beta := \int \lambda^n P(d\lambda) - \int \lambda^n P_\alpha(d\lambda)$. Since P_α is symmetric we have

$$\int \lambda^j \, P_\alpha(d\lambda) = 0 \qquad \text{for every odd number } j.$$

Hence the function $\phi_n : [1,\infty[\to \mathbb{R}$ defined by

$$\phi_n(x) := \int (x + \lambda\sqrt{x^2-1}\,)^n \, P_\alpha(d\lambda) \qquad (x \geq 1)$$

is a polynomial of degree less or equal n.

From the induction hypothesis and the binomial theorem we obtain

$$\psi_n(x) = \phi_n(x) + \beta\,(x^2-1)^{n/2} \qquad (x \geq 1) \qquad (3).$$

Since we have already proved that $\underset{P_\alpha}{\ast}$ is associative (2.7) we may apply the lemma both to P and P_α. This gives

$$\phi_n(x)^2 + 2\,\beta\,\phi_n(x)\,(x^2-1)^{n/2} + \beta^2\,(x^2-1)^n$$
$$= \psi_n(x)^2$$
$$= \int \psi_n(x^2 + \lambda(x^2-1))\; P(d\lambda)$$
$$= \int \phi_n(x^2 + \lambda(x^2-1))\, P_\alpha(d\lambda) + \int \phi_n(x^2+\lambda(x^2-1))\,(P-P_\alpha)(d\lambda) + \beta \int ((x^2+\lambda(x^2-1))^2-1)^{n/2} P(d\lambda)$$
$$= \phi_n(x)^2 + \beta\,\gamma\,(x^2-1)^n + \beta\,(x^2-1)^{n/2}.\int (\lambda+1)^{n/2}(x^2+1+\lambda(x^2-1))^{n/2} P(d\lambda)$$

where

$$\gamma := \int (1+\lambda)^n \, P_\alpha(d\lambda)$$

is the leading coefficient of ϕ_n. Thus we have proved

$$2\beta\phi_n(x) + \beta(\beta-\gamma)(x^2-1)^{n/2} = \beta \int (\lambda+1)^{n/2}\,(x^2+1+\lambda(x^2-1))^{n/2}\, P(d\lambda) \qquad (4).$$

3.3 First we consider the case that n is odd. The polynomial ϕ_n and the right hand side of (4) are $\frac{n+1}{2}$ times differentiable from the right at $x=1$. This is not the case with the function $x \to (x^2-1)^{n/2}$ on the left hand side and so $\beta(\beta-\gamma) = 0$, that is

$$2\,\beta\,\phi_n(x) = \beta \int (\lambda+1)^{n/2}\,(x^2 + 1 + \lambda(x^2-1))^{n/2}\; P(d\lambda) \qquad (5).$$

Analytic continuation onto the open right halfplane and $x \to 0$ implies

$$0 = 2\beta\,\phi_n(0) = \beta \int (\lambda+1)^{n/2}\,(1-\lambda)^{n/2}\; P(d\lambda) \qquad (6).$$

Now we claim that P cannot be concentrated on the set $\{-1,1\}$. Indeed, assume $P = a\varepsilon_1 + (1-a)\varepsilon_{-1}$ with $a \in [0,1[\setminus\{\frac{1}{2}\}$ (the cases $P = \varepsilon_1$ and $P = P_0$ are already excluded). Then n=1 gives $\beta = \int \lambda P(d\lambda) = 2a-1 \neq 0$ and (5) implies $2 = 2\phi_n(1) = \int (\lambda+1)^{1/2}.2^{1/2}(a\varepsilon_1+(1-a)\varepsilon_{-1})(d\lambda) = 2a$ in contradiction to $a\neq 1$.

Therefore we obtain $P(]-1,1[) > 0$ and $\int (1-\lambda^2)^{n/2} P(d\lambda) > 0$. From (6) we conclude $\beta=0$ and (2) is proved in this case.

3.4 Last we consider the case that n is even. We may assume $n\geq 4$ since (2) is trivially true in the case n=2 by the definition of α in (1).

Since n is even the integrand on the right hand side of (4) is a polynomial in λ and hence

$$\begin{aligned} &\int (1+\lambda)^{n/2} (x^2+1+\lambda(x^2-1))^{n/2} (P-P_\alpha)(d\lambda) \\ &= (x^2-1)^{n/2} \int \lambda^n (P-P_\alpha)(d\lambda) \\ &= \beta (x^2-1)^{n/2} \end{aligned}$$

by the induction hypothesis. Thus (4) implies

$$2\beta\,\phi_n(x) - \beta\gamma(x^2-1)^{n/2} = \beta \int (\lambda+1)^{n/2} (x^2+1+\lambda(x^2-1))^{n/2} P_\alpha(d\lambda).$$

For x=1 we obtain

$$\begin{aligned} 2\beta &= 2\beta\,\phi_n(1) \\ &= 2^{n/2}\beta \int (\lambda+1)^{n/2} P_\alpha(d\lambda) \\ &= 2^{n/2}\beta \sum_{j\leq n/4} \binom{n/2}{2j} \int \lambda^{2j} P_\alpha(d\lambda). \end{aligned}$$

But $n\geq 4$ and hence

$$2^{n/2} \sum_{j\leq n/4} \binom{n/2}{2j} \int \lambda^{2j} P_\alpha(d\lambda) \geq 2\left(1 + \binom{n/2}{2} \int \lambda^2 P_\alpha(d\lambda)\right) > 2.$$

This implies $\beta=0$ and (2) is proved. □

3.5 If $P = P_\alpha$ for some $\alpha\geq 0$ then $(\mathbb{R}_+, \underset{P}{*})$ is a hypergroup in the sense of [8]. This is an easy consequence from the fact that

$$\operatorname{supp}(\varepsilon_x \underset{P}{*} \varepsilon_y) = \{|x-y|, x+y\} \qquad (x,y \in \mathbb{R}_+)$$

if $\alpha=0$ and

$$\operatorname{supp}(\varepsilon_x \underset{P}{*} \varepsilon_y) = [|x-y|, x+y] \qquad (x,y \in \mathbb{R}_+)$$

if $\alpha>0$ (see also [11]).

It follows from 2.4 that a Haar measure of $(\mathbb{R}_+, \underset{P}{*})$ is given by

$$\omega := \operatorname{sh}^{2\alpha}\cdot\lambda_{\mathbb{R}_+}.$$

Furthermore from 2.6 one concludes that $\{\phi_\lambda : \lambda \geq 0\}$ is the dual of $(\mathbb{R}_+, \underset{p}{*})$.

In the case $\alpha=0$ it is clear that $(\mathbb{R}_+, \underset{p}{*})$ has a dual hypergroup with respect to pointwise multipli ation. However, if $\alpha>0$, the Plancherel measure has the Lebesgue density

$$\lambda \to 2^{-2\alpha}(\lambda-\alpha^2)^{-1/2}\,|\Gamma(\alpha+i\sqrt{\lambda-\alpha^2})|^2\,|\Gamma(\alpha+\tfrac{1}{2})\,\Gamma(i\sqrt{\lambda-\alpha^2})|^{-2}\,1_{]\alpha^2,\infty[}(\lambda).$$

This is a consequence of [5], (2.9). Therefore $(\mathbb{R}_+, \underset{p}{*})$ does not possess a hypergroup dual with respect to pointwise multiplication.

3.6 We conclude with some remarks about probabilistic applications of these hypergroups. It follows from [3] that for every convolution semigroup of probability measures $(\mu_t : t\geq 0)$ on $\mathbb{R}_+$ the Fourier transform of μ_t can be represented as

$$\int \phi_\lambda \, d\mu_t = \exp(-t\Psi)$$

where Ψ is of the form

$$\Psi(\lambda) = a\lambda + \int_{]0,\infty[} (1-\phi_\lambda)\, d\eta$$

with $a\geq 0$ and a Radon measure η on $]0,\infty[$ such that $\int \frac{x^2}{1+x^2}\,\eta(dx) < \infty$. It follows from [3], Theorème 8, that every hyperbolic hypergroup with $\alpha>0$ is transient in the sense that the potential kernel of every non trivial convolution semigroup vanishes at infinity. This result has recently been improved in [6].

4. Bibliography

[1] Bingham, N.H.: Random walks on spheres. Z. Wahrscheinlichkeitstheorie verw. Gebiete 22 (1972), 169-192

[2] Bloom, W.R., Heyer, H.: The Fourier transform for probability measures on hypergroups. Rendiconti di Matematika (2) 1982, Vol. 2, Serie VII, 315-334

[3] Chébli, H.: Opérateurs de translation généralisée et semi-groupes de convolution. In: Théorie du Potentiel et Analyse Harmonique. Edité par J. Faraut. Lecture Notes in Mathematics Vol. 404, Springer Berlin-Heidelberg-New York, 1974

[4] Faraut, J.: Analyse harmonique sur les paires de Guelfand et les espaces hyperboliques. Strasbourg

[5] Flensted-Jensen, M., Koornwinder, T.: The convolution structure for Jacobi function expansions. Arkiv för Matematik 11 (1973), 245-262

[6] Gallardo, L.: Random walks on hypergroups. Oral report, Oberwolfach, 1985

[7] Gasper, G.: Positivity and the convolution structure for Jacobi series. Annals of Math. 93 (1971), 112-118

[8] Heyer, H.: Probability theory on hypergroups: A survey. In: Probability Measures on Groups VII. Edited by H. Heyer. Lecture Notes in Mathematics Vol. 1064, Springer Berlin-Heidelberg-New York-Tokyo, 1984

[9] Jewett, R.I.: Spaces with an abstract convolution of measures. Advances in Math. 18 (1975), 1-101

[10] Kingman, J.F.C.: Random walks with spherical symmetry. Acta Math. 109 (1963), 11-53

[11] Zeuner, Hm.: One-dimensional hypergroups. To appear.

THEOREMES DE LA LIMITE CENTRALE POUR LES PRODUITS DE MATRICES EN DEPENDANCE MARKOVIENNE. RESULTATS RECENTS

par Philippe Bougerol

1. INTRODUCTION.

Considérons une chaine de Markov stationnaire $\{x_n, n \in \mathbb{N}\}$ sur un espace E. Si M est une application mesurable de E dans l'ensemble $G\ell(d,\mathbb{R})$ des matrices inversibles d'ordre d, on pose

$$M_n = M(x_n)\, M(x_{n-1}) \dots M(x_1) \ .$$

Notre but est d'exposer quelques résultats récents concernant la convergence en loi de la suite des matrices M_n , $n = 1,2, \dots$ convenablement normalisées. Citons quelques situations où intervient ce problème.

a) L'étude de la conductance dans une chaine désordonnée (O'Connor [17], Verheggen [25]).

b) L'étude de la transmission dans un guide d'onde radio (cf. Tutubalin [24] et sa bibliographie).

c) Le théorème de la limite centrale pour des produits d'éléments indépendants et de même loi d'un groupe de Lie (Raugi [20]).

d) L'étude de la stabilité des équations différentielles stochastiques linéaires (Bougerol [3], [4],[6]).

Le premier travail important sur cette question remonte à 1960 et est dû à Furstenberg et Kesten [8] . Il fut développé dans ce que nous appelerons le cas indépendant (i.e. celui où les x_n sont indépendants et de même loi) d'abord par Tutubalin et ses élèves dans les années 65-75 puis, plus récemment, par des membres de l'université de Rennes (Guivarc'h, Le Page, Raugi).

Nous nous intéressons à trois situations extrêmes :

Cas 1 : Les matrices $M(x)$ sont dans un sous-groupe compact de $G\ell(d,\mathbb{R})$.

Cas 2 : Les matrices $M(x)$ sont dans le groupe $N(d)$ des matrices triangulaires supérieures dont les termes diagonaux sont égaux à 1.

Cas 3 : L'ensemble des matrices $\{M(x), x \in E\}$ opère de façon irréductible sur $\mathbb{R}^d$.

En fait, pratiquement rien n'est connu hors de ces cas et de nombreux problèmes restent ouverts.

2. CADRE ET DEFINITICNS

Dans tout cet article nous ferons toujours l'hypothèse suivante :

Condition (c) :

Le noyau de transition P *de la chaine de Markov* $\{x_n, n \in \mathbb{N}\}$ *sur* E *admet une densité* p *par rapport à une probabilité invariante* π *, possédant les propriétés suivantes :*

a) $\pi \times \pi$ *presque partout,* p *est strictement positif.*

b) $\sup_{x \in E} p(x,y)$ *est dans* $L^1(\pi)$

Cette condition est extrémement forte et pourrait être affaiblie au moins dans l'étude des cas 1 et 2. D'un autre côté, même pour l'étude de sommes de variables aléatoires réelles du type

$$f(x_1) + f(x_2) + \ldots + f(x_n)$$

quelque chose ressemblant plus ou moins à cette condition est nécessaire si l'on veut obtenir, sans trop de difficultés, un théorème de la limite centrale non dégénéré (cf. par exemple Gordin, Lifschitz [9]).

Sous la condition (c) la chaine de Markov admet une seule probabilité invariante et $\| P^n - \pi \|_\infty$ tend exponentiellement vite vers zéro, pour la norme d'opérateurs sur $L^\infty(E, d\pi)$.

Notation

Etant donnéeune application mesurable M *de* E *dans* $G\ell(d,\mathbb{R})$ *on notera* $M(\pi)$ *l'image de la probabilité* π *par* M *et* S_π *le support de* $M(\pi)$.

Afin d'éviter des répétitions inutiles posons, si $M_n = M(x_n) \ldots M(x_1)$,

Définition

Soient $F_1, F_2, \ldots$ *des applications mesurables de* $G\ell(d,\mathbb{R})$ *dans un espace topologique* V *et* β *une probabilité sur* $E \times V$.

On dit que

"$(x_n, F_n(M_n))$ *converge en loi vers* β"

si pour toute fonction ϕ *de* $L^\infty(E, d\pi)$ *et toute fonction* $\psi : V \to \mathbb{R}$ *continue à support compact*

$$\lim_{n\to\infty} \mathbb{E}_x \{\phi(x_n)\, \psi(F_n(M_n))\} = \int \phi(x)\psi(u)\, d\beta(x,u)$$

pour π *presque tout* x *de* E .

Dans cet énoncé, $\mathbb{E}_x$ désigne l'espérance suivant la loi du processus $\{x_n, n \geq 0\}$ sachant que $x_0 = x$. Il n'est pas difficile de voir que dans les résultats qui suivent on peut remplacer la convergence pour π presque tout x par la convergence pour tout x de E.

Un théorème de la limite centrale pour M_n est un énoncé qui permet de calculer de façon approchée la loi de M_n pour n grand. La situation idéale est celle où on peut trouver des homéomorphismes F_n du groupe engendré par S_π dans un espace topologique V tels que $(x_n, F_n(M_n))$ converge en loi vers une probabilité non dégénérée . Nous allons voir que ceci est vrai dans les deux premiers cas et "presque" vrai dans le troisième, sous des hypothèses convenables.

3. LE CAS COMPACT

Commençons, à titre d'introduction, par examiner la situation très simple où les matrices $M(x_i)$ sont portées par un groupe compact. Le cas indépendant a été traité dès 1940 par Kawada et Ito ([15]) et on en trouvera un exposé très complet dans Heyer [13] .

Dans le cas markovien, on a sous la condition (c) le théorème suivant. Nous en donnons rapidement l'essentiel de la démonstration, qui est élémentaire, car elle fait bien apparaître le rôle joué par cette condition. Cette condition intervient dans les résultats suivants pour des raisons fondamentalement analogues.

Théorème

Supposons que le plus petit sous groupe fermé de $G\ell(d,\mathbb{R})$ *contenant le support* S_π *de* $M(\pi)$ *soit un groupe compact* K *de mesure de Haar normalisée* m.

Si S_π *n'est pas contenu dans une classe latérale d'un sous-groupe distingué fermé propre de* K *, alors*

(x_n, M_n) *converge en loi vers* $\pi \otimes m$.

Démonstration

En utilisant par exemple la formule de Plancherel on voit qu'il suffit de montrer que pour toute représentation unitaire irréductible L de K et tout ϕ de $L^\infty(E)$,

$$\lim_{n\to\infty} \mathbb{E}_x \{\phi(x_n) L^*_{M_n}\} = \int \phi(x) d\pi(x) \int L^*_g \, dm(g)$$

(pour π presque tout x). Pour cela introduisons l'opérateur T sur $L^\infty(E;\mathbb{C}^p)$, où $p = \dim L$, défini par, si $R = L^*$,

$$T\Phi(x) = \mathbb{E}_x [R_{M_1} \{\Phi(x_1)\}] \quad , \qquad \Phi \in L^\infty(E;\mathbb{C}^p), \ x \in E .$$

La propriété de Markov entraîne que pour tout entier n ,

$$T^n\Phi(x) = \mathbb{E}_x [R_{M_n} \{\Phi(x_n)\}].$$

Le théorème sera donc démontré si l'on vérifie que T^n tend vers zéro lorsque L n'est pas la représentation triviale. A l'aide de la condition (c) on voit facilement

que T est un opérateur quasi compact de norme inférieure ou égale à un.
Il suffit donc d'examiner sous quelles conditions T admet une valeur propre de module un.

Supposons qu'il existe $\lambda \in \mathbb{C}$, $|\lambda| = 1$, et $\Phi \in L^{\infty}(E;\mathbb{C}^p)$ tels que $T\Phi = \lambda\Phi$.
Puisque R est unitaire, on a π p.p.

$$\|\Phi(x)\| = \|T\Phi(x)\| \leq \mathbb{E}_x \| R_{M_1} \Phi(x_1)\| \leq \mathbb{E}_x \|\Phi(x_1)\|,$$

donc, si $f(x) = \|\Phi(x)\|$, $f \leq Pf$. La condition (c) entraine que f est p.p. constante donc que

$$\mathbb{E}_x \| R_{M_1} \Phi(x_1)\| = \| \mathbb{E}_x (R_{M_1} \Phi(x_1))\| .$$

Il en résulte que sous $\mathbb{P}_x$, $R_{M_1} \Phi(x_1)$ est proportionnel à la constante $\|\Phi(x_1)\|$.
Reportant dans l'équation $T\Phi = \lambda\Phi$ on en déduit que,

pour π presque tout x, $P(x,dy)$ p.p. $R_{M(y)} \Phi(y) = \lambda \Phi(x)$.
On utilise alors la stricte positivité de la densité de P par rapport à π, donnée par la condition (c), pour en conclure que

$$R_{M(y)} \Phi(y) = \lambda \Phi(x), \qquad \pi \otimes \pi \quad \text{p.p.}$$

La fonction $\Phi(x)$ ne dépend donc pas de x. Notons u sa valeur. Par continuité on a, pour tout g de S_π,

$$L_g^* u = R_g u = \lambda u.$$

Comme S_π engendre K, l'irréductibilité de $\hat{L}$ entraîne que cette représentation est de dimension un. Il existe donc un caractère χ de K tel que $\chi(g) = \lambda^{-1}$ pour tout g de S_π. Soit $g_0 \in S_\pi$, $H = \{g \in K;\ \bar{\chi}(g) = 1\}$ est un sous-groupe distingué fermé de K et S_π est contenu dans la classe $g_0 H$ car λ est de module un. Par hypothèse ceci n'est possible que si $H = K$, c'est-à-dire si L est la représentation triviale.

4. LE CAS DES MATRICES UNIPOTENTES

Considérons maintenant le cas où les matrices $M(x_i)$ sont dans le groupe $N(d)$ des matrices triangulaires supérieures dont les termes diagonaux sont égaux à un.
Le cas indépendant a été traité successivement par Tutubalin [21a] Virtser [26], Crépel et Raugi [7], Raugi [19]. Le cas markovien a été complètement résolu dans l'important article [21] de Raugi dont nous allons exposer quelques résultats.
Comme ailleurs, nous supposerons la condition (c) vérifiée, renvoyant à [21] pour une situation plus générale.

Notre présentation diffère un peu de celle de [21] mais tous les résultats que nous donnons dans ce paragraphe s'y trouvent ou en sont des conséquences immédiates.

Le théorème de la limite centrale pour les produits de matrices de N(d) apparaît comme conséquence d'un énoncé plus général (mais moins précis...) concernant les produits d'éléments aléatoires d'un groupe de Lie nilpotent simplement connexe que nous allons d'abord donner.

Considérons une algèbre de Lie nilpotente $(\mathcal{N},[\ ])$ et un idéal $\mathfrak{J}$ de $\mathcal{N}$. On leur associe une suite décroissante d'idéaux $\{\mathfrak{J}_\ell,\ \ell \geq 1\}$ de $\mathcal{N}$ construits par récurrence de la façon suivante :

$$\mathfrak{J}_1 = \mathcal{N},\quad \mathfrak{J}_2 = \mathfrak{J} \text{ et pour } r \geq 3\ ,\quad \mathfrak{J}_r = \sum_{i+j=r} [\mathfrak{J}_i, \mathfrak{J}_j]\ .$$

Puisque $\mathcal{N}$ est nilpotente il existe un plus grand entier k tel que $\mathfrak{J}_k$ ne soit pas réduit à $\{0\}$.

Soit $\mathcal{M}_i$ un sous-espace supplémentaire de $\mathfrak{J}_{i+1}$ dans $\mathfrak{J}_i$. Pour tout X de $\mathcal{N}$ on note $X^{(i)}$ sa composante sur $\mathcal{M}_i$ dans la décomposition en somme directe

$$\mathcal{N} = \mathcal{M}_1 \oplus \mathcal{M}_2 \oplus \dots \oplus \mathcal{M}_k\ .$$

A l'aide de la formule de Campbell-Hausdorff on définit un produit o sur $\mathcal{N}$

$$X \circ Y = X + Y + \frac{1}{2}[X,Y] + \dots$$

tel que $(\mathcal{N},\circ)$ est un groupe de Lie simplement connexe d'algèbre de Lie $(\mathcal{N},[\ ,\ ])$, pour lesquels l'application exponentielle est l'application identique.

Enfin, associons à $\mathcal{N}$ et $\mathfrak{J}$ un nouveau crochet de Lie, noté $[\ ,\ ]'$, vérifiant

$$[X,Y]' = [X,Y]^{(i+j)} \quad \text{si} \quad X \in \mathcal{M}_i \text{ et } Y \in \mathcal{M}_j.$$

et un produit o' sur $\mathcal{N}$, défini comme au-dessus mais pour l'algèbre $(\mathcal{N},[\ ,\ ]')$.

A l'aide de ces notations on peut énoncer certains résultats de [21] sous la forme suivante (x_n est comme ailleurs une chaine de Markov sur E satisfaisant la condition (c)).

Préposition

Soit $\mathfrak{J}$ *un idéal d'une algèbre de Lie nilpotente* $\mathcal{N}$ *contenant* $[\mathcal{N},\mathcal{N}]$. *Considérons une application mesurable* $f : E \to \mathcal{N}$ *telle que*

(i) $\mathbb{E}_\pi \|f(x_1)\|^2 < \infty$,

(ii) le support de $f(\pi)$ *engendre algébriquement* $\mathcal{N}$,

(iii) $\bar{f} = \int f(x)d\pi(x)$ *est un élément de* $\mathfrak{J}$

Alors

$$(x_n, \sum_{i=1}^{k} n^{-i/2} \{f(x_n) \circ f(x_{n-1}) \circ \dots \circ f(x_1)\}^{(i)})$$

converge en loi vers $\pi \otimes \nu$ *, où* ν *est la loi au temps un d'un mouvement brownien sur* $(\mathcal{N},o')$.

De plus il existe un élément Z *de* $\mathfrak{J}$ *, de même image que* $\bar{f}^{(2)}$ *dans* $\mathcal{N}/_{[\mathcal{N},\mathcal{N}]}$ *tel que la projection de* ν *sur la sous algèbre* $\mathcal{A}$ *de* $(\mathcal{N},[\ ,\ ]')$ *engendrée par* $\{X; X \in \mathcal{M}_1\}$ *et* $\{[X,Z]'\ ,\ X \in \mathcal{M}_1\}$ *ait une densité* $\mathcal{C}^\infty$.

Sans entrer dans les détails de la démonstration qui est assez longue, donnons en l'idée directrice. Supposons pour simplifier que $\{f(x), x \in E\}$ soit borné . Considérons une base (e_i) de $\mathcal{N}$ adaptée à la décomposition $\mathcal{N} = \oplus \mathcal{M}_i$, et notons (x_i) la base duale. Soit p la dimension de $\mathcal{N}$. Pour $\alpha = (\alpha_1, \dots \alpha_p)$ dans $\mathbb{N}^p$ on pose

$$x^\alpha = x_1^{\alpha_1} x_2^{\alpha_2} \dots x_p^{\alpha_p}$$

et $$\deg x^\alpha = \sum_{i=1}^{p} \alpha_i d_i$$

si $d_i = j$ lorsque $e_i \in \mathcal{M}_j$. Notons

$$T_n = \sum_{i=1}^{k} n^{-i/2} \{f(x_n) \circ \dots \circ f(x_1)\}^{(i)}$$

On commence par montrer (Proposition 3.4 de [21]) qu'il existe des réels $\sigma(x^\alpha)$ tels que les "moments généralisés" $\mathbb{E}_x(x^\alpha(T_n))$ convergent vers $\sigma(x^\alpha)$ quand n tend vers l'infini, pour α pas trop grand.

On en déduit (Proposition 6.12 de [21]) que la limite de (x_n, T_n) est la même que celle obtenue lorsqu'on effectue les produits dans $(\mathcal{N},o')$. On est donc ramené à prouver la proposition dans $(\mathcal{N},[\ ,\ ]')$. Utilisant alors une variante de la méthode des moments et le fait que les applications

$$U_n : (\mathcal{N},o') \longrightarrow (\mathcal{N},o')$$

définies par $U_n(X) = \sum n^{-i/2} X^{(i)}$ sont alors des automorphismes, Raugi en déduit que (x_n, T_n) converge en loi vers une probabilité de la forme $\pi \otimes \nu$, où ν est la loi au temps 1 de la diffusion sur $(\mathcal{N},o')$ de générateur

$$\Delta = \sum_{\{i, d_i=2\}} \sigma(x_i)\, E_i + \frac{1}{2} \sum_{\{i,j: d_i=d_j=1\}} \sigma(x_i x_j)\, E_i E_j$$

où E_i est le champ de vecteurs invariant à gauche associé à e_i .

Soit $Z, X_1, X_2, \dots, X_r$ des champs de vecteurs invariants à gauche tels que $\Delta = Z + \frac{1}{2} \sum X_i^2$. En utilisant la condition (c) et en examinant les projections sur $\mathcal{N}/_{[\mathcal{N},\mathcal{N}]}$ on voit que $Z, X_1, \dots, X_r$ engendrent $(\mathcal{N},[\ ,\ ]')$ et que la projection de ν sur l'algèbre $\mathcal{A}'$ engendrée par $\{[Z,X_1]', \dots, [Z,X_r]', X_1, \dots, X_r\}$ a une densité $\mathcal{C}^\infty$. Il en est de même de la projection de ν sur $\mathcal{A}$ car $\mathcal{A}$ est contenue dans $\mathcal{A}'$.

La condition de moment que nous avons donnée n'est pas optimale (cf. [21]).

La proposition précédente n'est qu'une étape vers un théorème de la limite centrale satisfaisant car la probabilité ν qui y apparaît peut être dégénérée. Cela est dû au fait que l'on n'a pas "recentré" les variables aléatoires.

Voyons donc maintenant, toujours suivant Raugi, comment l'adapter, mais dans la situation des matrices triangulaires supérieures.

Considérons donc des matrices $M(x), x \in E$, dans $N(d)$.
On sait que l'application Log est un difféomorphisme de $N(d)$ sur l'algèbre de Lie $\mathcal{N}(d)$ des matrices triangulaires supérieures dont les éléments diagonaux sont nuls. De plus, avec les notations précédentes, Log AB = Log A $\circ$ Log B.

Posons

$\mathcal{N}_0$ = sous-algèbre de $\mathcal{N}(d)$ engendrée par les matrices du support de Log $M(\pi)$.

$\bar{X} = \int \text{Log } M(x) \, d\pi(x)$

$\mathcal{N}_1$ = Idéal de $\mathcal{N}_0$ engendré par le support de Log $M(\pi) \circ (-\bar{X})$.

Lorsque $\bar{X}$ est dans $[\mathcal{N}_0, \mathcal{N}_0]$ on applique directement la proposition précédente avec $\mathcal{N} = \mathcal{N}_0$ et $\mathrm{J} = [\mathcal{N}_0, \mathcal{N}_0]$. Dans ce cas $\mathcal{A}$ est égal à $\mathcal{N}_0$ et l'on obtient

<u>Théorème</u> ([21])

Si $\mathbb{E}_\pi \| \text{Log } M(x_1) \|^2$ *est fini et* $\bar{X} \in [\mathcal{N}_0, \mathcal{N}_0]$, *alors* $(x_n, \sum n^{-i/2} (\text{Log } M_n)^{(i)})$ *converge en loi vers* $\pi \otimes \nu$ *où* ν, *loi au temps un d'un mouvement brownien sur* $(\mathcal{N}_0, o')$, *a une densité* $\mathcal{C}^\infty$.

Lorsque $\bar{X}$ n'est pas dans $[\mathcal{N}_0, \mathcal{N}_0]$, les variables aléatoires doivent être recentrées :

On commence par remarquer que

$\text{Log } M_n \circ (-n\bar{X}) = g(x_n) \circ (\bar{X} \circ g(x_{n-1}) \circ -\bar{X}) \circ (2\bar{X} \circ g(x_{n-2}) \circ -2\bar{X}) \circ \ldots$
$$\ldots \circ ((n-1)\bar{X} \circ g(x_1) \circ (1-n)\bar{X})$$

où $g(x_i) = \text{Log } M(x_i) \circ (-\bar{X})$ est $\mathcal{N}_1$.

Autrement dit, si N est le produit semi-direct $\mathcal{N}_1 \times_\sigma \mathbb{R}$, où $\mathbb{R}$ opère sur $\mathcal{N}_1$ par

$\sigma(t)Y = t \bar{X} \circ Y \circ (-t\bar{X})$, $t \in \mathbb{R}$, $Y \in \mathcal{N}_1$

alors
$(\text{Log } M_n \circ (-n\bar{X}), n) = f(x_n) f(x_{n-1}) \ldots f(x_1)$

où $f(x_i) = (g(x_i), 1)$ est dans N, et le produit du terme de droite est effectué dans N.

L'algèbre de Lie $\mathcal{N}$ de N s'identifie à $\mathcal{N}_1 \oplus \mathbb{R}\, T$ où $[T,Y] = [\bar{X},Y]$ pour $Y \in \mathcal{N}_1$. On applique la proposition précédente à $\mathcal{N}$, $\mathfrak{I} = [\mathcal{N},\mathcal{N}] \oplus \mathbb{R}T$ et f. Après avoir montré que l'algèbre de Lie $\mathcal{A}$ qui y apparaît est égale à $\mathcal{N}_1$ on en déduit

Théorème ([21]) :

Si $\mathbb{E}_\pi(\|\mathrm{Log}\, M(x_1)\|^2)$ *est fini,* $(x_n, \sum_{i=1}^{k} n^{-i/2} (\mathrm{Log}\, M_n \circ -n\, \bar{X})^{(i)})$ *converge en loi vers* $\pi \otimes \nu$ *où* ν *a une densité sur* $\mathcal{N}_1$.
De plus, $\nu \otimes \delta_1$ *est la loi au temps un d'un mouvement brownien sur* $(\mathcal{N}, o')$.

Exemple : Cas des matrices d'ordre 3.

Ecrivons, lorsque $d = 3$,

$$M_n = \begin{pmatrix} 1 & a_n & c_n \\ 0 & 1 & b_n \\ 0 & 0 & 1 \end{pmatrix}$$

Cas 1 : Si $\mathbb{E}_\pi(a_1) = \mathbb{E}_\pi(b_1) = 0$ et (a_1,b_1) n'est pas porté par une droite, on est sous les hypothèses du premier théorème.
Dans ce cas $(\mathcal{N}_0, o) = (\mathcal{N}_0, o') = \mathcal{N}(3)$ et on obtient que $(x_n, n^{-1/2}a_n, n^{-1/2}b_n, n^{-1}c_n)$ converge vers une loi $\pi \otimes \nu$ où ν a une densité. Si, par exemple, les x_i sont indépendants et la matrice de covariance est l'identité, la loi de ν est celle de $(B_1^{(1)}, B_1^{(2)}, \int_0^1 B_1^{(1)}\, dB_s^{(2)} + \mathbb{E}_\pi(c_1))$ où $(B^{(1)}, B^{(2)})$ est un mouvement brownien standard sur $\mathbb{R}^2$.

Cas 2 : Si, par exemple, $\mathbb{E}_\pi(b_1) = 0$, $\mathbb{E}_\pi(a_1) = 1$ et (a_1,b_1) n'est pas porté par une droite, on doit appliquer le second théorème. Reprenons dans ce cas la construction de $(\mathcal{N}, [\,,\,]')$.

Soit $\{U, V, W\}$ la base de $\mathcal{N}(3)$ formée des matrices dont tous les termes sont nuls sauf U_{12}, V_{23} et W_{33} qui sont égaux à un. Ici $\bar{X} = \mathbb{E}_\pi(\mathrm{Log}\, M_1)$ est dans $\mathbb{R}U \oplus \mathbb{R}W$ et $\mathcal{N}_0 = \mathcal{N}_1 = \mathcal{N}(3)$. L'algèbre $(\mathcal{N}, [\,,\,])$ est $\mathcal{N}(3) \oplus \mathbb{R}T$ muni du crochet vérifiant

$$[U,V] = W \quad , \quad [T,V] = W$$

où les autres crochets entre U, V, W et T sont nuls. On a $\mathfrak{I} = \mathbb{R}T \oplus \mathbb{R}W$ et $\mathfrak{I}_3 = \mathbb{R}W$.

Donc $\mathcal{M}_1 = \mathbb{R}U \oplus \mathbb{R}V$, $\mathcal{M}_2 = \mathbb{R}T$, $\mathcal{M}_3 = \mathbb{R}W$ et l'on devra normaliser a_n

et b_n par $\sqrt{n}$ et c_n par $n^{3/2}$. Pour le nouveau crochet de Lie $[\ ,\ ]'$ on a $[T,V]' = W$ et tous les autres crochets entre U, V, W et T nuls.

On en déduit que $(x_n\ ,\ \frac{a_n - n}{\sqrt{n}}\ ,\ \frac{b_n}{\sqrt{n}}\ ,\ \frac{c_n}{n^{3/2}})$ converge en loi vers une probabilité $\pi \otimes \nu$ où ν a une densité sur N(3). Dans le cas indépendant par exemple, si la matrice de covariance de (a_1, b_1) est l'identité, ν est la loi de $(B_1^{(1)}, B_1^{(2)}, \int_0^1 t\, dB_1^{(2)})$.

Remarque : Considérons un produit semi-direct $G = N(d) \times_\sigma K$ de N(d) par un groupe compact K. Soient $(Y_1,k_1), (Y_2,k_2),\ldots$ des variables aléatoires indépendantes et de même loi à valeurs dans G.

On a, si $(M_n,K_n) = (Y_1,k_1)(Y_2,k_2)\ldots(Y_n,k_n)$,

$$K_n = k_1 k_2 \ldots k_n$$

$$M_n = Y_1(\sigma(K_1)Y_2)(\sigma(K_2)Y_3)\ \ldots\ (\sigma(K_{n-1})Y_n)\ .$$

Si on pose $x_n = (Y_n, K_{n-1})$ et $M(x_n) = \sigma(K_{n-1})Y_n$ on voit donc que

$$(M_n,K_n) = (M(x_1)M(x_2)\ \ldots\ M(x_n),K_n)\ .$$

Les résultats précédents permettent alors d'établir un théorème de la limite centrale pour (M_n,K_n) . Suivant cette démarche Raugi a étudié le cas général des groupes de type R (cf.[19],[20]). On peut aussi traiter, utilisant les résultats de la section 3, le cas où les variables (Y_n,k_n) sont en dépendance markovienne. Ceci permet d'étudier les produits de matrices "triangulaires supérieures par blocs" dont les termes diagonaux sont des blocs variant dans un groupe compact.

5. LE CAS IRREDUCTIBLE

Nous allons maintenant exposer quelques résultats concernant le cas où l'ensemble des matrices $\{M(x), x \in E\}$ opère de façon irréductible sur $\mathbb{R}^d$ et n'est pas dans l'homothétique d'un groupe compact. On est donc à l'opposé des situations précédentes. On trouvera les démonstrations et des énoncés plus généraux dans Bougerol [6] .
Les méthodes utilisées sont des adaptations de celles employées par Tutubalin [22], Kaijser [14], Guivarc'h et Raugi [12], Le Page [16] et Bougerol-Lacroix [5] pour l'étude du cas indépendant. Elles s'appuient de façon essentielle sur l'étude des exposants de Lyapounov des produits markoviens menée dans Guivarc'h [11] .
Rappelons en la définition :

Soit, comme avant, $M_n = M(x_n)\ \ldots\ M(x_1)$ où (x_n) est une chaine de Markov sur E vérifiant la condition (c), et M une application mesurable de E dans $G\ell(d,\mathbb{R})$. Lorsque $\mathbb{E}_\pi\ (\mathrm{Log}^+ \|M(x_1)\| + \mathrm{Log}^+ \|M(x_1)^{-1}\|\ )$ est fini, $(M_n^* M_n)^{1/2n}$ converge

$\mathbb{P}_\pi$ p.s. vers une matrice dont les logarithmes des valeurs propres $\gamma_1 \geq \ldots \geq \gamma_d$ sont non-aléatoires et appelés exposants de Lyapounov.

Nous utiliserons les définitions suivantes

Définitions

On dit que le système (x_n, M) *est*

(a) *fortement irréductible si il n'existe pas de sous-espaces propres* $V_1, \ldots, V_k$ *en nombre fini tels que, pour* π *presque tout* $x \in E$

$$M(x)\,(V_1 \cup V_2 \cup \ldots \cup V_k) = V_1 \cup \ldots \cup V_k$$

(b) *contractant si il existe une suite* $\{A_n, n \in \mathbb{N}\}$ *dans le plus petit semi-groupe fermé de* $G\ell(d,\mathbb{R})$ *contenant le support de* $M(\pi)$ *, telle que*

$$\frac{A_n}{\|A_n\|}$$ *converge vers une matrice de rang un.*

On sait (décomposition d'Iwasawa) que l'on peut écrire de façon unique

$$M_n = K_n\, A_n\, N_n$$

où K_n est une matrice orthogonale, A_n une matrice diagonale à coefficients positifs et N_n est dans $N(d)$.

Nous allons étudier la convergence en loi de M_n , donc de (K_n, A_n, N_n) , convenablement normalisé. La première étape consiste à regarder, si $\{e_1, \ldots, e_d\}$ est la base canonique de $\mathbb{R}^d$, les lois de $(K_n e_1, A_n e_1, N_n^* e_1)$. Si $A_n = \mathrm{diag}(a_n^1, a_n^2, \ldots, a_n^d)$ on a $M_n e_1 = a_n^1 K_n e_1$, ce qui nous amène à étudier le comportement de $M_n u$, pour tout $u \neq 0$ de $\mathbb{R}^d$.

Rappelons d'abord l'analogue suivant de la loi des grands nombres

Proposition (Guivarc'h [11]) :

Si le système (x_n, M) *est fortement irréductible, pour tout* u *non nul,*

$$\lim_{n\to\infty} \frac{1}{n} \operatorname{Log} \|M_n u\| = \gamma_1 \quad , \qquad \mathbb{P}_\pi \text{ p.s.}$$

Rappelons qu'on ne peut espérer que $K_n e_1 = \dfrac{M_n e_1}{\|M_n e_1\|}$ converge toujours en loi. Dans le cas en effet où les coefficients des matrices $M(x)$ sont négatifs, $K_n e_1$ est à coordonnées positives pour n pair, négatives pour n impair. Pour cette raison nous n'étudions que la direction de $K_n e_1$ (ou, plus généralement, celle de $M_n u$) .

Notations

Si u *est un vecteur non nul de* $\mathbb{R}^d$ *on note* $\bar{u}$ *sa direction, i.e. son image dans l'espace projectif* $P(\mathbb{R}^d)$. *Pour* $M \in G\ell(d,\mathbb{R})$ *on pose* $M.\bar{u} = \overline{Mu}$.

L'idée naturelle pour étudier la loi de $(x_n, M_n.\bar{u}, \|M_n u\|)$ est d'utiliser la transformée de Fourier. Pour cela définissons, de façon formelle, si ϕ est une fonction de $E \times P(\mathbb{R}^d)$ dans $\mathbb{C}$ et $z \in \mathbb{C}$,

$$T(z)\,\phi(x,\bar{u}) = \mathbb{E}_x\left[\left\{\frac{\|M_1\,u\|}{\|u\|}\right\}^z \phi(x_1, M_1.\bar{u})\right] \quad , \quad x \in E, \bar{u} \in P(\mathbb{R}^d) \ .$$

On vérifie à l'aide de la propriété de Markov que pour tout entier n,

$$T(z)^n\,\phi(x,\bar{u}) = \mathbb{E}_x\left[\left\{\frac{\|M_n u\|}{\|u\|}\right\}^z \phi(x_n, M_n.\bar{u})\right] \ .$$

Le théorème de la limite centrale résulte de l'étude précise des opérateurs $T(z)$, au moins pour z petit. Elle se fait, à l'aide de la théorie des perturbations, à partir de celle de $T(0)$.

Remarquons que $T(0)$ est l'opérateur de transition de la chaine de Markov $(x_n, M_n.\bar{u})$ sur $E \times P(\mathbb{R}^d)$. En l'absence d'hypothèse de régularité on ne peut espérer que cette chaine de Markov satisfasse aux conditions usuelles (Döblin, Harris) sous lesquelles l'étude de $T(0)$ est classique. On suit la démarche développée par Le Page [16] dans le cas indépendant. Pour cela considérons la distance δ sur $P(\mathbb{R}^d)$ définie par

$$\delta(\bar{u},\bar{v}) = |\sin(u,v)| \ .$$

On voit facilement à l'aide de la proposition précédente que, pour tout $\bar{u},\bar{v}$ dans $P(\mathbb{R}^d)$, $\mathbb{P}_\pi$ p.s.

$$\overline{\lim_{n\to\infty}}\ \frac{1}{n} \operatorname{Log} \delta(M_n.\bar{u}, M_n.\bar{v}) \le \gamma_2 - \gamma_1 \ .$$

Ceci nous amène, lorsque γ_2 est strictement inférieur à γ_1 , à travailler sur les fonctions sur $E \times P(\mathbb{R}^d)$ dépendant de façon höldérienne de la seconde coordonnée. Plus précisément, pour $\alpha > 0$, considérons l'espace de Banach $\mathcal{L}^\alpha$ défini ainsi :

$$\mathcal{L}^\alpha = \{\phi : E \times P(\mathbb{R}^d) \to \mathbb{C} \quad \text{telles que}$$

$$\|\phi\|_\alpha = \left\| \sup_{\bar{u},\bar{v}} \frac{|\phi(.,\bar{u}) - \phi(.,\bar{v})|}{\delta(\bar{u},\bar{v})^\alpha} + \sup_{\bar{u}} |\phi(.,\bar{u})| \right\|_{L^\infty(\pi)} \quad \text{soit fini}\}$$

Sous les hypothèses du théorème suivant, Guivarc'h montre dans [11] qu'effectivement $\gamma_1 > \gamma_2$. Etudiant les opérateurs $T(z)$, pour $|z|$ assez petit, définis sur $\mathcal{L}^\alpha$, pour α assez petit, on montre dans [6] que

Théorème

Supposons que le système (x_n, M) *soit fortement irréductible et contractant et qu'il existe* $\beta > 0$ *tel que*

$$\int \sup_x p(x,y)\left(\|M(y)\|^{\beta} + \|M(y)^{-1}\|^{\beta} \right) d\pi(y) \qquad (*)$$

soit fini.

Il existe alors $\sigma > 0$ *tel que, pour tout* u *non nul de* $\mathbb{R}^d$ *et tout* x *de* E

$$(x_n, M_n.\bar{u}, \frac{1}{\sigma\sqrt{n}} \{\mathrm{Log}\,\|M_n u\| - n\gamma_1\}, N_n^* e_1)$$

converge en loi vers $\nu \otimes \Gamma \otimes m_x$ *où* ν *est l'unique probabilité invariante de la chaine de Markov* $(x_n, M_n.\bar{v})$, Γ *la loi normale centrée réduite et* m_x *une probabilité sur* $\mathbb{R}^d$.

Corollaire : *Sous les hypothèses précédentes, si* $\sum_n$ *est la matrice de coefficients* $\frac{1}{\sigma\sqrt{n}} \{\mathrm{Log}|M_n(i,j)| - n\,\gamma_1\}$, $1 \le i,j \le d$, *alors* $(x_n, \sum_n)$ *converge en loi vers* $\pi \otimes \rho$, *où* ρ *est la loi d'une matrice aléatoire dont tous les coefficients sont égaux et suivent une loi normale centrée réduite.*

Pour $1 \le p < d$ associons à toute matrice M de $G\ell(d,\mathbb{R})$ l'opérateur $\wedge^p M$ sur $\wedge^p \mathbb{R}^d$ défini par

$$(\wedge^p M)(u_1 \wedge u_2 \wedge \ldots \wedge u_p) = Mu_1 \wedge Mu_2 \wedge \ldots \wedge Mu_p .$$

Le théorème suivant étudie les composantes K_n, A_n, N_n de M_n écrite dans la décomposition d'Iwasawa. D y désigne le groupe des matrices diagonales dont les termes diagonaux valent ± 1. On a aussi noté $a_n^1, \ldots, a_n^d$ les coefficients diagonaux de A_n.

Théorème

Supposons que pour tout p, $1 \le p < d$, *le système* $(x_n, \wedge^p M)$ *soit fortement irréductible et contractant et que la condition d'intégrabilité* $(*)$ *soit satisfaite.*

Alors, si $\bar{K}_n$ *est l'image de* K_n *dans* $O(d)_{/D}$, *pour tout* x *de* E,

$$(x_n, \bar{K}_n, \frac{1}{\sqrt{n}} (\mathrm{Log}\, a_n^i - n\,\gamma_i)_{i=1,\ldots,d}, N_n)$$

converge en loi sous $\mathbb{P}_x$ *vers une probabilité de la forme* $\nu \otimes \Gamma \otimes m_x$ *où*

(i) ν *est une probabilité sur* $E \times O(d)_{/D}$

(ii) Γ *est une loi gaussienne centrée sur* $\mathbb{R}^d$

(iii) m_x *est une probabilité sur* $N(d)$.

Les résultats de cette section peuvent être précisés : on peut estimer la vitesse de convergence dans le T.C.L. et montrer un théorème de grandes déviations.

6. GENERALISATIONS DANS LE CAS IRREDUCTIBLE

Les résultats de la section précédente peuvent être étendus dans plusieurs directions (cf. Bougerol [6]). On montre par exemple le résultat suivant qui ne fait aucune hypothèse de contraction.

Proposition :

Supposons que le système (x_n, M) *soit fortement irréductible et que la condition d'intégrabilité (*) soit satisfaite.*
Alors, pour tout u *non nul de* $\mathbb{R}^d$, $(x_n, \frac{1}{\sqrt{n}}(\text{Log} \|M_n u\| - n\,\gamma_1))$ *converge en loi vers* $\pi \otimes \Gamma_\sigma$, *où* Γ_σ *est la loi normale centrée de variance* σ^2. *De plus* $\sigma = 0$ *si et seulement si il existe un réel* $r > 0$ *et* $Q \in G\ell(d, \mathbb{R})$ *tels que, pour* π *presque tout* x, $r\,Q\,M(x)Q^{-1}$ *soit une matrice orthogonale.*

Il peut aussi être intéressant de considérer une situation plus générale que celle des fonctions d'une chaine de Markov.

Considérons un système $(\Omega, F_t, x_t, \theta_t, \mathbb{P}_x, t \in T, x \in E)$ où $T = \mathbb{N}$ ou $\mathbb{R}^+$, $(\mathcal{F}_t)$ est une famille croissante de tribus sur Ω, (x_t) adapté à valeurs dans E et

a) $x_t \circ \theta_s = x_{t+s}$, $\theta_{t+s} = \theta_t \circ \theta_s$, pour tous t, s de T,

b) $\mathbb{P}_x(x_0 = x) = 1$

c) Pour toute variable aléatoire Z bornée, $\bigvee_{t \in T} \mathcal{F}_t$ mesurable,

$$\mathbb{E}_x(Z \circ \theta_t / \mathcal{F}_t) = \mathbb{E}_{x_t}(Z), \qquad \forall x \in E, \quad \forall t \in T.$$

On peut étudier le théorème de la limite centrale (cf. [6]) pour les processus M_t à valeurs dans $G\ell(d, \mathbb{R})$ vérifiant

(i) M_t est $\mathcal{F}_t$ mesurable

(ii) $M_{t+s} = M_t \circ \theta_s$ pour tous $t, s \in T$.

lorsque le processus de Markov (x_t) satisfait à la condition (c).
Ce type de situation intervient par exemple quand on regarde le flot dérivé du flot associé à une diffusion elliptique sur une variété compacte.

Les conditions de validité d'un théorème de la limite centrale sont un peu longues à écrire. Citons seulement l'analogue suivant de la proposition précédente, qui intervient dans la théorie de la stabilité des équations différentielles stochastiques linéaires (cf. Arnold, Oeljeklaus, Pardoux [1]), Baxendale [2], Bougerol [3], [4], [5]).

Proposition (6) :

Soit $\{x_t,\ t \in \mathbb{R}^+\}$ *une diffusion elliptique sur une variété compacte* E . *Considérons l'équation différentielle stochastique sur*

$$dZ_t^u = \sum_{i=1}^{k} A_i(x_t)Z_t^u \ dB_t^i + A_0(x_t)Z_t^u \ dt, \qquad Z_0^u = u$$

où

1) $(B_t^1,\dots,B_t^k)$ *est un* k*-mouvement brownien indépendant du processus* (x_t)

2) $A_0, A_1, \dots, A_k$ *sont des applications continues de* E *dans l'ensemble des matrices carrées d'ordre* d .

Supposons qu'il n'existe pas de sous-espace propre V *de* $\mathbb{R}^d$ *tel que*

$A_i(x)V \subset V$ *pour tout* x *de* E *et tout* $i = 0,1,\dots,d.$

Alors, il existe un réel γ *tel que pour tout* u *non nul de* $\mathbb{R}^d$ *et tout* $x \in E$

$$\lim_{n\to\infty} \frac{1}{t} \operatorname{Log} \|Z_t^u\| = \gamma \qquad , \qquad \mathbb{P}_x \quad \text{p.s}$$

et, sous $\mathbb{P}_x$, $\frac{1}{\sqrt{t}}\{\operatorname{Log}\|Z_t^u\| - t\gamma\}$ *converge en loi vers une loi normale centrée dont la variance n'est nulle que si il existe une matrice inversible* Q *pour laquelle* $Q(A_0(x) - \gamma Id)Q^{-1}$ et $Q\,A_i(x)Q^{-1}$, $i = 1,\dots,d$ *sont, pour tout* x *de* E, *antisymétriques.*

7. PROBLEMES OUVERTS

En dehors des situations examinées précédemment,et même dans le cas indépendant, peu de choses sont connues. Les résultats les plus complets sont annoncés dans Raugi [20].

L'exemple clef qui mérite d'être étudié est celui des matrices triangulaires à coefficients diagonaux arbitraires. On ne connait essentiellement que le cas des matrices d'ordre deux indépendantes (cf. Grincevicius [10]).

BIBLIOGRAPHIE

[1] Arnold, L., Oeljeklaus, E., Pardoux, E. (1986) Almost sure stability and moment stability for linear Ito equations. A paraître.

[2] Baxendale, P. (1986). Moment stability and large deviations for linear stochastic differential equations. A paraître.

[3] Bougerol, Ph. (1984). Stabilité en probabilité des équations différentielles stochastiques linéaires et convergence de produits de matrices aléa-

toires C.R. Acad. Sc. Paris, (299), Série 1. 631-634.

[4] Bougerol, Ph. (1984). Tightness of products of random matrices and stability of linear stochastic systems. A paraître dans Annals of Probability.

[5] Bougerol, Ph. Lacroix, J. (1985). Products of random matrices with applications to Schrödinger operators. Birkhäuser, P.P.M., Boston, Basel, Stuttgart.

[6] Bougerol, Ph. (1986). Théorèmes limites pour les systèmes linéaires à coefficients markoviens. (En préparation).

[7] Crépel, P., Raugi A. (1975). Théorème central limite sur les groupes nilpotents. C.R.A.S. (281), 605-608.

[8] Furstenberg, H. Kesten, H. (1960). Products of random matrices. Ann. Math. Statist. (31), 457-469.

[9] Gordin, M.I; Lifschitz, B.A. (1978). The central limit theorem for stationary Markov processes. Soviet Math. Dokl. (19), n°2, 392-394.

[10] Grincevicius A.K. (1974). A central limit theorem for the group of linear transformations of the real axis. Soviet Math Dokl. (15), 1512-1515.

[11] Guivarc'h, Y. (1984). Exposants caractéristiques des produits de matrices aléatoires en dépendance markovienne. In "Probability measures on groups 7", ed. H. Heyer. Lecture Notes in Math n° 1064. Springer Verlag. Berlin, Heidelberg, New-York, 161-181.

[12] Guivarc'h, Y., Raugi, A. (1985). Products of random matrices : convergence theorems. A paraître dans Symposium on pure and applied Maths.

[13] Heyer, H. (1977). Probability measures on locally compact groups. Springer Verlag. Berlin, Heidelberg, New-York.

[14] Kaijser, T. (1978). A limit theorem for markov chains in compact metric spaces with applications to products of random matrices. Duke Math. Journ. (45), 311-349.

[15] Kawada, Y., Ito, K. (1940). On the probability distribution on a compact group, I. Proc. Phys. Math. Soc. Japan (22), 977-999.

[16] Le Page, E. (1982). Théorèmes limites pour les produits de matrices aléatoires. Dans "Probability measures on groups", ed. H. Heyer. Lecture Notes in Math n° 928. Springer Verlag. Berlin, Heidelberg, New-York, 258-303.

[17] O'Connor, (1975). Disordered harmonic chain. Comm. Math. Phys. (45), 63-77.

[18] Raugi, A. (1977). Fonctions harmoniques et théorèmes limites pour les marches aléatoires sur les groupes. Bull. Soc. Math. France, Mémoire 54, 127p.

[19] Raugi, A. (1979). Théorème de la limite centrale pour un produit semi-direct d'un groupe de Lie résoluble simplement connexe de type rigide par un groupe compact. Dans "Probability measures on groups", ed. H. Heyer. Lecture Notes in Math. N° 706. Springer Verlag. Berlin, Heidelberg, New-York, 257-324.

[20] Raugi, A. (1980). Quelques remarques sur le théorème de la limite centrale sur un groupe de Lie C.R. Acad. Sc. Paris, (290), 103-106.

[21] Raugi, A. (1980). Théorème de la limite centrale sur les groupes de Lie nilpotents, pour des chaines semimarkoviennes. Séminaire de Probabilités. Université de Rennes.

[21a] Tutubalin, V.N. (1964). Composition of measures on the simplest nilpotent group. Theor. Proba. Appl. (9), 479-487.

[22] Tutubalin, V.N. (1965). On limit theorems for products of random matrices. Theor. Proba. Appl. (10),(10), 15-27.

[23] Tutubalin, V.N. (1969). Some theroems of the type of the strong law of large numbers. Theor. Proba. Appl. (14), 313,319.

[24] Tutubalin, V.N. (1977). The central limit theorem for products of random matrices and some of its applications. Symposia Math. (21), 101-116.

[25] Verheggen, T. (1979). Transmission coefficient and heat conduction of a harmonic chain with random masses. Comm. Math. Phys. (68), 69-82.

[26] Virtser, A.D. (1974). Limit theorems for composition of distributions on certain nilpotent Lie groups. Theor. Proba. Appl. (19), 86-105.

Philippe Bougerol

U.E.R. de Mathématiques
Université Paris VII
2, Place Jussieu

75251 - PARIS Cedex 05

ENTROPIE, THEOREMES LIMITE ET MARCHES ALEATOIRES

Y. DERRIENNIC
UNIVERSITE DE BRETAGNE OCCIDENTALE
BREST, FRANCE

La notion d'entropie d'une distribution de probabilité a pour origine la formule $H = -\sum p \log p$ de Boltzmann. Comme on le sait, elle est fondamentale dans de nombreux sujets. Le but poursuivi ici est d'exposer le rôle de cette notion dans l'étude des sommes de variables aléatoires indépendantes équidistribuées à valeurs réelles, vectorielles ou encore à valeurs dans un groupe localement compact.

Une suite de v.a. indépendantes X_n, équidistribuées sur un groupe localement compact G, engendre la marche aléatoire $S_1 = X_1$, $S_2 = X_1 X_2$, ... $S_n = X_1 \ldots X_n$ (ou $S_n = X_1 + \ldots\ldots + X_n$ si G est abélien). Le comportement asymptotique de cette marche est lié aux propriétés de la distribution des X_n et à celles du groupe G. Pour une part importante ce comportement est traduit par l'ordre de grandeur, quand n devient grand, de l'entropie $H(S_n)$ de la v.a. S_n. En effet, dans de nombreuses situations, il apparait que $H(S_n)$ a un développement asymptotique de la forme $H(S_n) \sim hn + c \log n$ dont la signification probabiliste est remarquable. Le premier terme hn, dans lequel figure la constante h, appelée dans la suite entropie asymptotique de la marche aléatoire, est lié à la loi des grands nombres ; sa nullité caractérise les marches aléatoires ne présentant qu'un seul comportement asymptotique distinguable, ou encore n'admettant pas d'autres fonctions harmoniques que les constantes. Le second terme $c \log n$ est lié au théorème limite central. Découvrir les liens entre la distribution des X_n, la structure de G et l'ordre de grandeur de l'entropie $H(S_n)$ est donc le sujet traité ici. Ce sujet n'est pas clos. La détermination de toutes les situations où $H(S_n)$ a le développement asymptotique donné ci-dessus, est loin d'être achevée.

Nous nous sommes efforcés de tenir compte de tous les résultats en notre connaissance (certains travaux très récents mis à part [27] [35]

[36]), mais bon nombre d'entre eux sont exposés sans démonstration complète, afin de ne pas répéter inutilement d'autres publications aisément accessibles. Le cas des marches aléatoires sur les groupes discrets a fait l'objet d'une publication récente très riche, comportant une bibliographie étoffée ([25]). Nous avons essayé de traiter le cas général ou au moins le cas où la loi de la marche aléatoire a une densité et les résultats de ce travail paraîssent ici pour la première fois sous forme détaillée (un résumé, contenant quelques erreurs et imprécisions, est paru dans [10]).

Voici, pour conclure cette introduction, le plan de l'article :

I

PRELIMINAIRES. ENTROPIE ET CONVOLUTION

Un groupe localement compact à base dénombrable G et une mesure de probabilité μ définie sur la tribu borélienne de G sont donnés. Une suite (X_n) de v.a. indépendantes de loi μ engendre alors la suite des produits, suivant la loi de groupe :

$$S_n = X_1 \dots X_n,$$

qui constitue la marche aléatoire droite définie par μ, issue de e, l'élément neutre de G. Si G est abélien, en particulier $\mathbb{R}^d$ ou $\mathbb{Z}^d$, la loi de groupe est notée, comme d'habitude, additivement et

$$S_n = X_1 + \dots + X_n.$$

La distribution de probabilité de la v.a. S_n est la $n^{\text{ième}}$ puissance de convolution μ^n de μ :

$$\int_G f(s)d\mu^n(s) = \int_{G^n} f(x_1 \dots x_n)\, d\mu(x_1) \dots d\mu(x_n).$$

La loi de probabilité de la marche aléatoire (S_n) est la loi pour laquelle les X_n sont indépendantes, de même distribution μ. Il est parfois utile de fixer les idées en considérant que les X_n sont les projections canoniques sur G de l'espace $\Omega = G^{\mathbb{N}}$, muni de sa tribu borélienne et de la mesure produit infini $\mu^{\otimes \mathbb{N}}$.

Une mesure de Haar à gauche sur G, notée m, est fixée. Si μ a une densité par rapport à m, elle est notée φ. Alors μ^n a une densité, notée φ_n, donnée par :

$$\varphi_n(y) = \int_G \varphi(x)\, \varphi_{n-1}(x^{-1}y)dm(x)$$

Dans le cas où G est un groupe discret dénombrable, m est la mesure de dénombrement ; on peut alors identifier μ et φ. L'entropie de S_n est définie par :

$$H(S_n) = H(\mu^n) = - \sum_{x \in G} \mu^n(x) \log \mu^n(x) \; ;$$

c'est l'entropie absolue de la distribution μ^n.

Dans le cas absolument continu, μ ayant la densité φ, l'entropie différentielle de S_n est définie par :

$$\tilde{H}(S_n) = \tilde{H}(\varphi_n) = -\int_G \varphi_n(x) \log \varphi_n(x)\, dm(x).$$

La première définition peut être considérée comme un cas particulier de la seconde.

Pour comprendre le sens de ces quantités, il faut aussi considérer l'information mutuelle des v.a. S_1 et S_n :

$$I(S_1, S_n) = H(\lambda(S_1, S_n) ; \lambda(S_1) \otimes \lambda(S_n))$$

qui est l'entropie relative de la loi du couple $\lambda(S_1, S_n)$ par rapport à la loi produit des marginales $\lambda(S_1) \otimes \lambda(S_n)$. Dans un appendice sont résumées les propriétés de l'entropie H et de l'information I. On se propose d'indiquer, dans cette première partie, quelques propriétés liant l'entropie, l'information et la convolution.

Considérons d'abord le cas discret. Si la série $-\sum \mu(x) \log \mu(x)$ converge, c'est à dire si $H(\mu)$ est finie, il en est de même pour μ^n et $H(\mu^n) \leq n\, H(\mu)$. En effet μ^n est image de la mesure produit $\mu \otimes \ldots\ldots \otimes \mu$ (n facteurs) dont l'entropie est évidemment $nH(\mu)$ (voir l'appendice). Le même argument montre que $H(\mu^n)$ est une suite sous-additive, c'est à dire $H(\mu^{n+k}) \leq H(\mu^n) + H(\mu^k)$. Ceci a été prouvé, avec un argument différent, par Avez [1], puis par Kaimanovich-Vershik [25]. Le rôle de l'information mutuelle $I(S_1, S_n)$ apparaît dans la proposition suivante.

Proposition : *Dans le cas discret, si* $H(\mu) < \infty$, *alors*

$$I(S_1, S_n) = H(\mu^n) - H(\mu^{n-1}).$$

Démonstration : Par définition

$$I(S_1, S_n) = \sum_{x,y \in G} \left(\log \frac{\mu(x)\, \mu^{n-1}(x^{-1}y)}{\mu(x)\, \mu^n(y)} \right) \mu(x)\, \mu^{n-1}(x^{-1}y).$$

Après simplification, par la convergence des séries :

$$I(S_1, S_n) = \sum_{x,y} (\log \mu^{n-1}(x^{-1}y))\, \mu(x)\, \mu^{n-1}(x^{-1}y) - \sum_y \mu^n(y) \log \mu^n(y).$$

$$= H(\mu^n) - H(\mu^{n-1}).$$

Si l'entropie $H(\mu)$ est infinie, la quantité $I(S_1, S_n)$ peut être soit infinie, soit finie ; les deux cas peuvent se présenter. Cependant la loi conjointe $\lambda(S_1, S_n)$ est toujours absolument continue par rapport à la loi produit $\lambda(S_1) \otimes \lambda(S_n)$.

Le cas absolument continu est plus compliqué. Mais si l'on suppose que toutes les intégrales $\int_G \varphi_n(x) \,|\log \varphi_n(x)|\, dm(x)$ convergent, la proposition précédente est encore valide.

Proposition : *Dans le cas absolument continu, si pour tout* n $\varphi_n \log \varphi_n$ *est intégrable, alors*

$$I(S_1, S_n) = \tilde{H}(\varphi_n) - \tilde{H}(\varphi_{n-1}).$$

Démonstration : La densité de la loi conjointe $\lambda(S_1, S_n)$ par rapport à $m \otimes m$ est $\varphi(x)\,\varphi_{n-1}(x^{-1}y)$; celle de la loi produit $\lambda(S_1) \otimes \lambda(S_n)$ est $\varphi(x)\,\varphi_n(y)$. Comme $\varphi_n(y) = 0$ si et seulement si $\varphi(x)\,\varphi_{n-1}(x^{-1}y) = 0$ m p.p. en x, la fonction $\varphi(x)\,\varphi_{n-1}(x^{-1}y) = 0$ $m \otimes m$ p.p. sur l'ensemble où $\varphi(x)\,\varphi_n(y) = 0$. D'après le théorème de Gelfand-Yaglom-Perez rappelé en appendice, on trouve, après simplification :

$$I(S_1, S_n) = \int \log\left(\frac{\varphi_{n-1}(x^{-1}y)}{\varphi_n(y)}\right) \varphi(x)\,\varphi_{n-1}(x^{-1}y)\, dm(x)\, dm(y).$$

Cette formule est valide sans hypothèse sur φ ; l'intégrale est définie au sens large, prenant une valeur positive, finie ou infinie. Sous l'hypothèse faite sur φ on trouve, comme précédemment, par l'invariance à gauche de m :

$$I(S_1, S_n) = \int \varphi_{n-1}(y) \log \varphi_{n-1}(y)\, dm(y) - \int \varphi_n(y) \log \varphi_n(y)\, dm(y)$$
$$= \tilde{H}(\varphi_n) - \tilde{H}(\varphi_{n-1}).$$

Dans la suite des conditions de moments sur φ, suffisantes pour que les $\varphi_n \log \varphi_n$ soient intégrables, seront données. Observons déjà que cela est réalisé pour φ bornée et portée par un ensemble de mesure m finie. Les deux propositions qui viennent d'être démontrées unifient en quelque sorte, les notions différentes d'entropie absolue et d'entropie relative. Le cas discret apparaît comme un cas particulier du cas absolument continu. C'est un des avantages de la considération de l'information mutuelle $I(S_1, S_n)$.

Dans le cas général, μ ayant une partie singulière avec m, la quantité $I(S_1, S_n)$ conserve un sens ; elle prend une valeur positive finie ou infinie (voir l'appendice). C'est alors cette quantité dont il faut étudier la croissance avec n. Il est important de remarquer que le cas $I(S_1, S_n) = +\infty$ se scinde en deux. On peut avoir $I(S_1, S_n)=+\infty$ en raison de la divergence des intégrales ou des séries définissant l'information mutuelle, comme dans le cas absolument continu ou discret. Un exemple est facile à donner en prenant une mesure μ sur les entiers telle que, à l'infini, $\mu(n) \sim \frac{C}{n(\log n)^\gamma}$, $1 < \gamma < 2$. Mais si μ n'est pas étalée, c'est à dire si aucune de ses puissances μ^k n'a de partie absolument continue non nulle, si de plus μ n'a pas de partie discrète, il est possible que, pour tout n, les lois $\lambda(S_1, S_n)$ et $\lambda(S_1) \otimes \lambda(S_n)$ soient étrangères. Un exemple de ce phénomène est donné par la mesure μ sur $\mathbb{R}$, construite par Fourt [15]. Cette mesure est diffuse et portée par un ensemble A tel que $A \cap (A + x)$ a au plus un élément si $x \neq 0$. L'ensemble $B = \{(x,y) \in \mathbb{R}^2 ; y - x \in A\}$ est de mesure 1 pour la loi conjointe car :

$$\lambda(S_1, S_2)(B) = \int_A d\mu(x)\, \mu(A) = 1,$$

mais de mesure 0 pour le produit des marginales car :

$$\lambda(S_1) \otimes \lambda(S_2)(B) = \int_A d\mu(x)\; \mu^2(A+x)$$

$$= \int_A d\mu(x) \int_A d\mu(u)\; \mu(A+x-u) = 0$$

Il n'est pas surprenant que ce cas se révèle plus difficile que l'autre.

II

ENTROPIE ET THEOREME LIMITE CENTRAL

Le maximum de l'entropie d'une distribution de probabilité à support fini est réalisé par la distribution uniforme. Cela est bien connu depuis longtemps. Le travail de Shannon contient une observation du même type concernant les lois gaussiennes : parmi les densités de variance donnée, sur $\mathbb{R}$ ou $\mathbb{R}^d$, le maximum de l'entropie différentielle est réalisé par la densité gaussienne ([34], p.88). Ces deux propriétés "variationnelles" ont d'intéressantes connexions avec les théorèmes de convergence en loi. C'est le thème de cette seconde partie.

La démonstration la plus courante de ces deux résultats repose sur l'inégalité

$$H(f;g) = \int f(x) \log \frac{f(x)}{g(x)} dx \geqslant 0$$

toujours satisfaite par l'entropie relative d'une densité f par rapport à une densité g dominant f, dans laquelle l'inégalité n'a lieu que si $f = g$ (c'est un corollaire direct de l'inégalité $\log x < x-1$). En posant $g \equiv 1$ on trouve la première propriété. Plus généralement, pour une densité φ portée par un ensemble de mesure finie E, on a

$$\tilde{H}(\varphi) = - \int_E \varphi(x) \log \varphi(x) \, dm(x) \leqslant \log m(E),$$

mais $\tilde{H}(\varphi)$ peut valoir $-\infty$. En posant $g(x) = \Psi_\sigma(x)$ où

$$\Psi_\sigma(x) = \frac{1}{\sigma\sqrt{2\Pi}} e^{-x^2/2\sigma^2}$$

notation qui sera utilisée dans la suite, on trouve la seconde propriété, car pour f centrée et de variance σ^2 on a :

$$- \int f(x) \log \Psi_\sigma(x) \, dx = \frac{1}{2} \log (2\Pi e \sigma^2) = \tilde{H}(\Psi_\sigma).$$

Le centrage n'est pas une restriction : l'entropie différentielle est invariante par translation. Cela prouve aussi que toute densité bornée ayant une variance finie, sur $\mathbb{R}$ ou $\mathbb{R}^d$, a une entropie différentielle finie.

Il n'est peut-être pas inutile de rappeler l'argument original de Shannon. Le maximum de $-\int f(x) \log f(x) \, dx$ sous les conditions $\int f(x) \, dx = 1$ et $\int x^2 f(x) \, dx = \sigma^2$, ne peut être atteint que si

$$\frac{d}{df} \int (-f(x) \log f(x) + \alpha f(x) x^2 + \beta f(x)) \, dx = 0.$$

En dérivant sous le signe somme on trouve

$$\int (-1 - \log f(x) + \alpha x^2 + \beta) dx = 0.$$

En fixant les multiplicateurs de Lagrange α et β de façon à vérifier les conditions on trouve que $f(x) = \Psi_\sigma(x)$ vérifie cette égalité. On a reproché à cette preuve de manquer de rigueur ([30] note 2), mais il n'est pas difficile de la rendre précise, en calculant la dérivée par rapport à un paramètre bien choisi. Par exemple, si l'on pose $F(t) = \tilde{H}(t\Psi_\sigma + (1-t)f)$ pour $0 \leqslant t \leqslant 1$, on vérifie facilement $F'(0) = 0$ et $F''(t) < 0$, pour f permettant les dérivations sous le signe somme. Cela montre que F a un maximum strict en $t = 0$ et donc que $\tilde{H}(\Psi_\sigma) > \tilde{H}(f)$; on passe alors à des densités f quelconques par approximation.

Pour faire apparaître la relation avec les théorèmes limite, considérons d'abord le cas où G est un groupe compact et où μ a une densité φ par rapport à la mesure de Haar m normalisée par $m(G) = 1$. Par le premier principe variationnel $\tilde{H}(\varphi) \leqslant 0$, avec égalité seulement si $\varphi = 1$ m p.p. Calculons $\tilde{H}(\varphi_{n+1})$. Par convexité

$$-\varphi_{n+1}(y)\log\varphi_{n+1}(y) \geqslant \int_G (-\varphi_n(x^{-1}y)\log\varphi_n(x^{-1}y))\,\varphi(x)\,dm(x) \;;$$

en intégrant en y on trouve

$$\tilde{H}(\varphi_{n+1}) \geqslant \int_G \tilde{H}(\varphi_n)\,\varphi_n(x)\,dm(x) = \tilde{H}(\varphi_n).$$

La suite $\tilde{H}(\varphi_n)$, qui est donc croissante et négative, a une limite. Si φ est continue, la suite φ_n est équicontinue ; par un argument de compacité on montre alors que cette limite est 0, autrement dit :

$$\lim_n \tilde{H}(\varphi_n) = \lim_n H(\mu^n;m) = 0$$

ce qui implique

$$\lim_n \int_G |\varphi_n(x) - 1|\,dm(x) = 0$$

(d'après [31] p.20). Cet argument a été donné par Csiszar pour démontrer le théorème de Ito-Kawada selon lequel la suite μ^n converge faiblement vers m quand μ est strictement apériodique ([7]). Il montre que la tendance vers l'équirépartition n'est que la tendance vers le maximum de l'entropie.

Le travail de Csiszar s'inspirait d'un article de Linnik, dans lequel le théorème limite central était démontré à partir de l'étude de $\tilde{H}(\varphi_n)$, φ étant une densité sur $\mathbb{R}$, ([30]). Il ressort de l'article de Linnik que la convergence en loi de $\frac{1}{\sqrt{n}} S_n$ vers une loi gaussienne ne fait qu'exprimer la tendance vers le maximum de l'entropie, étant donné les variances. Commentant ce travail dans son livre, Renyi a écrit "ainsi le théorème limite central apparaît analogue au second principe de la thermodynamique". ([32] p 554).

Pour comprendre la relation entre la croissance de $\tilde{H}(\varphi_n)$, φ étant une densité sur $\mathbb{R}$, ou de $H(\mu^n)$, μ étant une probabilité sur $\mathbb{Z}$, et le théorème limite central, commençons par quelques observations simples.

Proposition *: Si* φ *est une densité sur* $\mathbb{R}$ *, bornée, de variance* σ^2 *, de fonction caractéristique intégrable, alors*

$$\lim_n \tilde{H}(\varphi_n) - \frac{1}{2}\log n\sigma^2 = \frac{1}{2}\log(2\Pi e)$$

i.e.

$$\lim_n (\tilde{H}(\varphi_n) - \tilde{H}(\Psi_{\sigma\sqrt{n}})) = 0$$

(Si φ *est centrée, ce qui ne change rien au problème le résultat s'écrit encore* $\lim_n H(\varphi_n ; \Psi_{\sigma\sqrt{n}}) = 0$).

Démonstration : Supposons φ centrée. Soit $\widehat{\varphi_n}(x) = \sigma\sqrt{n}\ \varphi_n(x\sigma\sqrt{n})$ qui est la densité de $\frac{1}{\sigma\sqrt{n}} S_n$. Sous l'hypothèse faite sur la fonction caractéristique le théorème limite local est valide :

$$\lim_n \widehat{\varphi_n}(x) = \Psi_1(x) \quad \text{uniformément sur } \mathbb{R} \quad [14]$$

Pour n assez grand, $\widehat{\varphi_n}(x) < 1$ donc on a

$$\tilde{H}(\widehat{\varphi_n}) \geqslant \int_{-L}^{+L} -\widehat{\varphi_n}(x) \log \widehat{\varphi_n}(x)\, dx, \quad \text{pour tout } L.$$

Cette intégrale tend avec n, vers $\int_{-L}^{L} - \Psi_1(x) \log \Psi_1(x)\, dx$.

Comme $\tilde{H}(\Psi_1) \geqslant \tilde{H}(\widehat{\varphi_n})$ et que L est arbitraire, on obtient :

$$\lim_n \tilde{H}(\widehat{\varphi_n}) = \tilde{H}(\Psi_1).$$

C'est le résultat cherché car $\tilde{H}(\widehat{\varphi_n}) = \tilde{H}(\varphi_n) - \frac{1}{2}\log n\,\sigma^2$.

Proposition *: Si* μ *est la loi de Bernoulli sur les entiers :* $\mu(1) = p$, $\mu(0) = 1-p$ $(0 < p < 1)$, *la suite* $H(\mu^{n+1}) - H(\mu^n)$ *est décroissante* , $H(\mu^{n+1}) - H(\mu^n) \sim \frac{1}{2} n$ *et* $H(\mu^n) \sim \frac{1}{2}\log n$.

Démonstration : En utilisant la formule de Kolmogorov (voir l'appendice on a :

$$H(S_{n+1}, X_{n+1}) = H(X_{n+1}) + H(S_{n+1} \mid X_{n+1}) = H(X_{n+1}) + H(S_n)$$

et d'autre part :

$$H(S_{n+1}, X_{n+1}) = H(S_{n+1}) + H(X_{n+1} \mid S_{n+1}).$$

Soit $g(x) = -x \log x - (1-x) \log (1-x)$, pour $0 < x < 1$. Sachant $S_{n+1} = k$, la loi de $(X_1, \ldots, X_{n+1})$ est équirépartie sur les suites de 0 ou 1 de longueur $n+1$, comportant k fois 1. La loi conditionnelle de X_{n+1} est donc

$$P(X_{n+1}=1 \mid S_{n+1} = k) = \frac{k}{n+1}.$$

Alors $$H(X_{n+1} \mid S_{n+1}) = \sum_{k=0}^{n+1} \binom{n+1}{k} p^k (1-p)^{n+1-k} \; g\left(\frac{k}{n+1}\right),$$

ce qui donne

$$H(\mu^{n+1}) - H(\mu^n) = H(S_{n+1}) - H(S_n) = g(p) - B_{n+1}g(p)$$

où $B_{n+1} g$ est le polynôme de Bernstein de g de degré $n+1$. Le résultat annoncé résulte alors directement du théorème de de Moivre-Laplace et de la concavité de g.

Ces propositions suggèrent que, μ étant une probabilité de variance finie sur $\mathbb{R}$ ou $\mathbb{Z}$, le théorème limite central implique l'ordre de grandeur $\frac{1}{2} \log n$ pour $\tilde{H}(\varphi_n)$ ou $H(\mu^n)$. (L'hypothèse minimale sur μ sous laquelle cette estimation est valide reste à préciser, en particulier dans le cas discret). Dans l'espace de dimension d l'estimation est $\frac{d}{2} \log n$. La démarche de Linnik consiste à contrôler de façon directe l'ordre de grandeur de $\tilde{H}(\varphi_n)$ afin d'en déduire le théorème limite central. Le point essentiel de son argumentation est l'inégalité suivante.

<u>Inégalité de Linnik</u> Soit φ une densité sur $\mathbb{R}$, centrée, de variance σ^2 et de la forme $\varphi = \Psi_{\delta^2} * \gamma$ où γ est une densité à support compact. Alors

$$(\tilde{H}(\varphi_{n+1}) - \frac{1}{2} \log (n+1)\, \sigma^2) - (\tilde{H}(\varphi_n) - \frac{1}{2} \log n\, \sigma^2)$$

$$= \frac{1}{2n} \left(n\sigma^2 \int_{-\infty}^{+\infty} \left(\frac{\varphi_n'}{\varphi_n}\right)^2 (x)\, \varphi_n(x)\, dx - 1\right) + O\left(\frac{1}{n} \left(\tau + \int_{|x| > \tau\sigma\sqrt{n}} x^2 \varphi(x) dx\right)\right)$$

(le dernier terme est O par rapport à $n > 1$ et $\tau > 0$).

Le premier membre s'écrit encore

$$H(\varphi_{n+1} \; ; \; \Psi_{(n+1)\sigma^2}) - H(\varphi_n \; ; \; \Psi_{n\sigma^2}).$$

L'expression $\int_{-\infty}^{+\infty} (\varphi'_n / \varphi_n)^2(x)\ \varphi_n(x)\ dx$ est l'information de Fisher de la densité φ_n. D'après l'inégalité de Cramer-Rao-Fréchet, elle est supérieure ou égale à $1/n\sigma^2$ avec égalité seulement si φ_n est gaussienne. La démonstration, assez difficile, de Linnik exploite à la fois les propriétés de H et celles de l'information de Fisher. Un article récent de Brown permet de la simplifier un peu, en particulier grâce à l'inégalité

$$\text{Inf. Fisher } (\varphi * \varphi) < 2 \text{ Inf. Fisher}(\varphi)$$

dans laquelle l'égalité n'est réalisée que si φ est gaussienne ([6]).

Le raisonnement que l'on vient de résumer semble assez peu commode. Cependant il conduit à plusieurs questions intéressantes. La propriété d'entropie maximum à variance donnée, caractérise-t-elle les lois gaussiennes sur un groupe de Lie autre que $\mathbb{R}^d$? Le théorème limite central prend-il alors le sens de tendance vers le maximum de l'entropie ? La considération de l'entropie n'élimine pas la difficulté liée à la définition de la notion de variance, mais elle libère l'énoncé du théorème limite du problème de la "normalisation" : en effet le théorème peut se formuler directement sous la forme $\lim_n H(\varphi_n ; \Psi_n) = 0$ (le même avantage est présenté par la considération de la distance en variation au lieu de l'entropie relative).

Pour une loi stable d'indice α , $0 < \alpha < 2$, sur $\mathbb{R}$ on vérifie immédiatement que

$$\tilde{H}(\varphi_n) \sim \frac{1}{\alpha}\log n$$

On peut aussi se demander si la convergence vers une loi stable autre que gaussienne correspond à un principe variationnel ? La notion d'entropie différentielle ne conduit pas simplement à un tel principe. En effet, pour que φ appartienne au domaine d'attraction "standard" d'une densité stable d'indice α , la condition nécessaire et suffisante ne porte que sur le comportement à l'infini de φ; il est donc clair que la densité stable ne maximise pas l'entropie différentielle dans son domaine d'attraction "standard".

III

ENTROPIE ASYMPTOTIQUE D'UNE MARCHE ALEATOIRE

Alors que le travail de Linnik est resté peu exploité, l'étude des fonctions harmoniques sur les groupes de Lie, qui s'est développée depuis vingt-cinq ans, a fait usage de la notion d'entropie suivant un point de vue différent, suggéré par la théorie de Kolmogorov-Sinai-Ornstein. Déjà dans le travail de Fürstenberg, la notion d'*entropie asymptotique* d'une marche aléatoire est considérée, quoique l'expression ne soit pas employée ([16]). Dans l'étude des fonctions harmoniques sur les groupes discrets l'idée a été employée avec un certain succès par Avez ([1] [2] [3]) puis par Kaimanovich et Vershik ([24] [25]). On se propose dans cette partie de définir l'entropie asymptotique et de donner ses premières propriétés, pour une marche aléatoire sur un groupe localement compact séparable G définie par une probabilité μ *quelconque*.

La loi de probabilité de la marche aléatoire issue de e, $S_n = X_1 \dots X_n$ où les X_n sont des v.a. indépendantes de même distribution μ, est l'image de la mesure produit infini $\mu^{\otimes \mathbb{N}}$ par l'application :

$$(x_1, \dots, x_n \dots) \mapsto (s_1 = x_1, \dots, s_n = (x_1 \dots x_n), \dots) ;$$

on la note P_e. Une v.a. Z fonction de la suite $(S_1, \dots, S_n \dots)$ est dite "asymptotique" si elle ne dépend que des coordonnées S_n pour $n \geqslant k$ et ceci pour tout k. Les événements asymptotiques pour (S_n) forment la tribu "asymptotique" ; on note $\mathcal{A} = \bigcap_k \sigma(S_n ; n \geqslant k)$ (ne pas confondre $\mathcal{A}$ avec la tribu asymptotique de la suite indépendante (X_n)). Une v.a. Y fonction de la suite $(S_1, \dots S_n \dots)$ est dite "invariante" si elle vérifie :

$$Y(S_1, \dots, S_n, \dots) = Y(S_2, \dots, S_{n+1} \dots)$$

autrement dit si elle est invariante sous l'opérateur décalage

$$\vartheta(S_1, \dots, S_n, \dots) = (S_2, \dots, S_{n+1}, \dots).$$

Les événements invariants forment la tribu "invariante" notée $\mathcal{J}$. L'inclusion $\mathcal{J} \subset \mathcal{A}$ est évidente (pour plus de détails voir l'appendice 2).

Définition : L'entropie asymptotique de la marche aléatoire définie par μ *, quelconque, est l'information mutuelle des tribus* $\sigma(S_1)$ *et* $\mathcal{A}$ *pour* P_e *On note*

$$h(\mu) = I(S_1, \mathcal{A}).$$

Si le comportement asymptotique de la marche, c'est à dire la tribu $\mathcal{A}$ mod. P_e, est connu, on peut, en principe, en déduire $h(\mu)$. Réciproquement, s'il est possible de connaître, a priori, $h(\mu)$ on espère en déduire des informations sur la tribu $\mathcal{A}$. Le principal moyen pour atteindre $h(\mu)$ est l'approximation par $I(S_1, S_n)$.

Proposition : $I(S_1, S_n) = I(S_1, \sigma(S_k ; k \geqslant n)$
La suite $I(S_1, S_n)$ *décroît, au sens large Elle peut être constante égale à* $+\infty$
S'il existe n *tel que* $I(S_1, S_n) < \infty$ *alors* $h(\mu) = \lim_n I(S_1, S_n)$.

Démonstration : La propriété de Markov signifie que, sachant S_n, S_1 et $(S_{n+1}\ldots\ldots)$ sont indépendantes. En vertu de la formule de Kolmogorov, cela entraîne la première égalité (voir [31] partie 3.4). La décroissance de $I(S_1, S_n)$ et l'égalité $h(\mu) = \lim_n I(S_1, S_n)$ dans le cas fini, résultent alors de la propriété 5 de l'information mutuelle, rappelée en appendice.

Les calculs effectués dans la partie I donnent l'énoncé suivant :

Proposition : Dans le cas discret, si $H(\mu) < \infty$ *, on a*

$$h(\mu) = \lim_n (H(\mu^n) - H(\mu^{n-1})) = \lim_n \frac{1}{n} H(\mu^n).$$

Dans le cas absolument continu, si $\tilde{H}(\varphi_n)$ *est finie pour tout* n*, on a*

$$h(\mu) = \lim_n (\tilde{H}(\varphi_n) - \tilde{H}(\varphi_{n-1})) = \lim_n \frac{1}{n} \tilde{H}(\varphi_n).$$

Si les quantités $I(S_1, S_n)$ sont toutes infinies, $h(\mu)$ est plus difficile à atteindre.

Proposition : En désignant par $\mathcal{C}$ *une partition finie mesurable, arbitraire, de* G *et* $S_1^{-1}(\mathcal{C})$ *la sous-tribu de* $\sigma(S_1)$*engendrée par* $\mathcal{C}$ *, on a, pour* μ *quelconque :*

$$h(\mu) = \sup_{\mathcal{C}} I(S_1^{-1}(\mathcal{C}), \mathcal{A})$$
$$= \sup_{\mathcal{C}} \lim_n I(S_1^{-1}(\mathcal{C}), S_n).$$

Démonstration : La première égalité répète la définition de l'information mutuelle $I(S_1, \mathcal{A})$. Pour la seconde, on observe que

$I(S_1^{-1}(\mathscr{C}), \mathscr{A}) = \lim_n I(S_1^{-1}(\mathscr{C}), S_n)$; ceci résulte, comme précédemment, de la propriété de Markov qui implique la monotonie de la suite qui est dans ce cas finie.

Dans le travail d'Avez la quantité $h(\mu)$ est définie, dans le cas discret, par l'égalité

$$h(\mu) = \lim_n \frac{1}{n} H(\mu^n) \ ;$$

dans le cas absolument continu, la convergence de $\frac{1}{n} \tilde{H}(\varphi_n)$ n'est pas prouvée et seule $\limsup_n \frac{1}{n} \tilde{H}(\varphi_n)$ est considérée. Avez appelle $h(\mu)$ l'entropie de la marche aléatoire. La définition, plus générale, donnée ici est suggérée par le travail de Kaimanovich et Vershik, bien que seul le cas discret y soit envisagé. Il semble préférable de nommer $h(\mu)$ "entropie asymptotique" de la marche aléatoire. En effet l'entropie absolue de la loi P_e est infinie, sauf dans le cas déterministe ; d'autre part le gain d'entropie par unité de temps est celui du schéma de Bernoulli de base μ et vaut donc $H(\mu)$. La quantité $h(\mu)$ ne représente que la part d'entropie de la marche restant dans le comportement asymptotique $\mathscr{A}$, quand on connait la position S_1. Si les quantités considérées sont finies on a :

$$h(\mu) = H_{P_e}(\mathscr{A}) - H_{P_e}(\mathscr{A}/S_1).$$

La définition de $h(\mu)$ fait intervenir $\mathscr{A}$, la tribu asymptotique. On peut se demander quel rôle joue exactement la tribu invariante $\mathscr{I}$.

Théorème : *Les tribus asymptotique $\mathscr{A}$ et invariante $\mathscr{I}$ de la marche aléatoire coïncident modulo la loi de la marche aléatoire P_e. L'entropie asymptotique vérifie* $h(\mu) = I(S_1, \mathscr{A}) = I(S_1, \mathscr{I})$.

La démonstration qui sort un peu du sujet traité, est repoussée en appendice. Rappelons seulement ici que, dans le cas où μ^n et μ^{n+1} ne sont pas étrangères pour au moins un n, l'égalité $\mathscr{A} = \mathscr{I}$ mod P_e résulte de la loi "zéro ou deux" et est bien connue ([8]). La motivation d'Avez était l'étude des fonctions μ-harmoniques en relation avec la structure du groupe.

Définition : *Une fonction réelle, mesurable* g *définie sur* G *est dite* μ*-harmonique à droite si, pour tout* $x \in G$

$$g(x) = \int_G g(xy)\, d\mu(y).$$

(dans la suite on omet la précision "à droite").

Le lien entre fonctions µ-harmoniques et entropie asymptotique apparaît tout d'abord dans l'énoncé général suivant

Théorème : La tribu invariante $\mathcal{J}$ (ou la tribu asymptotique $\mathcal{A}$) est triviale mod P_e si et seulement si $h(\mu) = 0$.
   Si µ est adaptée (i.e. si G est le plus petit sous-groupe fermé contenant le support de µ, $\mathrm{Supp}(\mu)$), si la tribu $\mathcal{J}$ est triviale mod P_e, les fonctions µ-harmoniques bornées, continues, sont constantes. Réciproquement, si µ est étalée, si les fonctions µ-harmoniques, bornées, continues sont constantes alors $\mathcal{J}$ est triviale mod $P_{\dot{e}}$.

La démonstration du premier point repose sur le lemme suivant.

Lemme : Si $\mathcal{D}$ est une sous-tribu de $\mathcal{A}$, stable sous l'opérateur décalage ϑ, $\vartheta^{-1}\mathcal{D} = \mathcal{D}$, alors pour tout k, $I(S_k, \mathcal{D}) = kI(S_1, \mathcal{D})$.

<u>Démonstration</u> : D'après la propriété de Markov et la formule de Kolmogorov :

$$I(S_k, \mathcal{D}) = I((S_1, \ldots, S_k), \mathcal{D})$$
$$= I((S_1 \ldots S_{k-1}), \mathcal{D}) + EI(S_k, \mathcal{D}/S_1 \ldots S_{k-1}).$$

Par stationnarité $EI(S_k, \mathcal{D}/S_1 \ldots S_{k-1}) = I(S_1, \mathcal{D})$, donc $I(S_k, \mathcal{D}) = kI(S_1, \mathcal{D})$.

<u>Démonstration du théorème</u>: Si $h(\mu) = I(S_1, \mathcal{J}) = I(S_1, \mathcal{A}) = 0$ alors d'après le lemme $I(S_k, \mathcal{A}) = I(S_k, \mathcal{J}) = 0$ pour tout k.
D'après la propriété 2 de l'information mutuelle (voir l'appendice), cela signifie que, pour tout k, les tribus $\sigma(S_1, \ldots, S_k)$ et $\mathcal{A}$ (ou $\mathcal{J}$) sont indépendantes pour P_e. La tribu $\mathcal{A}$ (ou $\mathcal{J}$) étant une sous-tribu de $\sigma(S_1, \ldots, S_k \ldots)$ cela prouve que $\mathcal{A}$ (ou $\mathcal{J}$) est triviale mod. P_e. La réciproque est évidente.

La démonstration du second point repose sur la correspondance bien connue entre v.a. invariantes et fonctions harmoniques d'une chaîne de Markov. Soit g une fonction µ-harmonique bornée continue. La suite $g(S_n)$ forme une martingale pour P_e, qui converge vers une v.a. invariante Z. Si $\mathcal{J}$ est triviale mod P_e, alors

$$Z = E_e(Z) = g(e) \quad P_e \text{ p.s.}$$

et $g(S_1) = E_e(Z/S_1) = g(e)$ P_e p.s.
Ceci donne $g(y) = g(e)$ µ p.p. en y. La fonction g étant continue

on obtient $g(y) = g(e)$ pour tout $y \in \mathrm{Supp}(\mu)$. On trouve le même résultat pour la translatée à gauche $g_a(x) = g(ax)$ qui est aussi μ-harmonique. Donc le sous-groupe fermé des périodes à droite de g contient $\mathrm{Supp}(\mu)$. L'hypothèse d'adaptation implique que g est constante.

Réciproquement si μ est étalée, toutes les fonctions μ-harmoniques bornées sont continues donc ici constantes. Il en résulte que $\mathfrak{J}$ est triviale mod P_e, et même mod. la loi de la marche aléatoire de distribution initiale arbitraire sur G ([**8**] p. 119).

La constance des fonctions μ-harmoniques sous l'hypothèse $h(\mu) = 0$, a été démontrée, dans le cas discret ou absolument continu avec $\tilde{H}(\varphi_n)$ fini par Avez, avec un argument direct, élégant ([**2**], [**4**]). La réciproque a été prouvée, dans le cas discret dans [**9**] et [**24**]. La démonstration générale donnée ici s'inspire de celle de [**25**].

Dans la réciproque, l'hypothèse d'étalement est indispensable. Sinon il est possible qu'existent des fonctions μ-harmoniques constantes m.p.p. mais non constantes relativement à la mesure $\sum_0^\infty \mu^n/2^{n+1}$. Alors la tribu $\mathfrak{J}$ est triviale pour la loi de la marche aléatoire si la distribution initiale est absolument continue, mais n'est pas triviale pour P_e et on a $h(\mu) > 0$. Un exemple est le suivant : dans $SO(3, \mathbb{R})$ il existe un sous-groupe à deux générateurs a et b qui est libre ([**18**] p. 9) ; la moyenne des mesures de Dirac $\mu = \frac{1}{2}(\delta_a + \delta_b)$ est adaptée au sous-groupe *fermé* G engendré par a et b, qui est compact ; les fonctions μ-harmoniques continues sur G sont donc constantes, mais sur le sous-groupe libre engendré par a et b, il existe des fonctions μ-harmoniques non constantes, bornées, donc $h(\mu) > 0$. Sur un groupe abélien ce phénomène est impossible en vertu de la loi "zéro ou un" de Hewitt et Savage.

Examinons, pour conclure cette partie, le problème de la stabilité de l'entropie asymptotique $h(\mu)$: une suite de probabilités μ_j est donnée sur G, qui converge vers μ étroitement. Existe-t-il une relation, valide en général, entre $h(\mu)$ et $h(\mu_j)$? La réponse est négative et c'est une des difficultés du sujet. Si $\lim_j \mu_j = \mu$ alors $\lim_j \mu_j^n = \mu^n$ et $\lim_j \lambda_j(S_1, S_n) = \lambda(S_1, S_n)$ (loi conjointe de S_1 et S_n). Par un théorème de Dobrushin ([**11**], [**31**] p. 9) on a

$$I(S_1, S_n) = \sup_{\mathfrak{E}} \sum_{E,F} \log\left(\frac{\int_E d\mu(x)\, \mu^{n-1}(xF)}{\mu(E)\, \mu^n(F)}\right) \int_E d\mu(x)\, \mu^{n-1}(xF),$$

où $\mathcal{E}$ désigne la collection des partitions finies de $G \times G$ formées d'ensembles $E \times F$ boréliens, de frontière négligeable pour $\lambda(S_1) \otimes \lambda(S_n) = \mu \otimes \mu^n$ et pour $\lambda(S_1, S_n)$. La même égalité a lieu pour μ_j ; en passant à la limite dans la somme finie on trouve :

$$\liminf_j I_j(S_1, S_n) \geqslant I(S_1, S_n) \qquad ([\mathbf{31}]\ p13).$$

Quand on passe à la limite en n, même si toutes les quantités considérées sont finies, cette inégalité n'a pas de raison d'être préservée. Kaimanovich et Vershik ont donné à ce sujet deux exemples remarquables. Sur $\mathfrak{S}_\infty$ le groupe des permutations d'ordre fini d'un ensemble infini dénombrable, il existe une probabilité symétrique μ pour laquelle $H(\mu) < \infty$ et $H(\mu) > 0$ alors que, tout élément de $\mathfrak{S}_\infty$ étant d'ordre fini, toute probabilité à support fini, a une entropie asymptotique nulle ([**25**] p. 484). D'autre part sur le groupe des configurations finies de $\mathbb{Z}^3$, $F_0(\mathbb{Z}^3, \mathbb{Z}/2\mathbb{Z})$ décrit dans la partie IV, il existe une probabilité μ telle que $h(\mu) = 0$, alors que pour toute probabilité à support fini, adaptée, l'entropie asymptotique est non nulle. Il n'y a donc aucun moyen général d'approximer, ni de majorer ou minorer, l'entropie asymptotique d'une probabilité μ par l'entropie asymptotique de "restrictions" finies de μ convergeant vers μ.

IV

CROISSANCE, LOI DES GRANDS NOMBRES ET ENTROPIE ASYMPTOTIQUE NULLE

Comme on vient de le voir, la nullité de $h(\mu)$ équivaut à la constance des fonctions μ-harmoniques continues bornées, au moins dans le cas étalé. Alors la marche aléatoire ne présente qu' *un* comportement limite distinguable avec probabilité positive. La partie IV est consacrée aux problèmes suivants :

- caractériser les groupes G sur lesquels il existe μ adaptée telle que $h(\mu) = 0$;
- caractériser les groupes G tels que pour toute probabilité μ, adaptée, $h(\mu) = 0$;
- dans le cas intermédiaire, décrire les probabilités adaptées telles que $h(\mu) = 0$.

Pour que les problèmes soient bien posés on suppose toujours, dans cette partie, que μ est adaptée (i.e. G est le plus petit sous-groupe fermé portant μ).

Le premier problème est résolu par le théorème suivant.

Théorème *: Une condition nécessaire et suffisante pour qu'il existe sur* G *une probabilité* μ *, adaptée, telle que* $h(\mu) = 0$ *est que* G *soit moyennable.*

Il est connu depuis longtemps que l'existence sur G d'une probabilité μ pour laquelle les fonctions μ-harmoniques continues, bornées, sont constantes implique la moyennabilité. ([5] p. 43). Cela prouve la nécessité. La suffisance a été prouvée par Rosenblatt et indépendamment par Kaimanovich et Vershik ([33], [24], [25]) : sur tout groupe moyennable on peut construire une probabilité μ symétrique, absolument continue, pour laquelle les μ-harmoniques bornées sont constantes et donc $h(\mu) = 0$.

Le second problème n'a pas encore de solution complète. D'après la loi "zéro ou un" de Hewitt et Savage, qui implique le théorème de Choquet-Deny, si G est abélien $h(\mu) = 0$ pour tout μ. Il est aussi connu que, si G est nilpotent de degré 2, les fonctions μ-harmoniques bornées continues sont constantes, quelle que soit μ ; en degré supérieur le résultat reste vrai si μ est étalée, ou a un "moment" ([19]). Pour décrire complètement la classe des groupes G pour lesquels $h(\mu) = 0$, quelle que soit μ adaptée, il faudrait au moins savoir si cette classe contient les groupes à croissance non exponentielle discrets ou non. C'est dans l'étude de ce problème que se présente le résultat le plus remarquable d'Avez.

L'énoncé demande quelques rappels sur la croissance.

On suppose que G est à génération compacte, non compact. A un voisinage V de e, symétrique, compact, qui engendre G, on associe la *"longueur"* :

$$L_V(x) = \inf \{n \in \mathbb{N} \ ; \ x \in V^n \};$$

V^n est l'ensemble des produits $x_1 \dots x_n$ où $x_i \in V$. On a

$$L_V(xy) \leqslant L_V(x) + L_V(y).$$

En posant $C_V(n) = m(V^n) = m \ \{L_V \leqslant n\}$ on a

$$\log C_V(n+k) \leqslant \log C_V(n) + \log C_V(k)$$

et donc, par sous-additivité :

$$\lim_n \frac{1}{n} \log C_V(n) = \inf_n \frac{1}{n} \log C_V(n)$$

Le groupe G est dit à croissance *non exponentielle* si $\lim_n \frac{1}{n} \log C_V(n) = 0$
Il est dit à croissance *polynomiale* si $C_V(n) = O(n^d)$ pour un $d > 0$. Ces deux propriétés ne dépendent pas de l'ensemble V choisi ; elles ne dépendent que de G ([19]).

Théorème *: (Avez* [4]*) Si* G *est à croissance non exponentielle, si* μ *a une densité bornée à support compact, alors* $h(\mu) = 0$ *(i.e. les fonctions* μ*-harmoniques, bornées sont constantes).*

En raffinant les arguments on peut prouver un peu plus. Commençons par un lemme dans lequel les notations sont celles introduites ci-dessus

Lemme *: Si* μ *a une densité bornée* φ *et si*

$$\sum_{k=1}^{\infty} \mu\{L_V = k\} \log C_V(k) = E[\log C_V(L_V(X_1))] < \infty$$

alors $\tilde{H}(\varphi_n) < \infty$ *pour tout* n *et*

$$\tilde{H}(\varphi_n) \leqslant E[\log C_V(L_V(S_n))] - \sum_{k=1}^{\infty} \mu^n\{L_V = k\} \log \mu^n\{L_V = k\} .$$

<u>Démonstration</u> : Sur une partie B de mesure $m(B) < \infty$ le maximum de l'entropie différentielle est réalisé par la densité uniforme (voir partie II). Donc

$$-\int_{\{L_V=k\}} \varphi_n(x) \log \varphi_n(x)\, dm(x) \leqslant \mu^n\{L_V=k\} \log \frac{m\{L_V=k\}}{\mu^n\{L_V=k\}}$$

En sommant sur k on obtient :

$$\tilde{H}(\varphi_n) \leqslant E[\log C_V(L_V(S_n))] - \sum_{k=1}^{\infty} \mu^n\{L_V=k\} \log \mu^n\{L_V=k\} .$$

Comme $\log C_V(L_V(S_n)) \leqslant \sum_1^n \log C_V(L_V(X_i))$, l'hypothèse donne $E[\log C_V(L_V(S_n))] < \infty$. D'autre part

$$\sum_{k=1}^{\infty} \mu^n\{L_V=k\} \log k = E[\log L_V(S_n)] \leqslant E[\log C_V(L_V(S_n))] < \infty$$ car

$k = O(C_V(k))$. Cela implique $- \sum_{k=1}^{\infty} \mu^n\{L_V=k\} \log \mu^n\{L_V=k\} < \infty$

(voir partie VII)

Théorème : *Si* G *est un groupe à croissance non exponentielle, si* μ *a une densité* φ *bornée et vérifie :* $E[\log C_V(L_V(X_1))] < \infty$ *et* $E(L_V(X_1)) < \infty$ *alors* $\lim_n \frac{1}{n} \tilde{H}(\varphi_n) = h(\mu) = 0$ *(les notations sont celles introduites ci-dessus les deux hypothèses de "moments" finis ne dépendent pas de l'ensemble* V *choisi).*

Plus précisément si $E[\log C_V(L_V(X_1))] < \infty$ *est vérifié pour un choix de* V*, tel que* $\lim_n \frac{C_V(n)}{C_V(n-1)} = 1$ *alors* $\lim_n \frac{1}{n} \tilde{H}(\varphi_n) = h(\mu) = 0$ *; en particulier si,* G *est à croissance polynômiale, l'hypothèse* $E[\log (L_V(X_1))] < \infty$ *suffit pour que* $\lim_n \frac{1}{n} \tilde{H}(\varphi_n) = h(\mu) = 0$.

Démonstration : Soit $\mathcal{V}$ la partition de G en "couronnes" $\{L_V=k\}$, $k = 1, 2, \ldots$. Alors

$$- \sum_{k=1}^{\infty} \mu^n \{L_V=k\} \log \mu^n \{L_V=k\} = H(S_n^{-1}(\mathcal{V})).$$

Posons $T_n = \sum_{i=1}^{n} L_V(X_i)$. On a $L_V(S_n) \leqslant T_n$. Les propriétés de H, rappelées en appendice, permettent d'écrire :

$H(S_n^{-1}(\mathcal{V})) \leqslant H(S_n^{-1}(\mathcal{V}), T_n) = H(T_n) + EH(S_n^{-1}(\mathcal{V})/T_n)$. Comme T_n est une marche aléatoire sur $\mathbb{Z}$, avec $H(T_1) < \infty$ car $\sum_{k=1}^{\infty} \mu\{T_1=k\} \log k < \infty$, on a $\lim_n \frac{1}{n} H(T_n) = 0$. (Voir partie VII). Sachant $T_n = k$, S_n ne peut se trouver que dans, au plus, k atomes de $\mathcal{V}$; donc $H(S_n^{-1}(\mathcal{V})/T_n=k) \leqslant \log k$. Cela donne $EH(S_n^{-1}(\mathcal{V})/T_n) \leqslant E(\log T_n)$.

En vertu du lemme, on obtient :

$$\tilde{H}(\varphi_n) \leqslant E[\log C_V(L_V(S_n))] + E(\log T_n) + o(n).$$
$$\leqslant 2E[\log C_V(T_n)] + o(n).$$

La suite sous-additive $\log C_V(T_n)$ est intégrable et positive, donc $\frac{1}{n} \log C_V(T_n)$ converge p.s. et en moyenne d'après le théorème ergodique sous-additif. D'après l'hypothèse $E(L_V(X_1)) = E(T_1) < \infty$, $T_n = O(n)$ p.s. Le groupe G étant à croissance non exponentielle, on obtient $\lim_n \frac{1}{n} \log C_V(T_n) = 0$ p.s. et donc $\lim_n \frac{1}{n} E(\log C_V(T_n)) = 0$.

La première partie du théorème est ainsi démontrée.

Supposons que $\lim \frac{C_V(n)}{C_V(n-1)} = 1$; c'est évidemment le cas si G est à croissance polynomiale.

Alors $\log C_V(T_n) - \log C_V(T_n - T_1) = \log \frac{C_V(T_n)}{C_V(T_n - T_1)} \xrightarrow[n\to\infty]{} 0$ p.s. D'après la méthode de [9], cela implique $\lim_n \frac{1}{n} E(\log C_V(T_n)) = 0$, sans l'hypothèse $E(T_1) < \infty$. Si G est à croissance polynomiale $C_V(n) \simeq n^d$, donc $E[\log L_V(X_1)] < \infty$ implique $E[\log C_V(L_V(X_1))] < \infty$ et le théorème est complètement démontré.

L'énoncé précédent conduit à la question naturelle, mais difficile, suivante :

"Sur un groupe G *à croissance non exponentielle, a-t-on* $h(\mu)=0$, *quelle que soit* μ *(absolument continue) ?"*. i.e. les hypothèses de "moments" finis posées dans le théorème sont-elles indispensables ? Cette question se pose d'autant plus qu'on sait que pour G discret à croissance polynomiale la réponse est positive ; dans ce cas, l'hypothèse de moment logarithmique fini pour μ, posée dans le théorème, sous laquelle on a vu que $\lim_n \frac{1}{n} \tilde{H}(\varphi_n) = h(\mu) = 0$, n'est pas indispensable. En effet, Gromov a démontré que tout groupe discret à croissance polynomiale est une extension finie d'un groupe nilpotent [**19**] ; sur un groupe nilpotent discret les fonctions μ-harmoniques bornées sont constantes, comme on l'a déjà mentionné, et cette propriété est conservée par extension finie. Cela ne suffit pas à résoudre complètement la question, car Grigorchuk a démontré qu'il existe des groupes discrets à croissance non exponentielle qui ne sont pas à croissance polynomiale [**20**].

Concernant le problème de la description des probabilités μ, adaptées sur G moyennable, telles que $h(\mu) = 0$, le résultat général le plus frappant a été obtenu par Guivarc'h, dans le cadre de son étude de la loi des grands nombres sur les groupes de Lie [**22**]. Guivarc'h a introduit la notion de *croissance* d'une probabilité μ dans G, qui est la suivante dans le cas où μ a une densité continue.

Définition *: On appelle croissance dans* G *d'une probabilité* μ *de densité continue* φ *sur* G *, la borne inférieure* $\mathcal{C}(\mu)$ *des réels* $\alpha>0$ *tels qu'il existe une suite croissante exhaustive de boréliens* A_n *dans* G *vérifiant*

$$\lim_n \mu^n(A_n) = 1 \quad \text{et} \quad \limsup_n \frac{1}{n} \log m(A_n) \leqslant \alpha.$$

Cette notion apparaît comme plus primitive que la notion d'entropie asymptotique elle-même.

Proposition *: Si* μ *a une densité continue* φ *à support compact* V *, on a* $h(\mu) \leqslant \mathcal{C}(\mu)$.

<u>Démonstration</u> : On peut alors prendre les A_n de la définition précédente vérifiant $A_n \subset V^n$. On a

$$\tilde{H}(\varphi_n) = -\int_{A_n} \varphi_n(x) \log \varphi_n(x)\, dm(x) - \int_{V^n \setminus A^n} \varphi_n(x) \log \varphi_n(x)\, dm(x).$$

Le maximum de l'entropie différentielle étant réalisé par la densité uniforme, on obtient :

$$\tilde{H}(\varphi_n) \leqslant \mu^n(A_n) \log \frac{m(A_n)}{\mu^n(A_n)} + \mu^n(V^n \setminus A_n) \log \frac{m(V^n \setminus A_n)}{\mu^n(V^n \setminus A_n)}$$, d'où on tire

$$\tilde{H}(\varphi_n) \leqslant \log m(A_n) + (1-\mu^n(A_n)) \log m(V^n) - \mu^n(A_n) \log \mu^n(A_n) - \mu^n(V^n \setminus A_n) \log \mu^n(V^n \setminus A_n)$$

Comme $\frac{1}{n} \log m(V^n)$ est borné, on obtient

$$\lim_n \frac{1}{n} \tilde{H}(\varphi_n) = h(\mu) \leqslant \limsup_n \frac{1}{n} \log m(A_n),$$ ce qui prouve la proposition.

Dans le cas où G est un groupe de Lie connexe moyennable, la version de la loi forte des grands nombres de [22], s'énonce ainsi : $\lim_n \frac{1}{n} L_V(S_n) = 0$ p.s. si μ est *centrée*, c'est à dire si $\int_G \gamma(x)\, d\mu(x) = 0$ pour tout homomorphisme de groupe γ, de G dans $\mathbb{R}$; cela implique $\mathscr{C}(\mu) = 0$. On arrive ainsi au théorème suivant ([22] p. 75).

<u>Théorème</u> : *Si* G *est un groupe de Lie connexe moyennable, si* μ *est une probabilité centrée, de densité* φ *continue à support compact, alors* $\lim_n \frac{1}{n} \tilde{H}(\varphi_n) = h(\mu) = 0$ *(i.e. les fonctions* μ*-harmoniques bornées sont constantes).*

Pour conclure cette partie, décrivons la classe des groupes des *"configurations finies"* $F_o(\mathbb{Z}^k, \mathbb{Z}/2\mathbb{Z})$, introduite par Kaimanovich et Vershik pour résoudre la question suivante d'Avez [3] :

"Sur un groupe, discret, à croissance exponentielle, peut-il exister une probabilité à support fini μ, adaptée, telle que $h(\mu) = 0$?".

Le groupe $F_o(\mathbb{Z}^k, \mathbb{Z}/2\mathbb{Z})$ est le produit semi-direct de $\mathbb{Z}^k$, avec le groupe des applications à support fini dans $\mathbb{Z}^k$, à valeurs dans le groupe des entiers modulo 2, l'action de $\mathbb{Z}^k$ sur la seconde composante étant définie par les tranlations. Un élément de ce groupe a la forme (i, χ_A) où $i \in \mathbb{Z}^k$ et A est une partie finie de $\mathbb{Z}^k$ (χ désignant l'indicatrice); la loi de groupe s'écrit

$$(i, \chi_A)(j, \chi_B) = (i+j, \chi_A + \chi_{B+i} \bmod 2).$$

Ces groupes ($k \geqslant 1$) sont résolubles de degré 2 et à croissance exponentielle (donc ils ne sont pas nilpotents). Kaimanovich et Vershik ont montré que la mesure μ symétrique, élémentaire sur $F_o(\mathbb{Z}, \mathbb{Z}/2\mathbb{Z})$, définie par $\mu(1,0) = \mu(-1, 0) = 1/4$ et $\mu(0,\chi_{\{0\}}) = 1/2$, a une croissance $\mathcal{C}(\mu)$ nulle dans ce groupe [25]. D'après ce qu'on a vu ci-dessus, cela donne $h(\mu) = 0$, ce qui répond positivement à la question posée. De façon plus précise, ces auteurs démontrent le résultat suivant :

Proposition : *Soit* μ *à support fini sur* $F_o(\mathbb{Z}^k, \mathbb{Z}/2\mathbb{Z})$. *Une condition nécessaire et suffisante pour que les fonctions* μ*-harmoniques bornées soient constantes (i.e.* $h(\mu) = 0$*) est que la première composante de* S_n *, qui forme une marche aléatoire sur* $\mathbb{Z}^k$ *, soit récurrente. Si de plus* μ *est symétrique, cela est réalisé seulement pour* $k=1$ et $k=2$ ([25], p. 482).

L'idée de la démonstration est que, si la première composante de S_n est transitoire sur $\mathbb{Z}^k$, alors la seconde composante de S_n a une "limite" p.s. non constante, qui engendre des fonctions μ-harmoniques bornées non constantes.

Chacun de ces groupes est moyennable, donc porte une mesure μ adaptée telle que $h(\mu) = 0$. Kaimanovich et Vershik affirment aussi que, pour $k \geqslant 3$, cette mesure μ vérifie nécessairement $H(\mu) = \infty$. (voir fin de la partie III). ([25] p. 481).

V

FRONTIERE, ENTROPIE ASYMPTOTIQUE ET FORMULE DE FÜRSTENBERG

La notion d'entropie asymptotique $h(\mu)$ étudiée dans les deux parties précédentes est apparue pour la première fois sous une forme quelque peu différente, dans le travail de Fürstenberg sur les produits de matrices unimodulaires indépendantes ([**16**]). Considérons $G = S\ell(2,\mathbb{R})$ et μ une probabilité adaptée sur G, suffisamment régulière. S'il existe sur l'espace projectif $\mathbb{P}_1$ une mesure Π, équivalente à la mesur de Lebesgue, telle que $\mu * \Pi = \Pi$, la convolution étant déduite de l'action canonique de $S\ell(2,\mathbb{R})$ sur $\mathbb{P}_1$, Fürstenberg démontre que

$$\lim_n \frac{1}{n} \log \|S_n\| = -\frac{1}{2} \iint \log \frac{dx^{-1}\Pi}{d\Pi}(y)\, d\mu(x)\, d\Pi(y) \text{ p.s.}$$

($\|\cdot\|$ désigne la norme euclidienne d'une matrice de $S\ell(2,\mathbb{R})$). Il apparaît, a posteriori, que cette quantité n'est autre que l'entropie asymptotique de la marche aléatoire S_n. C'est ce qu'on se propose d'expliquer, en général, dans cette partie.

On reprend les notations des parties précédentes. On considère de plus l'action mesurable de G sur l'espace Ω des trajectoires $(s_1, \ldots, s_n \ldots)$ de la marche aléatoire, définie par :

$$(x, (s_1, \ldots, s_n, \ldots)) \mapsto (xs_1, \ldots, xs_n, \ldots).$$

Cette action commute avec l'opérateur de décalage ϑ. Elle engendre un produit de convolution $\rho * \nu$, pour une mesure ρ sur G et ν sur Ω. La convolution $\delta_x * P_e$, $x \in G$, représente la loi de la marche aléatoire issue de x, $(x\, S_n)_n$. On a donc $\vartheta P_e = \mu * P_e$.

La loi conjointe de S_1 et S_{n+1} s'écrit :

$$P_e(S_1 \in B_1, S_{n+1} \in B_{n+1}) = \int_{B_1} d\mu(x)\, (\delta_x * \mu^n)(B_{n+1})$$

$$= \int_{B_1} d\mu(x)\, (\delta_x * P_e)(S_n \in B_{n+1})$$

La "loi conjointe" de S_1 et $\mathfrak{J}$ s'écrit :

$$P_e((S_1 \in B_1) \cap \mathfrak{J}) = \int_{B_1} d\mu(x)\, (\delta_x * P_e)(\mathfrak{J})$$

où $J \in \mathfrak{J}$. Le calcul de l'information mutuelle $I(S_1, S_n)$ ou $I(S_1, \mathfrak{J})$ obéit à un principe général exprimé par la proposition suivante :

Proposition : *Soit* $(E, \mathcal{E}, m)$ *un espace probabilisé. Soit* $T(x,B)$ *une probabilité de transition, définie sur* E *, à valeurs dans* $(F, \mathfrak{F})$ *, la tribu* $\mathfrak{F}$ *étant à base dénombrable. Soit* $m' = Tm$ *la probabilité image sur* $(F, \mathfrak{F})$:

$m'(B) = \int_E T(x,B)\, dm(x)$. *Soit* ρ *la mesure définie sur l'espace produit* $E \times F$, $\mathcal{E} \otimes \mathfrak{F}$, *par* $\rho(A \times B) = \int_A T(x, B)\, dm(x)$.

Alors ρ *est absolument continue par rapport à* $m \otimes m'$ *si et seulement si, pour* m *presque tout* x, $T(x, .)$ *est absolument continue par rapport à* m' *et alors*

$$\frac{d\rho}{dm \otimes dm'}(x,y) = \frac{dT(x,.)}{dm'}(y) \quad m \otimes m' \text{ p.p.}$$

L'entropie relative $H(\rho \,;\, m \otimes m')$ *est finie si et seulement si l'intégrale*

$$\iint \left(\log \frac{dT(x,.)}{dm'}(y)\right) dT(x,.)(y)\, dm(x)$$

existe et alors elles sont égales.

Démonstration : Elle repose sur les arguments ordinaires de la différentiation, aussi on ne fait que l'esquisser. Soit $\mathfrak{F}_k$ une suite croissante de sous-tribus finies de $\mathfrak{F}$, formées d'ensembles de m'-mesure > 0 et telles que $\mathfrak{F}$ soit engendrée par $\bigcup_k \mathfrak{F}_k$ mod m. La fonction $\frac{T(x, F_k(y))}{m'(F_k(y))}$, où $F_k(y)$ désigne l'atome de $\mathfrak{F}_k$ contenant y, est une version de la densité de ρ par rapport à $m \otimes m'$ en restriction à $\mathcal{E} \otimes \mathfrak{F}_k$; ρ est absolument continue par rapport à $m \otimes m'$ si et seulement si cette suite de fonctions, qui forme une martingale, converge dans $L^1(m \otimes m')$; $T(x,.)$ est absolument continue par rapport à m', si et seulement si cette suite converge, à x fixé, dans $L^1(m')$. Cela démontre la première partie de la proposition. L'assertion concernant l'entropie relative résulte de la première partie, en vertu du théorème de Gelfand-Yaglom - Perez rappelé dans l'appendice 1.

Il faut souligner que cette proposition peut être en défaut si la tribu $\mathfrak{F}$ n'est pas à base dénombrable et ceci complique certains des énoncés qui vont suivre. En prenant $(E, \mathcal{E}) = (G, \mathcal{B})$, $(F, \mathfrak{F}) = (\Omega, \sigma(S_n, \ldots))$, $T(x, B) = \delta_x * P_e(B)$ on trouve tout d'abord la formule suivante pour $I(S_1, S_{n+1})$:

$$I(S_1, S_{n+1}) = \iint (\log \frac{d\delta_x * \mu^n}{d\mu^{n+1}}(y))\, d(\delta_x * \mu^n)(y)\, d\mu(x)$$

$$= \iint (\log \frac{d\delta_x * P_e}{d\vartheta P_e}\Big|_{\sigma(S_n\dots)}(\omega))\, d\delta_x * P_e(\omega)\, d\mu(x).$$

Dans la seconde la densité de $\delta_x * P_e$ par rapport à ϑP_e est prise en restriction à la tribu $\sigma(S_n,\dots)$. Le groupe G étant supposé à base dénombrable, l'hypothèse de la proposition est bien remplie.

Considérons maintenant $(F, \mathfrak{F}) = (\Omega, \mathfrak{I})$. Sur $\mathfrak{I}$ on a $P_e = \vartheta P_e = \int_G \delta_x * P_e\, d\mu(x)$. Mais il n'est plus possible, même dans les cas usuels, de considérer $\mathfrak{I}$ à base dénombrable : par exemple, pour une rotation ergodique sur le cercle la tribu des invariants n'est pas à base dénombrable. Cependant l'espace mesuré $(\Omega, \mathfrak{I}, P_e)$ est "séparable" en tant que sous-espace de $(\Omega, \mathcal{B}^\infty, P_e)$. Il existe donc au moins une sous-tribu $\mathfrak{I}'$ de $\mathfrak{I}$ qui soit à base dénombrable et qui soit égale à $\mathfrak{I}$ mod. P_e. On a évidemment $h(\mu) = I(S_1, \mathfrak{I}) = I(S_1, \mathfrak{I}')$. La proposition précédente peut alors être appliquée à $\mathfrak{I}'$.

Théorème : *Soit* $\mathfrak{I}'$ *une sous tribu à base dénombrable de la tribu* $\mathfrak{I}$ *telle que* $\mathfrak{I}' = \mathfrak{I}$ mod. P_e *L'entropie asymptotique* $h(\mu)$ *de la marche aléatoire s'écrit :*

$$h(\mu) = \iint (\log \frac{d\delta_x * P_e}{dP_e}\Big|_{\mathfrak{I}'}(\omega))\, d\delta_x * P_e(\omega)\, d\mu(x) \; ; \textit{ si } \mathfrak{I}' \textit{ est stable sous } G$$

$$h(\mu) = -\int_G d\mu(x) \int_\Omega (\log \frac{d\delta_{x^{-1}} * P_e}{dP_e}\Big|_{\mathfrak{I}'}(\omega))\, dP_e(\omega).$$

Cet énoncé signifie que $h(\mu)$ est finie si et seulement si l'intégrale considérée existe et alors ces deux quantités sont égales. On passe de la 1ère intégrale à la 2ème par le changement de variables $\omega' = x\omega$ (action de G sur Ω).

Quand $h(\mu) > 0$, l'espace des fonctions μ-harmoniques bornées est non trivial. Sa description complète est un problème difficile. Pour ce faire on introduit la notion de μ-frontière.

Soit $(M, \mathcal{M})$ un G-espace mesurable, la tribu $\mathcal{M}$ étant à base dénombrable. Soit ν une probabilité μ-invariante sur M :

$$\mu * \nu = \int_G \delta_x * \nu\, d\mu = \nu ,$$

la convolution étant déduite de l'action de G sur M.

Pour toute fonction f réelle, $\mathcal{M}$-mesurable, bornée sur M on définit une fonction μ-harmonique bornée sur G, $g = Rf$ par la formule :

$$g(x) = \int_M f(xz)\, d\nu(z) = \int_M f(z)\, d\, \delta_x * \nu(z).$$

L'opérateur R est linéaire, contractant pour la norme sup.

Définition : *L'espace mesuré* $(M, \mathcal{M}, \nu)$ *où* $\mathcal{M}$ *est à base dénombrable et où* ν *est* μ*-invariante, est une* μ*-frontière si l'opérateur* R *défini ci-dessus possède les deux propriétés suivantes :*

1) R *est injectif i.e.* $Rf = 0$ *sur* G *si et seulement si* $f = 0\ \delta_x * \nu$ p.p. *pour tout* $x \in G$.

2) R *est multiplicatif pour le produit propre* $\square$ *des fonctions* μ*-harmoniques bornées :* $R(ff') = Rf \square Rf'$.

Le produit propre de deux fonctions μ-harmoniques bornées g et g' est défini par :

$$g''(x) = g \square g'(x) = \lim_n \int_G g(xy)\, g'(xy)\, d\mu^n(y)$$

$$= \int_\Omega \lim_n g(x\, S_n)\, g'(x\, S_n)\, dP_e.$$

Il introduit une structure d'algèbre sur l'espace des fonctions μ-harmoniques bornées, isomorphe à celle de l'algèbre des v.a. invariantes, considérées modulo $\delta_x * P_e$, pour tout $x \in G$.

Définition *: La* μ*-frontière* $(M, \mathcal{M}, \nu)$ *est appelée frontière de Poisson, associée à* (G, μ) *, si de plus l'opérateur* R *est surjectif.*

A l'origine les notions de μ-frontière et de frontière de Poisson, ont été introduites par Fürstenberg de façon un peu différente ([**17**]). Fürstenberg définit une μ-frontière comme un G-espace compact M, métrisable, portant une probabilité ν , μ-invariante, de telle sorte que $\delta_{S_n} * \nu$ converge p.s. vers une mesure de Dirac sur M. On vérifie aussitôt qu'alors les deux conditions de la définition donnée ci-dessus sont vérifiées.

La relation avec l'entropie asymptotique apparaît dans l'énoncé suivant :

<u>*Théorème*</u> : *Si* $(M, \mathcal{M}, \nu)$ *est une* μ*-frontière, on a l'inégalité :*

$$h(\mu) \geq - \int_G d\mu(x) \int_M \left(\log \frac{d\delta_{x^{-1}} * \nu}{d\nu}(z)\right) d\nu(z).$$

C'est une égalité si et seulement si $(M, \mathcal{M}, \nu)$ *est frontière de Poisson, dans le cas où* $h(\mu) < \infty$.

<u>Démonstration</u> : D'après la proposition donnée au début de cette partie, l'intégrale apparaissant ici est l'entropie relative de la mesure, sur $G \times M$, $\int_B d\mu(x)\, \delta_x * \nu(N)$, par rapport à la mesure $\mu \otimes \nu\ (B \times N)$.
D'après la définition de μ-frontière, la tribu $\mathcal{M}$ munie de la famille des probabilités $\delta_x * \nu$, $x \in G$, est représentée par une sous-tribu $\mathcal{J}$ de $\mathfrak{I}$, stable sous l'action de G, à base dénombrable, munie de la famille $\delta_x * P_e$, $x \in G$. En effet à $N \in \mathcal{M}$ correspond la fonction μ-harmonique $R\chi_N$; à cette fonction correspond une classe mod. $\delta_x * P_e$, pour tout x, de v.a. invariante. Comme R est multiplicatif cette v.a. invariante est l'indicatrice d'un événement invariant. L'intégrale considérée ci-dessus n'est autre que l'information mutuelle $I(S_1, \mathcal{J})$. L'inégalité avec $h(\mu)$ est alors évidente.

Si c'est une égalité $h(\mu) = I(S_1, \mathfrak{I}) = I(S_1, \mathcal{J})$; par le lemme de la partie III on trouve aussi $I((S_1 \dots S_k), \mathcal{J}) = I(S_k, \mathcal{J}) = kI(S_1, \mathcal{J}) = I((S_1, \dots S_k), \mathfrak{I})$. Ceci étant vrai pour tout k, les tribus $\mathfrak{I}$ et $\mathcal{J}$ sont égales mod P_e. De même on montre leur égalité mod $\delta_x * P_e$, pour tout $x \in G$. Alors toutes les fonctions μ-harmoniques bornées sont représentées sur Ω par une v.a. $\mathcal{J}$ mesurable, mod $\delta_x * P_e$ pour tout x. Cela prouve la surjectivité de l'opérateur R.

Ce théorème donne un critère qui permet de reconnaitre la frontière de Poisson parmi les μ-frontières éventuelles ; autrement dit, on peut ainsi reconnaître si la μ-frontière considérée permet la représentation de toutes les fonctions μ-harmoniques bornées.

Dans le cas discret Kaimanovich et Vershik ont développé le point précédent de façon plus simple et introduit la notion de transformée de Radon-Nikodym [25]. Ils ont montré, en particulier, que l'inégalité du théorème précédent est valide pour tout G-espace muni d'une probabilité μ-invariante, même si ce n'est pas une μ-frontière, mais sous l'hypothèse $H(\mu) < \infty$. Il serait intéressant de savoir si cela reste vrai en général.

Pour illustrer le principe qu'on vient de décrire, considérons l'exemple de la marche aléatoire élémentaire sur le groupe libre F_2 à deux générateurs a et b. La probabilité μ est alors $\mu = \frac{1}{4}(\delta_a + \delta_b + \delta_{a^{-1}} + \delta_{b^{-1}})$. Calculons, tout d'abord, de façon directe l'entropie asymptotique $h(\mu)$. En notant $L(x)$ la longueur de $x \in F_2$, c'est à dire le nombre de symboles de son écriture réduite, on a :

$$H(S_n) = H(S_n, L(S_n)) = H(L(S_n)) + EH(S_n/L(S_n)).$$

Sachant $L(S_n) = k$, S_n est équirépartie sur les $4(3^{k-1})$ éléments de F_2, de longueur k. Donc :

$$EH(S_n/L(S_n)) = \sum_{k\geq 1} \log\,(4(3^{k-1}))\quad P(L(S_n) = k)$$

$$= (\log 4) + (\log 3)\ E(L(S_n) - 1).$$

Pour n grand, $L(S_n)$ se comporte comme la marche aléatoire sur $\mathbb{Z}$ définie par $\frac{3}{4}\delta_1 + \frac{1}{4}\delta_{-1}$. Cela donne

$$\lim_n \frac{1}{n} H(L(S_n)) = 0,$$

$$\lim_n \frac{1}{n} E(L(S_n)) = \frac{1}{2}$$

d'où
$$\lim_n \frac{1}{n} H(S_n) = h(\mu) = \frac{1}{2}\log 3.$$

Considérons maintenant la μ-frontière formée par l'espace M des mots réduits infinis écrits avec les quatre symboles a, b, a^{-1}, b^{-1}, et ν la probabilité sur M obtenue comme la loi de la chaîne de Markov à quatre états, de transition :

	a	b	a^{-1}	b^{-1}
a	1/3	1/3	0	1/3
b	1/3	1/3	1/3	0
a^{-1}	0	1/3	1/3	1/3
b^{-1}	1/3	0	1/3	1/3

et de distribution initiale uniforme ($\frac{1}{4}$ sur chaque état). Alors

$$\frac{d\delta_a * \nu}{d\nu}(m) = 3 \quad \text{pour}\quad m \in \{m_0 = a \text{ ou } b \text{ ou } b^{-1}\}$$

$$= \frac{1}{3} \quad \text{pour}\quad m \in \{m_0 = a^{-1}\},$$

d'où
$$\int \log \frac{d\delta_a * \nu}{d\nu}(m)\ d\nu(m) = \frac{3}{4}\log 3 - \frac{1}{4}\log 3 = \frac{1}{2}\log 3$$

On obtient finalement :

$$\int_G d\mu(x) \int_M \log \frac{d\delta_{x^{-1}} * \nu}{d\nu}(m) \; d\nu(m) = \frac{1}{2} \log 3.$$

Le théorème précédent nous permet de conclure que (M, ν) est l'espace de Poisson de (F_2, μ). Autrement dit l'ensemble des fonctions μ-harmoniques bornées est exhaustivement représenté par l'espace des fonctions mesurables bornées sur M, par l'intermédiaire de la formule :

$$g(x) = \int_M f(x\, m)\, d\nu(m).$$

Ce résultat a été prouvé pour la première fois, par un argument direct différent, par Dynkin et Maliutov ([**12**]). Le calcul direct de l'entropie effectué ci-dessus, montre aussi que le développement au "2ème ordre" de $H(S_n)$ donne un terme de l'ordre de $\frac{1}{2} \log n$, comme sur $\mathbb{Z}$ (voir partie II).

Dans [**28**] et [**29**], Ledrappier a utilisé le même principe "entropique" pour obtenir la frontière de Poisson de certains groupes discrets de matrices.

VI

ENTROPIE ET SYMETRIE

On se propose, dans cette partie, de comparer les fonctions harmoniques et les entropies des marches aléatoires droite $S_n = X_1 \dots X_n$ et gauche $S'_n = X_n \dots X_1$ définies par une probabilité μ sur G. Par l'automorphisme $x \mapsto x^{-1}$ la marche aléatoire droite $S_n = X_1 \dots X_n$ est changée en la marche aléatoire gauche $S_n^{-1} = X_n^{-1} \dots X_1^{-1}$ définie par la probabilité $\check{\mu}$, image de μ par l'automorphisme inverse. L'étude de la marche aléatoire gauche définie par μ se ramène ainsi à celle de la marche aléatoire droite définie par $\check{\mu}$. On se propose donc de comparer les fonctions μ-harmoniques et $\check{\mu}$-harmoniques, ainsi que les entropies asymptotiques $h(\mu)$ et $h(\check{\mu})$ en ne considérant, comme précédemment, que des marches aléatoires droites.

La fonction module du groupe G est notée Δ :

$$\int_G f(x)\, dm(x) = \int_G f(x^{-1})\, \Delta(x)^{-1}\, dm(x)$$

(m désigne comme précédemment la mesure de Haar gauche). Si μ^n a une densité φ_n par rapport à m, alors $(\check{\mu})^n$ a une densité, notée Ψ_n donnée par $\Psi_n(x) = \varphi_n(x^{-1})\ \Delta(x)^{-1}$.

En effet, $X_1 \ldots . X_n$ et $X_n \ldots . X_1$ ayant même loi μ^n on trouve :

$$\int_G f(x)\ d(\check{\mu})^n(x) = E\ [f(X_1^{-1} \ldots X_n^{-1})] = E\ [f(X_1 \ldots . X_n)^{-1}]$$

$$= \int_G f(x^{-1})\ d\mu^n(x) = \int_G f(x^{-1})\varphi_n(x)\ dm(x) = \int_G f(x)\varphi_n(x^{-1})\Delta(x)^{-1} dm(x)$$

Il est alors facile de comparer les entropies différentielles $\tilde{H}(\varphi_n)$ et $\tilde{H}(\Psi_n)$.

Proposition : *Si l'intégrale* $\int_G \log \Delta(x)\ d\mu(x) = E(\log \Delta(X_1))$ *converge, l'entropie différentielle* $\tilde{H}(\mu^n) = \tilde{H}(\varphi_n)$ *est finie si et seulement si l'entropie différentielle* $\tilde{H}(\check{\mu}^n) = \tilde{H}(\Psi_n)$ *l'est. Alors* $\tilde{H}(\Psi_n) = \tilde{H}(\varphi_n) - n\ E(\log\Delta\ (X_1))$.

Démonstration : C'est un calcul direct :

$$\tilde{H}(\Psi_n) = -\int \log\ (\varphi_n(x^{-1})\ \Delta(x)^{-1})\varphi_n(x^{-1})\ \Delta(x)^{-1}\ dm(x)$$

$$= -\int \log\ (\varphi_n(x)\ \Delta(x))\ \varphi_n(x)\ dm(x)$$

$$= \tilde{H}(\varphi_n) - \int (\log\ \Delta(x))\ \varphi_n(x)\ dm(x)\ ;$$

Δ étant multiplicatif, $\int (\log\ \Delta(x))\ \varphi_n(x)\ dm(x) = E(\log\ \Delta\ (S_n))$ $= n\ E(\log\ \Delta(X_1))$.

Théorème : *Dans le cas absolument continu, les entropies différentielles* $\tilde{H}(\mu^n) = \tilde{H}(\varphi_n)$ *étant finies, si l'intégrale* $E(\log \Delta(X_1)) = \int_G \log\Delta\ (x)\ d\mu(x)$ *converge, l'entropie asymptotique vérifie l'inégalité :*

$$h(\mu) \geqslant E(\log \Delta(X_1)).$$

De plus $\quad h(\check{\mu}) = h(\mu) - E(\log \Delta(X_1))$.

Démonstration : D'après la partie III, $h(\mu) = \lim_n \frac{1}{n} \tilde{H}(\varphi_n)$, et $h(\check{\mu}) = \lim_n \frac{1}{n} \tilde{H}(\Psi_n)$. La proposition donne alors

$$h(\check{\mu}) = h(\mu) - E(\log\ \Delta(X_1)).$$

La quantité $h(\check{\mu})$ est positive, donc $h(\mu) \geqslant E(\log\ \Delta(X_1))$.

Ce théorème a plusieurs corollaires remarquables.

<u>Corollaire</u>: Si G est unimodulaire (i.e. $\Delta \equiv 1$), si μ est absolument continue avec les entropies différentielles $\tilde{H}(\varphi_n)$ finies, alors $h(\check{\mu}) = h(\mu)$. En particulier, si toutes les fonctions μ-harmoniques bornées sont constantes il en est de même pour $\check{\mu}$.

La seconde assertion du corollaire résulte de la première en vertu de la partie II. On peut se demander si l'égalité $h(\check{\mu}) = h(\mu)$ est vraie, pour G unimodulaire, sans restriction sur μ. D'après Kaimanovich et Vershik la réponse à cette question est négative, même sur un groupe discret. Dans le cas discret, le corollaire précédent dit que $h(\mu)=h(\check{\mu})$ dès que $H(\mu) < \infty$. Mais si $H(\mu) = \infty$, des fonctions bornées non constantes harmoniques pour μ peuvent exister alors qu'il n'en existe pas pour $\check{\mu}$ ([25] p. 483). Cela montre, de nouveau, qu'il n'y a pas en général de principe d'approximation de $h(\mu)$ par des mesures à support fini. (voir la fin de la partie III).

<u>Corollaire</u> : Si G n'est pas unimodulaire, si μ est absolument continue avec les entropies différentielles $\tilde{H}(\varphi_n)$ finies, si $\int_G \log\Delta(x)\, d\mu(x) > 0$, alors il existe des fonctions μ-harmoniques bornées non constantes.

Considérons l'exemple du groupe affine de la droite réelle : $G = \left\{\begin{pmatrix} a & b \\ 0 & 1\end{pmatrix} ;\ a \neq 0,\ a, b, \in \mathbb{R}\right\}$. La mesure de Haar gauche est donnée par $dm(a,b) = \frac{1}{a^2}\, da db$; la fonction module est $\Delta\begin{pmatrix} a & b \\ 0 & 1\end{pmatrix} = 1/a$. Le corollaire précédent nous dit que pour μ à densité bornée à support compact (ou telle que les entropies différentielles $\tilde{H}(\varphi_n)$ soient finies), il existe des fonctions μ-harmoniques bornées non constantes dès que $\int_G \log a(x)\, d\mu(x) < 0$ ($a(x)$ désigne le 1er coefficient de la matrice $x \in G$). On retrouve ainsi, pour une part, un énoncé d'Azencott ([5] ; voir aussi [13]). On sait aussi, dans ce cas, que si $\int_G \log a(x) d\mu(x) > 0$ les fonctions μ-harmoniques bornées sont nécessairement constantes, donc $h(\mu) = 0$. Le théorème précédent donne alors l'énoncé suivant.

<u>Corollaire</u> : Sur le groupe affine de la droite réelle $G = \left\{\begin{pmatrix} a & b \\ 0 & 1\end{pmatrix} ;\ a \neq 0,\ a,b \in \mathbb{R}\right\}$, si μ a une densité bornée à support

compact, alors

$$h(\mu) = - \int \log a(x)\, d\mu(x) = -E(\log a(X_1))$$

si cette quantité est positive, et 0 sinon.

Sous ces hypothèses, la méthode de la partie V s'applique alors pour décrire l'espace de Poisson de la marche aléatoire définie par μ.

VII

QUELQUES COMPLEMENTS

A) L'entropie, au sens de Kolmogorov-Sinai, d'un processus stationnaire peut être obtenue à partir de la connaissance d'une trajectoire typique, par un passage à la limite. C'est le contenu du théorème de Shannon-Mc Millan-Breiman. Dans [**1**], Avez a demandé si un énoncé analogue était valide pour l'entropie asymptotique d'une marche aléatoire. Dans le cas discret, avec $H(\mu) < \infty$, la réponse est positive.

Théorème *: Sur un groupe* G *discret, si* $H(\mu) < \infty$ *, alors* $\lim_n - \frac{1}{n} \log \mu^n(S_n) = h(\mu)$ P_e p.s. *et dans* $L^1(P_e)$.

Ce théorème a été démontré comme un corollaire du théorème ergodique sous-additif dans [**9**]. Une démonstration plus directe, plus proche de la démonstration habituelle du théorème de Shannon-Mc Millan-Breiman, a été donnée indépendamment dans [**25**]. On peut se demander quel énoncé du même type est valide dans le cas non discret.

B) La démonstration du lemme de la partie IV utilise la proposition suivante

Proposition *: Soit* $(\rho_n)_{n \in \mathbb{N}}$ *une distribution de probabilité discrète. L'entropie* $H(\rho) = - \sum_{n=1}^{\infty} \rho_n \log \rho_n$ *est finie si et seulement si la série* $\sum_{n=1}^{\infty} \rho'_n \log n$ *converge, où* (ρ'_n) *désigne la réarrangée décroissante de la suite* ρ_n. *La convergence de la série* $\sum_{n=1}^{\infty} \rho_n \log n$ *est donc une condition suffisante, impossible à améliorer, pour que l'entropie* $H(\rho)$ *soit finie.*

Démonstration : L'entropie est invariante par réarrangement: $H(\rho) = H(\rho')$.

Supposons $\sum_{n=1}^{\infty} \rho_n \log n < \infty$. Soit $A = \{ n \; ; \; -\log \rho_n < 2 \log n \}$. Alors $- \sum_{n\in A} \rho_n \log \rho_n$ converge. Pour $n \notin A$, $\rho_n < \frac{1}{n^2}$; la convergence de la série $\sum_{n=1}^{\infty} \frac{1}{n^2} \log n^2$ implique donc celle de $- \sum_{n\notin A} \rho_n \log \rho_n$.

Réciproquement, supposons (ρ_n) décroissante et $- \sum_{n=1}^{\infty} \rho_n \log \rho_n < \infty$. Soit $B = \{n; \rho_n > \frac{1}{n} \}$. Si B était infini on construirait la suite croissante t_k : $t_1 = 1$ et $t_{k+1} = \inf \; \{n > t_k \; ; \; \rho_n > \frac{1}{n} \}$. On aurait $\rho_n > \frac{1}{t_{k+1}}$ pour $t_{k+1} \geqslant n > t_k$ en raison de la décroissance de ρ_n. La convergence de $- \sum_{n=1}^{\infty} \rho_n \log \rho_n$ donnerait alors celle de $\sum_{k=1}^{\infty} \frac{t_{k+1} - t_k}{t_{k+1}} \log t_{k+1}$. Or la série $\sum_{k=1}^{\infty} \frac{t_{k+1} - t_k}{t_{k+1}}$ diverge nécessairement. Donc l'ensemble B est fini. Comme $\sum_{n\in B^c} \rho_n \log n \leqslant \sum_{n\in B^c} \rho_n \log \rho_n$, l'hypothèse $H(\rho) < \infty$ implique $\sum_{n=1}^{\infty} \rho_n \log n < \infty$.

Un énoncé analogue est valide pour une densité bornée φ sur $\mathbb{R}$ ou $\mathbb{R}^d$. L'entropie différentielle $\tilde{H}(\varphi) = - \int \varphi(x) \log \varphi(x) \, dx$ est finie si et seulement si le moment logarithmique $\int_1^{+\infty} \bar{\varphi}(x) \log x \, dx$ est fini pour la réarrangée décroissante $\bar{\varphi}$ de φ. D'après le lemme de la partie IV, quand le groupe G est à croissance polynomiale, le moment logarithmique fini pour φ, densité bornée, est une condition suffisante pour que toutes les $\tilde{H}(\varphi_n)$ soient finies. La proposition précédente montre que cette condition de moment ne peut pas être affaiblie.

C) D'après le théorème de la partie IV, si μ est une probabilité sur $\mathbb{Z}$ ayant un moment logarithmique fini, $\lim_n \frac{1}{n} H(\mu^n) = h(\mu) = 0$. L'un des arguments de la démonstration est le suivant :

Proposition *: Si* μ *est une probabilité sur* $\mathbb{Z}$ *telle que* $\sum \mu(n) \log |n| < \infty$, *alors* $\lim_n \frac{1}{n} \log S_n = 0$ p.s. *et en moyenne.*

Pour μ quelconque sur $\mathbb{Z}$, il résulte du théorème de Choquet-Deny que $h(\mu) = 0$. Il serait intéressant de produire une démonstration directe de ce fait. Cela fournirait probablement de nouveaux arguments conduisant à des versions du théorème de Choquet-Deny pour d'autres situations.

Si μ a un moment d'ordre 2, l'ordre de grandeur attendu pour $H(\mu^n)$ est $\frac{1}{2}\log n$; si μ appartient au domaine d'attraction d'une loi stable d'indice α , c'est $\frac{1}{\alpha}\log n$ (voir la partie II). En considérant des exemples où $\mu(n) \sim \dfrac{1}{n(\log n)^a}$, $a > 2$, on peut se convaincre que $H(\mu^n)$ peut être de l'ordre de $n^{1-\varepsilon}$, quel que soit $\varepsilon > 0$. L'énoncé $\lim_n \frac{1}{n} H(\mu^n) = 0$, pour $H(\mu) < \infty$, semble donc le meilleur possible, même sur $\mathbb{Z}$.

D) On peut aussi se demander si la croissance $\mathscr{C}(\mu)$ dans $\mathbb{Z}$ ou $\mathbb{R}$ d'une probabilité μ, de densité continue φ , est une quantité accessible. Plus précisément, en suivant la définition de $\mathscr{C}(\mu)$ donnée dans la partie IV, peut-on préciser l'ordre de grandeur de la suite des mesures de Lebesgue $m(A_n)$, pour une suite croissante de boréliens A_n telle que $\int_{A_n} \varphi_n(x)\,dx > 1-\varepsilon$? Il apparaît qu'on peut avoir $\mathscr{C}(\mu) = \infty$ et même que la suite $m(A_n)$ peut croître à une vitesse arbitrairement grande. La démonstration de ce fait repose sur un théorème de Polya : "Toute fonction réelle f, paire, telle que $f(0) = 1$, convexe sur $[0, +\infty[$, est la fonction caractéristique d'une mesure de probabilité sur $\mathbb{R}$" ([**14**] p. 509). Indiquons rapidement cette démonstration. Pour une densité continue φ , paire, décroissante sur $[0, +\infty[$, la croissance de $m(A_n)$ est celle de la suite

$$\inf \{a > 0 \; ; \int_{-a}^{+a} \varphi_n(x)\,dx > 1-\varepsilon\}$$

ou encore celle de

$$\inf \{a > 0 \; ; \frac{1}{\sqrt{2\Pi}} \int_{-\infty}^{+\infty} e^{-u^2/2} \varphi_n(ua)\,a\,du > 1-\varepsilon\}.$$

Par l'identité de Parseval et après troncation de l'intégrale on est ramené à considérer

$$\inf \{a > 0 \; ; \frac{2}{\sqrt{2\Pi}} \int_0^2 e^{-u^2/2} (\hat{\varphi})^n(\tfrac{u}{a})\,du > 1-\varepsilon\}.$$

où $\hat{\varphi}$ désigne la fonction caractéristique de φ . En prenant $\hat{\varphi}$ comme dans le théorème de Polya on a

$$1 + (\hat{\varphi})^n \left(\frac{2}{a}\right) > \int_0^2 (\hat{\varphi})^n \left(\frac{u}{a}\right) du$$

et l'ordre de grandeur cherché est minoré par celui de

$\inf \{a > 0 \; ; \; (\hat{\varphi})^n (\frac{2}{a}) > \delta\}$ pour un δ, $0 < \delta < 1$. La pente de $\hat{\varphi}$ au voisinage de 0 pouvant être arbitraire, cette quantité peut croître avec n aussi vite que l'on veut.

APPENDICE 1

QUELQUES GENERALITES SUR LES NOTIONS D'ENTROPIE ET D'INFORMATION

Cet appendice ne contient que des rappels, sans démonstration, des résultats, plus ou moins classiques, qui ont été utilisés dans l'exposé précédent. La référence principale est le livre de Pinsker [31].

L'entropie d'une distribution de probabilité discrète (p_k) est définie par la formule de Boltzmann-Shannon :

$$H(p) = -\sum p_k \log p_k .$$

Grâce à la convexité de la fonction $-t \log t$ il est facile de voir que ce nombre est toujours positif, éventuellement $+\infty$, et n'est nul que si (p_k) est une distribution de Dirac. Si X est une v.a. de distribution (p_k) on dit que $H(p) = H(X)$ est l'entropie de X. Une permutation des p_k ne change pas $H(p)$ donc $H(X)$ ne dépend que de la distribution réarrangée de X.

Pour définir l'entropie d'une loi de probabilité ayant une densité φ sur $\mathbb{R}$ ou sur G, on introduit suivant Shannon ([34]) la quantité

$$\tilde{H}(\varphi) = -\int \varphi(x) \log \varphi(x)\, dx.$$

Malgré la similarité avec la formule précédente, cette quantité n'est pas en général positive ; elle peut prendre toute valeur de $-\infty$ à $+\infty$ inclus ou même ne pas être définie. De plus elle est modifiée par tout changement de variable de déterminant jacobien différent de 1. Pour ces raisons quand l'intégrale $-\int \varphi(x) \log \varphi(x)\, dx$ a un sens, on l'appelle *entropie différentielle* de la densité φ, et on la note $\tilde{H}(\varphi)$. Si φ est la loi d'une v.a. X, on dit que $\tilde{H}(\varphi) = \tilde{H}(X)$ est l'entropie différentielle de X. Cette quantité ne dépend que de la densité réarrangée de X.

Pour faire apparaître le lien entre l'entropie absolue et l'entropie différentielle et atteindre une généralité suffisante, il est nécessaire d'introduire la notion d'*entropie relative* suivante. Etant données deux probabilités P et Q sur un espace mesurable $(\Omega, \mathcal{B})$, l'entropie de

Q relative à P est définie par

$$H(Q\ ;\ P) = \sup_{\mathfrak{F}} \sum_{F\in\mathfrak{F}} \left(\log \frac{Q(F)}{P(F)}\right) Q(F)$$

où la somme est prise sur l'ensemble des parties, de P-mesure positive, F d'une partition finie mesurable $\mathfrak{F}$ et où le sup. est pris sur l'ensemble de toutes ces partitions. C'est une quantité positive, éventuellement infinie, nulle seulement si $P = Q$. Le théorème suivant dû à Gelfand, Yaglom et Perez ([11] ; voir aussi [31]) est alors fondamental.

Théorème : *Si* $H(Q\ ;\ P) < \infty$ *alors* Q *est absolument continue par rapport à* P.

Si Q *est absolument continue par rapport à* P *alors*

$$H(Q\ ;\ P) = \int \left(\log \frac{dQ}{dP}\right) dQ$$

(intégrale éventuellement infinie).

L'intégrale apparaissant ici est aussi appelée "information de Kullback". L'entropie relative $H(Q\ ;\ P)$ est donc infinie dans deux cas : si Q n'est pas absolument continue par rapport à P, ou si l'intégrale $\int \left(\log \frac{dQ}{dP}\right) dQ$ est infinie.

La notion d'*information mutuelle* de deux v.a. X et Y est déduite de celle d'entropie relative par la formule :

$$I(X,Y) = H(\lambda(X,Y)\ ;\ \lambda(X) \otimes \lambda(Y))$$

où $\lambda(X,Y)$ est la loi conjointe et $\lambda(X) \otimes \lambda(Y)$ la loi produit des marginales du couple (X,Y). Cette définition ne dépend que des tribus $\sigma(X)$ et $\sigma(Y)$ engendrées par les v.a. X et Y. Dans le cas où X et Y ne prennent qu'un nombre fini de valeurs on a :

$$I(X,Y) = H(X,Y) - H(X) - H(Y) = H(X) - H(X/Y)$$

formule d'usage constant dans la théorie de Kolmogorov-Sinai-Ornstein. Grâce à la concavité de la fonction log on démontre alors les propriétés suivantes ([31] chap II).

Propriétés de l'information mutuelle

1) $I(X,Y) \geq 0$
2) $I(X,Y) = 0$ si et seulement si X et Y sont indépendantes.
3) $I(X, f(Y)) \leq I(X,Y)$, f étant une fonction mesurable quelconque.

4) $\lim_n \uparrow I(X,(Y_1 \ldots Y_n)) = I(X, (Y_1 \ldots Y_n \ldots))$.

5) s'il existe n tel que $I(X, (Y_n, Y_{n+1} \ldots))$ soit fini,
$\lim_n \downarrow I(X, (Y_n, Y_{n+1}, \ldots)) = I(X, \mathcal{C})$ où $\mathcal{C} = \bigcap_n \sigma(Y_n, Y_{n+1} \ldots))$.

Pour retrouver dans ce cadre la première formule de Shannon on observe que $I(X,X) < \infty$ si et seulement si X est discrète de distribution (p_k) vérifiant $-\sum p_k \log p_k < \infty$; alors $I(X,X) = -\sum p_k \log p_k$. On pose donc, en général :

$$H(X) = I(X,X).$$

On a l'inégalité $I(X,Y) \leqslant H(X)$.

Pour retrouver la seconde formule de Shannon, on considère deux densités de probabilité φ et Ψ , sur $\mathbb{R}$ ou G, avec $\varphi(x)/\Psi(x) < \infty$ p.p. Alors

$$H(\varphi;\Psi) = \int (\log \frac{\varphi(x)}{\Psi(x)}\ \varphi(x)\ dx = \int (\log \varphi(x))\ \varphi(x)\ dx - \int (\log \Psi(x))\ \varphi(x) dx.$$

La notion d'entropie ou information *conditionnelle* définie et étudiée par Dobrushin ([11] [31]) est essentielle. Etant données trois v.a. X, Y, Z, les deux premières prenant leurs valeurs dans un espace "séparable" afin que les versions régulières des probabilités conditionnelles $\lambda(X,Y/Z=z)$, $\lambda(X/Z=z)$, $\lambda(Y/Z=z)$ existent, on définit l'information mutuelle de X et Y sachant Z par :

$$EI(X,Y/Z) = \int I(X,Y/Z=z) d\ \lambda_Z(z).$$

($I(X,Y/Z=z)$ étant l'entropie de $\lambda(X,Y/Z=z)$ relative à $\lambda(X/Z=z) \otimes \lambda(Y/Z=z)$). On a alors l'importante formule

$$I((X,Z),Y) = I(Y,Z) + EI\ (X,Y/Z),$$

dite formule de Kolmogorov.

APPENDICE 2
SUR LES TRIBUS ASYMPOTIQUE ET INVARIANTE

On donne ici quelques indications supplémentaires sur les tribus asymptotique et invariante d'une marche aléatoire et la démonstration du théorème de la partie III.

Considérons $\Omega = G^{\mathbb{N}}$ muni de la tribu borélienne produit $\mathcal{B}^{\infty}$. Notons S_n les coordonnées canoniques. Munissons $(\Omega, \mathcal{B}^{\infty})$ de la probabilité P_e pour laquelle $(S_n)_n$ est une réalisation de la marche aléatoire, définie par μ, issue de e ; P_e est l'image de la probabilité produit $\mu^{\otimes \mathbb{N}}$ par l'application :

$$(x_1, \ldots, x_n \ldots) \longmapsto (s_1 = x_1,\ s_2 = x_1 x_2, \ldots,\ s_n = (x_1 \ldots x_n), \ldots).$$

C'est aussi la loi de la chaîne de Markov de transition

$$P(S_{n+1} \in B \ / \ S_n = x) = \mu(x^{-1}B)$$

et de distribution initiale δ_e. L'application décalage ϑ sur Ω est définie par

$$\vartheta(s_1, s_2, \ldots, s_n \ldots) = (s_2, s_3, \ldots, s_{n+1} \ldots).$$

C'est une application $\mathcal{B}^{\infty}$-mesurable, surjective sur Ω.

Le groupe G agit de façon mesurable sur Ω par

$$(x, (s_1, \ldots, s_n)) \longmapsto (xs_1, xs_2, \ldots, xs_n, \ldots).$$

Cette action permet de définir la "convolution" d'une mesure ρ sur G avec une mesure ν sur Ω:

$$\int_{\Omega} f(s_1, \ldots, s_n \ldots)\, d\, \rho * \nu\, (s_1, \ldots, s_n)$$
$$= \int_G d\rho(x) \int_{\Omega} f(xs_1, \ldots, xs_n, \ldots)\, d\nu(s_1, \ldots, s_n, \ldots).$$

On a alors la relation :

$$\vartheta P_e = \mu * P_e.$$

La tribu asymptotique de la marche aléatoire, notée $\mathcal{A}$, est définie par :

$$\mathcal{A} = \bigcap_{n \geq 1} \vartheta^{-n} \mathcal{B}^{\infty}.$$

La tribu invariante, notée $\mathcal{J}$, est définie par

$$\mathcal{J} = \{B \in \mathcal{B}^{\infty};\ \vartheta^{-1} B = B\}.$$

L'inclusion $\mathcal{J} \subset \mathcal{A}$ est évidente.

Proposition *: L'action de ϑ sur $\mathcal{A}$ est celle d'un automorphisme de tribu.*

Démonstration : La surjectivité de ϑ est essentielle. Par ϑ^{-1} on définit un endomorphisme surjectif de la tribu $\vartheta^{-n} \mathcal{B}^{\infty}$ sur $\vartheta^{-n-1}\mathcal{B}^{\infty}$. En passant à la limite on voit que ϑ^{-1} est un endomorphisme surjectif de $\mathcal{A}$. Comme ϑ est surjective, $\vartheta\,\vartheta^{-1}F = F$ pour tout $F \subset \Omega$, donc ϑ est inverse à gauche de l'endomorphisme surjectif ϑ^{-1} de $\mathcal{A}$; ϑ est aussi inverse à droite et cela prouve la proposition.

Théorème *: Quelle que soit μ, les tribus $\mathcal{A}$ et $\mathcal{J}$ sont égales* mod P_e *; c'est à dire que, pour tout $A \in \mathcal{A}$ il existe $J \in \mathcal{J}$ vérifiant*

$$A = J \quad P_e \text{ p.s.}$$

Démonstration : La démonstration se scinde en deux, suivant qu'il existe deux entiers n et n' tels que μ^n et $\mu^{n'}$ ne soient pas étrangères ou non.

Considérons le 1er cas. Soit alors k le plus petit des entiers $j \geq 1$ tels qu'il existe $n \geq 0$ avec $\|\mu^{n+j}-\mu^n\| < 2$ ($\|\ \ \|$ note la variation totale des mesures ; μ^{n+j} et μ^n ne sont pas étrangères si et seulement si $\|\mu^{n+j}-\mu^n\| < 2$). Alors, d'après la loi "zéro ou deux" ([8] p. 120) $\lim_i \|\mu^{(i+1)k} - \mu^{ik}\| = 0$ et $\mathcal{A} = \mathcal{J}_k$ mod P_e où $\mathcal{J}_k = \{B \in \mathcal{B}^{\infty};\ \vartheta^{-k}B = B\}$ est la tribu invariante de ϑ^k. Si $k = 1$ c'est le résultat annoncé. Sinon on doit démontrer que $\mathcal{J}_k = \mathcal{J}$ mod P_e. L'inclusion $\mathcal{J} \subset \mathcal{J}_k$ est évidente. Pour alléger prenons $k=2$, le cas général étant similaire. Alors μ^{2i} et μ^{2j+1} sont étrangères pour tout i et j. Comme $\vartheta^{2i} P_e = \mu^{2i} * P_e$ est portée par tout ensemble de la forme $[S_1 \in C]$ où C est un borélien de G portant μ^{2i+1}, les mesures $\vartheta^{2i}P_e$ et $\vartheta^{2j+1} P_e$ sont étrangères pour tout i et j. Posons

$\xi = \sum_{i=0}^{\infty} \vartheta^{2i} P_e/2^{i+1}$ et $\rho = \vartheta\xi$; ξ et ρ sont étrangères sur Ω . Soient

S et T deux ensembles de $\mathcal{B}^\infty$, disjoints, tels que $\xi(S) = \rho(T) = 1$. Les ensembles $S_1 = S\cap \bar{\vartheta}^1 T$ et $S_2 = \bar{\vartheta}^1 S_1$ sont disjoints ; le premier porte ξ et le second ρ. Soit $F\in \mathcal{J}_2$. Posons $\bar{F} = (F\cap S_1)\cup(\bar{\vartheta}^1 F\cap S_2)$; alors $\bar{\vartheta}^1\bar{F} = (\bar{\vartheta}^1 F\cap S_2)\cup(F\cap\bar{\vartheta}^1 S_2)$.

Comme $\vartheta^2\xi << \xi$ on a $\xi(\bar{\vartheta}^2 S_1) = \xi(\bar{\vartheta}^1 S_2) = 1$ et $F\cap S_1 = F\cap\bar{\vartheta}^1 S_2$ mod ξ d'où $\bar{F} = \bar{\vartheta}^1 F$ mod $(\frac{\xi+\rho}{2})$. Comme $\vartheta\ (\frac{\xi+\rho}{2}) << \frac{\xi+\rho}{2}$, il existe $F'\in \mathcal{J}$ tel que $\bar{F} = F'$ mod $(\frac{\xi+\rho}{2})$ ([8] p. 114). On arrive ainsi à $F = \bar{F} = F'$ mod ξ donc aussi mod P_e.

Considérons le 2^d cas. Pour tout i et $j, i \neq j, \|\mu^i - \mu^j\| = 2$ donc $\vartheta^i P_e$ et $\vartheta^j P_e$ sont étrangères. Tout d'abord il est facile de construire une suite d'ensembles mesurables S_n, disjoints deux à deux, tels que $\vartheta^n P_e(S_n) = 1$. Posons $T_n^k = S_n \cap \bar{\vartheta}^1 S_{n+1} \cap \ldots \cap \bar{\vartheta}^k S_{n+k}$. Pout tout n et k, $\vartheta^n P_e(T_n^k) = 1$, donc $\vartheta^n P_e(T_n) = 1$ pour $T_n = \lim_k T_k^n$. On a aussi $T_n = S_n \cap \bar{\vartheta}^1 T_{n+1}$. Soit $F \in \mathcal{A}$. Posons $F_n = \vartheta^n F\cap T_n$ et $\bar{F} = \bigcup_{n\geq 0} F_n$; ces ensembles sont $\mathcal{B}^\infty$-mesurables, d'après la proposition. Par surjectivité de ϑ on a $\vartheta T_n \subset T_{n+1}$, donc $\vartheta\bar{F} \subset \bar{F}$ et $\bar{F} \subset \bar{\vartheta}^1\vartheta\bar{F} \subset \bar{\vartheta}^1\bar{F}$. La suite $\bar{\vartheta}^n\bar{F}$ est donc croissante avec pour limite $F'\in \mathcal{J}$. Comme ϑ est un automorphisme de $\mathcal{A}$ on a aussi, $F_n^c \cap T_n = \vartheta^n(F^c)\cap T_n$ d'où $\bar{\vartheta}^n\bar{F}\cap T_o = F\cap T_o$ et à la limite $F'\cap T_o = F\cap T_o$. Ceci prouve le résultat cherché : $F = F'$ P_e p.s.

BIBLIOGRAPHIE

[1] **AVEZ A.** (1972) - Entropie des groupes de type fini C.R. Acad. Sc. Paris 275 A 1363-1366

[2] **AVEZ A.** (1974) - Théorème de Choquet-Deny pour les groupes à croissance non exponentielle. C.R. Acad. Sc. Paris 279 A, 25-28

[3] **AVEZ A.** (1976) - Croissance des groupes de type fini et fonctions harmoniques. L.N. in Math. Springer n° 532, 35-49.

[4] **AVEZ A.** (1976) - Harmonic functions on groups. Diff. Geom. and Relativity 27-32, Reidel (Holland)

[5] **AZENCOTT R.** (1970) - Espaces de Poisson des groupes localement compacts. L.N. in Math. Springer n°148

[6] **BROWN L.** (1982) - A proof of the central limit theorem motivated by the Carmer-Rao inequality. Statistics and Proba : essays in honor of C.R. Rao. North. Holland 141-148.

[7] **CSISZAR I.** (1964) - A note on limiting distributions on topological groups. Publ. Math. Inst. Hung. Acad. Sc. Vol 9, 595-598.

[8] **DERRIENNIC Y.** (1976) - Lois "zéro ou deux" pour les processus de Markov. Ann. Inst. H. Poincaré, Sect. B, 12, 111-129.

[9] **DERRIENNIC Y.** (1980) - Quelques applications du théorème ergodique sous-additif. Astérisque 74, 183-201

[10] **DERRIENNIC Y.** (1985) - Entropie et frontière d'une marche aléatoire : le cas général. Probabilités sur les structures géométriques. Publications de l'Université Paul Sabatier, Toulouse.

[11] **DOBRUSHIN R.L.** (1959) - General formulation of Shannon's basic theorems of the theory of information. Usp. Math. Nauk, 14, n5 6, 3-104. Traduction allemande : VEB Deutscher Verlag der Wissenschaften Berlin 1963.

[12] **DYNKIN E.B. et MALJUTOV M.B.** (1961) - Random walks on groups with a finite number of generators. Soviet Math. Dokl. 2, 399-402.

[13] **ELIE L.** (1978) - Fonctions harmoniques positives sur le groupe affine. L.N. in Math. Springer n° 706 96-110.

[14] **FELLER W.** (1966) - An introduction to probability theory Vol II (Wiley)

[15] **FOURT G.** (1972) - Existence de mesures à puissances singulières à toutes leurs translatées. C.R. Acad. Sc. Paris, 274 A, 648-650

[16] **FÜRSTENBERG H.** (1963) - Non communicating random products. T.A.M.S. 108, 377-428.

[17] **FÜRSTENBERG H.** (1971) - Random walks and discrete subgroups of Lie groups. Adv. Proba. Vol 1, 3-63 (Dekker).

[18] **GREENLEAF F.P.** (1969) - Invariant means on topological groups. Van Nostrand.

[19] **GRIGORCHUK R.** (1983) - On Milnor's problem of group growth. Soviet Math. Dokl. 28, n° 1, 23-26.

[20] **GROMOV M.** (1981) - Groups of polynomial growth and expanding maps. Publ. Math. IHES, 53, 53-78.

[21] **GUIVARC'H Y.** (1973) - Croissance polynomiale et période des fonctions harmoniques. Bull. SMF,101, 333-379.

[22] **GUIVARC'H Y.** (1980) - Sur la loi des grands nombres. Astérisque, 74, 47-98.

[23] **GUIVARC'H Y.** (1980) - Quelques propriétés asymptotiques des produits de matrices aléatoires. L.N. in Math, Springer, n° 774, 177-250.

[24] **KAIMANOVICH V.A. et VERSHIK A.M.** (1979) - Random walks on groups : boundary, entropy, uniform distribution. Soviet Math. Dokl. 20, 1170-1173.

[25] **KAIMANOVICH V.A. et VERSHIK A.M.** (1983) - Random walks on discrete groups : boundary and entropy. The annals of Proba. Vol 11, n° 3, 457-490.

[26] **KAIMANOVICH V.A.** (1982) - The differential entropy of the boundary of a random walk on a group. Russian Math. Surveys

[27] **KAIMANOVICH V.A.** (1985) - An entropy criterion for maximality of the boundary of random walks on discrete groups. Soviet Math. Dokl. Vol 31, n° 1, 193-197.

[28] **LEDRAPPIER F.** (1983) - Une relation entre entropie, dimension et exposant pour certaines marches aléatoires. C.R. Acad. Sc. Paris 296 A, 369-372.

[29] **LEDRAPPIER F.** (1985) - Poisson boundaries of discrete groups of matrices. Israel J. of Math. Vol 50, n° 4, 319-336.

[30] **LINNIK Yu. V.** (1959) - An information-theoretic proof of the central limit theorem with the Lindeberg condition. Theory of Probability and its applications. Vol IV, n° 3, 288-299.

[31] **PINSKER M.S.** (1964) - Information and information stability of random variables and processes. Holden-Day.

[32] **RENYI A.** (1967) - Calcul des probabilités, DUNOD.

[33] **ROSENBLATT J.** (1981) - Ergodic and mixing random walks on locally compact groups. Math. Ann. 257, 31-42.

[34] **SHANNON C.E.** (1949) - The mathematical theory of communication. The University of Illinois Press.

[35] **VAROPOULOS N.T.** (1986) - Théorie du potentiel sur des groupes et des variétés. C.R. Acad. Sc. Paris t. 302, série I, n° 6, 203-205. (et les références du même auteur qui y sont indiquées).

[36] **BARRON A.R.** (1986) - Entropy and the central limit theorem. The Annals of Probability Vol 14, n° 1, 336-342.

Yves DERRIENNIC
Faculté des Sciences
29287 BREST CEDEX
FRANCE

RANDOM WALKS ON GRAPHS

by Peter GERL (Salzburg)

In this survey we will discuss some recent ideas and developments in the theory of Markov chains. We only consider discrete parameter Markov chains with a countable number of states and we will find it convenient to represent the state space with the possible transitions on a graph (the transition probabilities are stationary); this gives a very intuitive picture of what is going on and allows us to use the language of graph theory. We will try to explain the underlying ideas and not always give the most general results with all the technical details.

0. INTRODUCTION ([3], [14])

In this paper we consider graphs $\Gamma = (V,E)$ with vertex set = V, edge set = E with the following properties:

(i) V is (at most) countable

(ii) Γ is connected

(iii) Γ is locally finite, i.e. $d_x < \infty$ for all vertices x where

d_x = degree of the vertex x = # (edges through x).

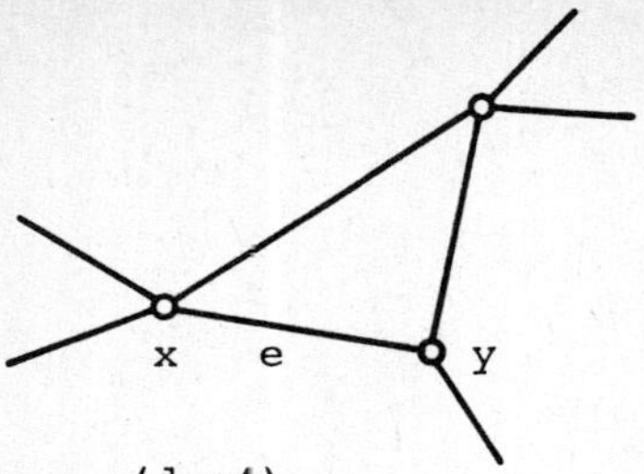

$(d_x=4)$

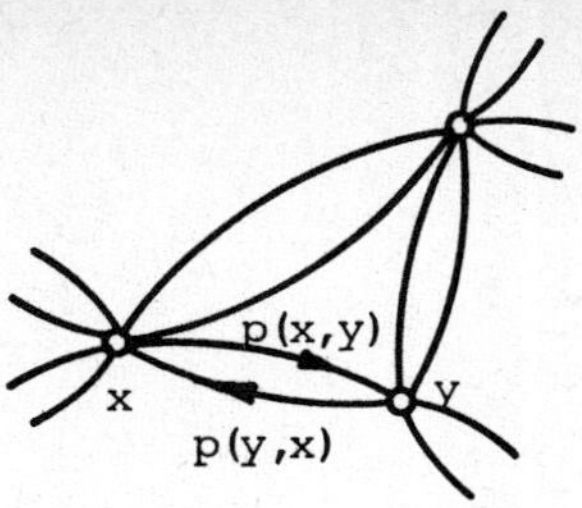

$x \sim y$ means that the vertices x,y are connected by an edge (are neighbours)

We now replace every edge of Γ by two oppositly oriented edges xy, yx (this set is again denoted by E) and define (one step) <u>transition probabilities</u> by

$$p: \quad E \longrightarrow \mathbb{R}$$
$$xy \longmapsto p(x,y)$$

with $p(x,y) \geqq 0$ and $\sum_y p(x,y) = 1$. We will always asssume that p is <u>irreducible</u>, i.e. that

for all $x,y \in V$ there is a path $x = x_0, x_1, \ldots, x_n = y$ with

$$p(x_{i-1}, x_i) > 0 \qquad \text{for } i = 1,\ldots,n.$$

The pair (Γ,p) is then called a <u>random walk on Γ.</u>

An important special case is the so called <u>simple random walk</u> (Γ,p), where

$$p(x,y) = \begin{cases} \frac{1}{d_x} & \text{if } xy \in E \\ 0 & \text{otherwise.} \end{cases}$$

This is well defined because of (iii) and irreducible by property (ii).

In general the transition probability p of a random walk (Γ,p) is not symmetric ($p(x,y)$ and $p(y,x)$ are not always equal). But often it is useful (or even necessary) to consider transition probabilities which are close to symmetric ones. To make this precise we say that p is reversible if there is a function

$$\lambda : V \longrightarrow \mathbb{R}$$

such that (a) $\lambda_x > 0$ for all $x \in V$

(b) $\lambda_x p(x,y) = \lambda_y p(y,x)$ for all $x,y \in V$.

It is then clear that

- the simple random walk is reversible ($\lambda_x = d_x$)
- a symmetric random walk ($p(x,y) = p(y,x)$) is reversible ($\lambda_x \equiv 1$)
- every random walk on a tree is reversible.

We are interested to find relations between

geometric properties of Γ and probabilistic properties of random walks on Γ.

We introduce some more notation:

$$p^n(x,y) = \text{Prob}\left(x \xrightarrow[\text{(in n steps)}]{\text{along n edges}} y \right) \ (n \geq 1), \quad p^0(x,y) = \delta(x,y)$$

$$f^n(x,y) = \text{Prob}\left(x \xrightarrow[\text{for the first time}]{\text{along n edges}} y \right) \ (n \geq 1), \quad f^0(x,y) = 0.$$

Then

$$F(x,y) = \sum_{n\geq 1} f^n(x,y) = \text{Prob}\ (x \xrightarrow[\text{once}]{\text{at least}} y)$$

and

$$G(x,y) = \sum_{n\geq 0} p^n(x,y) = \delta(x,y) + \frac{F(x,y)}{1-F(y,y)} = \text{Green function.}$$

The <u>random walk</u> (Γ,p) is called

<u>recurrent</u> if $F(x,y) = 1$ (or $G(x,y) = \infty$)

<u>transient</u> if $F(x,y) < 1$ (or $G(x,y) < \infty$)

(this property is independent of the vertices x,y).

The <u>graph</u> Γ is called

<u>recurrent</u> / <u>transient</u> if the simple random walk on Γ is recurrent / transient

The first main problems are:

<u>Question 1</u>: Which graphs are recurrent? (= TYPE problem)

<u>Question 2</u>: Which random walks on a (recurrent) graph are recurrent?

A recurrent random walk (Γ,p) is either

<u>positive recurrent</u> $(\limsup_n p^n(x,y) > 0)$ or

<u>null recurrent</u> $(\lim_n p^n(x,y) = 0)$

(this property is independent of the vertices x,y).

The most satisfying characterization is then ([10]):

finite graph = positive recurrent graph (i.e. the simple random walk on it is positive recurrent).

<u>Remark:</u> Random walks on <u>discrete groups</u> can be considered in this setting by looking at the random walk on the associated

Cayley graph; there are two important special features:

1) the underlying graph is very regular (i.e. to arbitrary vertices x,y there is an automorphism sending x into y)

2) the transition probability is (automorphism) group invariant.

1. FLOWS ON A GRAPH ([5], [11], [13])

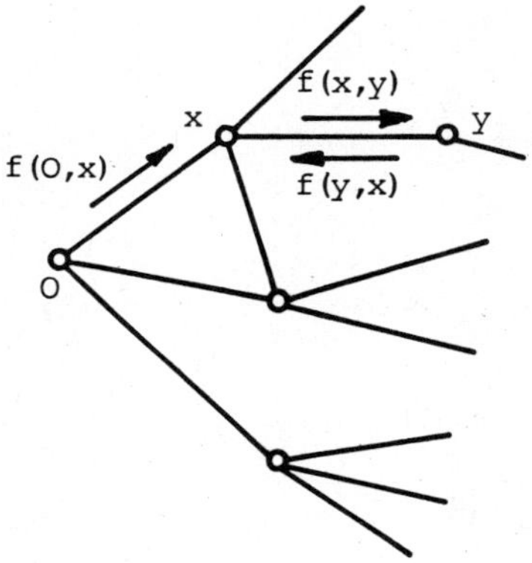

Let $\Gamma = (V,E)$ be an infinite graph with the properties (i),(ii),(iii) as in the introduction and fix a reference vertex O (= source).

A flow (from O to ∞) is a function

$$f: E \longrightarrow \mathbb{R}$$

with the property

(i) $f(x,y) = -f(y,x)$ for all edges xy

(ii) $\sum_{y(\sim x)} f(x,y) = 0$ for all vertices $x \neq 0$

($\sum_{y(\sim x)}$ denotes summation over all edges through x).
Property (ii) means that nothing is lost. For convenience we let $f(x,y) = 0$ if xy is not an edge.

We call

$f_O = \sum_{y(\sim O)} f(O,y) =$ value of the flow f and we say that f is a unit flow if $f_O = 1$.

It is sometimes useful to consider the following physical interpretation ([11]): Let the graph be represented by a system of pipes,

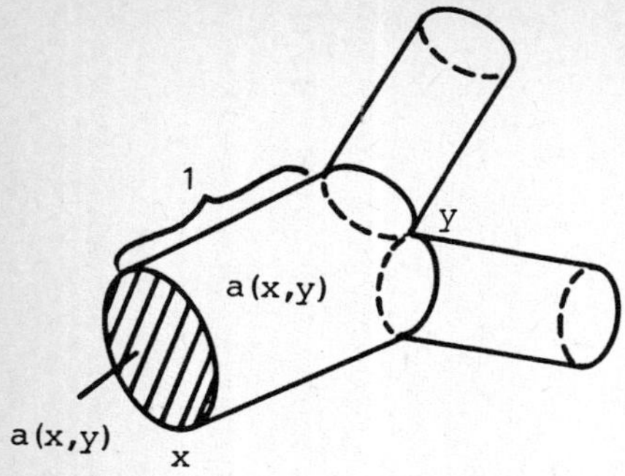

i.e. every edge xy is a pipe of length 1 with cross-sectional area

$$a(x,y) = a(y,x)$$

(these a(x,y) determine a reversible random walk, see below). The pipes are full with a liquid and form a closed system except at the vertex O: there the liquid enters at a constant rate. Then

f(x,y) can be considered as the volume rate at which the fluid flows from x to y ($xy \in E$)

a(x,y) as the mass of the fluid in the pipe xy.

Therefore

$$\frac{f(x,y)}{a(x,y)} = \text{velocity of the fluid in the pipe xy}$$

and so

$$a(x,y) \cdot \left(\frac{f(x,y)}{a(x,y)}\right)^2 = \frac{f^2(x,y)}{a(x,y)} = \text{kinetic energy.}$$

With this in mind we define the

(total kinetic) <u>energy of the flow f</u> $= E(f) = \sum_{xy\in E} \frac{f^2(x,y)}{a(x,y)} =$

$$= \frac{1}{2} \sum_{x\sim y} \frac{f^2(x,y)}{a(x,y)} .$$

Now a positive symmetric function

$$a: \; E \longrightarrow \mathbb{R}$$

($a(x,y) \geq 0$ and $a(x,y) = a(y,x)$ for all $xy \in E$) with

$$a_x = \sum_{y(\sim x)} a(x,y) > 0 \quad \text{for all vertices } x$$

determines a reversible random walk on Γ through

$$p(x,y) = \frac{a(x,y)}{a_x} \quad (\geq 0),$$

because then $\sum_y p(x,y) = 1$ and $a(x,y) = a_x \; p(x,y) = a_y\, p(y,x) =$ $= a(y,x)$ (and the correspondence $a \leftrightarrow p$ can be made one-to-one). The particular function $a(x,y) \equiv 1$ for all edges xy defines the simple random walk on Γ $(a_x = d_x)$.

Furthermore every reversible random walk (Γ,p) defines a flow i [e], the so called harmonic [unit] flow on Γ through the following construction:

Since p is reversible we get the symmetric function a by $a(x,y) = \lambda_x p(x,y) = \lambda_y p(y,x) = a(y,x)$ for all edges xy.

Let

$$h_x = \text{Prob}\,\left(x \xrightarrow[\text{for the first time}]{\text{along} \geq 0 \text{ edges}} 0\right) = \begin{cases} 1 & \text{for } x = 0 \\ F(x,0) & \text{for } x \neq 0 . \end{cases}$$

Then h is a harmonic function for $x \neq 0$ (i.e.

$$h: V \to \mathbb{R}$$

and $\sum_y p(x,y)\, h(y) = h(x)$ for $0 \neq x \in V$).

Therefore

$$i(x,y) = (h_x - h_y)\, a(x,y) = (h_x - h_y)\, \lambda_x\, p(x,y)$$

for all edges xy defines a flow on Γ with value

$$i_0 = a_0(1 - F(0,0) = \lambda_0(1 - F(0,0)) ,$$

the harmonic flow of (Γ,p). If $i_0 > 0$ then

$$e(x,y) = \frac{i(x,y)}{i_0}$$

defines the harmonic unit flow of (Γ,p). Its energy is given by

$$E(e) = \begin{cases} \infty & \text{for } i_0 = 0 \\ \frac{1}{2} \sum_{x\sim y} \lambda_x p(x,y)(h_x - h_y)^2 = \frac{1}{i_0} & \text{for } i_0 > 0 \end{cases}$$

Some important properties and consequences are:

Let (Γ,p) be a reversible random walk and

$$a(x,y) = \lambda_x p(x,y), \qquad a_0 = \sum_y a(0,y) = \lambda_0.$$

Then we have

1) <u>Thomson-principle</u>: For every unit flow f on Γ is

$$E(e) \leqq E(f).$$

2) $E(e) = \frac{1}{\lambda_0} G(0,0).$

Since $G(0,0) < \infty$ is equivalent to the transience of (Γ,p), we obtain

3) (Γ,p) is transient iff there exists a unit flow f on Γ with finite energy $E(f)$ ([5], [11]).

4) If two reversible random walks (Γ,p), $(\Gamma,\overline{p})$ on Γ satisfy

$$\overline{a}(x,y) = \overline{\lambda}_x \overline{p}(x,y) \leqq \lambda_x p(x,y) = a(x,y) \text{ for all edges } xy,$$

then

$$E(e) \leqq E(\overline{e}).$$

This gives for the case of simple random walks:

5) A graph with loops is recurrent iff the same graph without loops is recurrent.

6) Every subgraph of a recurrent graph is recurrent.
Every graph containing a transient one is transient.

It is also interesting to consider the particular case of simple random walks on trees: To every tree Γ with a reference vertex 0 (= root) and without dead ends ($d_x \geqq 2$ for all vertices $x \neq 0$) we associate a sequence of squares. We start

Γ:

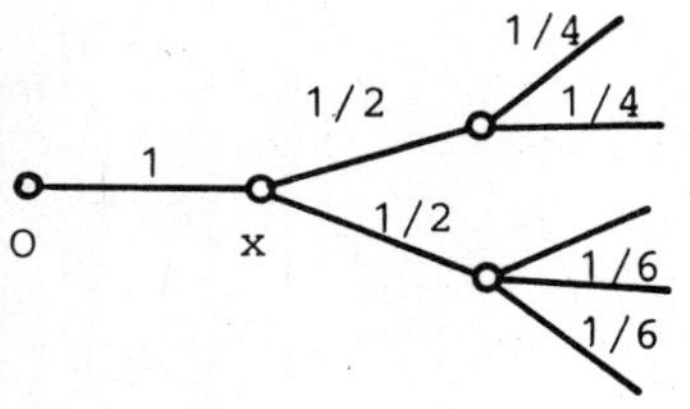

$V(\Gamma)$:

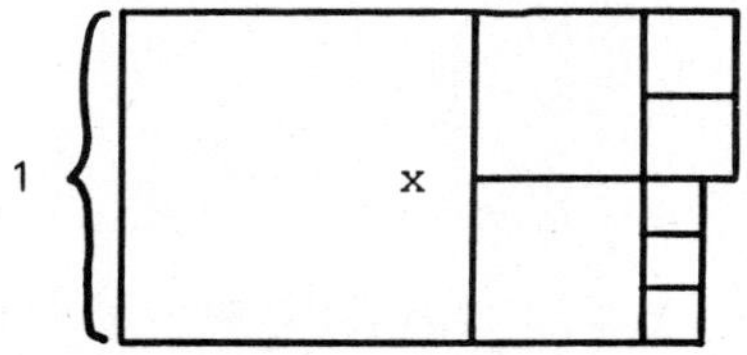

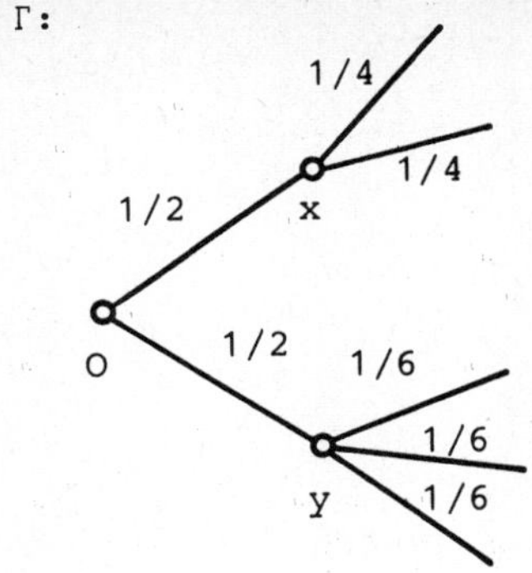

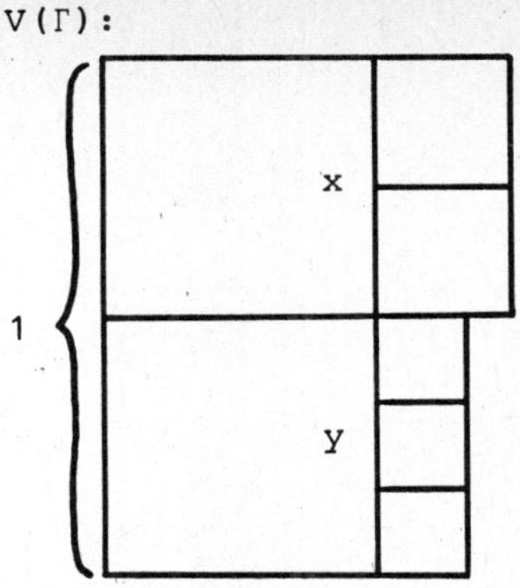

with d_o squares of side length $\frac{1}{d_o}$, then for every $x \sim O$ we subdivide one such square into $d_x - 1$ smaller ones of side length $\frac{1}{d_o}\frac{1}{d_x-1}$ and so on. We define the <u>volume $V(\Gamma)$ of the tree Γ</u> by

$$V(\Gamma) = \Sigma \quad \text{(areas of these squares)}.$$

The volume of a tree can be identified with the energy of a flow s, the <u>simple unit flow on Γ</u>:

$$s(O,x) = \frac{1}{d_o} \quad \text{for } x \sim O$$

$$s(y,z) = \frac{1}{d_y-1} \quad \text{for } O \neq y \sim z \text{ and } y \text{ closer to } O \text{ than } z$$

(as in the figure). Then we have

$$E(s) = V(\Gamma).$$

From properties 1) and 2) above we infer that

$$E(e) = \frac{1}{d_o} G(O,O) \leq E(s) = V(\Gamma),$$

so that $V(\Gamma) < \infty$ implies the transience of Γ. I do not know if the converse is also true, this is

<u>Question 3</u>: Does the transience of Γ (= tree without dead ends) imply that $V(\Gamma) < \infty$? ([7])

One could think that for deciding the type of a graph (Question 1) it would be sufficient to study subtrees. But this is not true as there exists a transient graph without transient subtree (P. Doyle).

2. ANOTHER CHARACTERIZATION OF TRANSIENCE ([16])

Let (Γ,p) be a reversible random walk on the graph $\Gamma = (V,E)$ and write

$$a(x,y) = \lambda_x p(x,y) = a(y,x) \quad (\geq 0).$$

Let

$c_o(V)$ = all real valued functions on V with finite support

and introduce the notation

$$(f,g) = \sum_{x\varepsilon V} \lambda_x f(x)\, g(x)$$

$$(f,g)_D = \frac{1}{2} \sum_{x\sim y} a(x,y)(f(x)-f(y))(g(x)-g(y))$$

$$||f||_D^2 = (f,f)_D$$

(this makes certainly sense for f or g in $c_o(V)$).

Then we have the

Theorem: The following properties are equivalent:

(i) (Γ,p) is transient

(ii) to every $x \varepsilon V$ there is a constant $C_x > 0$ such that

$$|f(x)| \leq C_x\, ||f||_D \qquad \text{for all } f \varepsilon c_o(V)$$

(iii) there exist $x_o \varepsilon V$ and $C > 0$ such that

$$|f(x_o)| \leq C\, ||f||_D \qquad \text{for all } f \varepsilon c_o(V).$$

Proof: If we write as usual

$$Pf(x) = \sum_{y\varepsilon V} p(x,y)\, f(y)$$

then the following relation holds:

$$(f,g)_D = (f, (I-P)g).$$

<u>(i) ⟹ (ii)</u> (as in [17], App.): If (Γ,p) is transient then $G(x,y) = \sum_{n\geq 0} p^n(x,y) < \infty$ for all $x,y \in V$. Furthermore

$$(I-P)\,G(y,x) = \delta_x(y)$$

and

$$\|G(.,x)\|_D^2 = (G(.,x), (I-P)\,G(.,x)) = (G(.,x), \delta_x) =$$
$$= \lambda_x G(x,x) < \infty .$$

This implies for $f \in c_o(V)$ and a fixed $x \in V$:

$$\lambda_x |f(x)| = |(f,\delta_x)| = |(f, (I-P)\,G(.,x))| =$$
$$= |(f,G(.,x))_D| \leq \|G(.,x)\|_D \; \|f\|_D = (\lambda_x G(x,x))^{1/2} \|f\|_D$$

and this is (ii) with $C_x = \lambda_x^{-1/2} G^{1/2}(x,x) > 0$.

<u>(iii) ⟹ (i)</u>: For any integer N we write

$$G_N(x,y) = \sum_{n=0}^{N} p^n(x,y).$$

Then for a fixed x_o we have $G_N(.,x_o) \in c_o(V)$.

Since

$$(I-P)\,G_N(x,x_o) = \delta_{x_o}(x) - p^{N+1}(x,x_o)$$

we therefore get from (iii) for $f(x) = G_N(x,x_o)$:

$$|G_N(x_o,x_o)|^2 \leq C^2 \, \|G_N(.,x_o)\|_D^2 = C^2 (G_N(.,x_o), (I-P)G_N(.,x_o))$$
$$\leq C^2 (G_N(.,x_o), \delta_{x_o}) = C^2 \lambda_{x_o} G_N(x_o,x_o).$$

But this gives

$$G(x_o,x_o) = \lim_{N\to\infty} G_N(x_o,x_o) \leq C^2 \lambda_{x_o} < \infty ,$$

so (i) holds.

<u>Remark</u>: Other proofs can be found in [16] (using Dirichlet-spaces) or [17], Appendix (using Hilbert-spaces).

3. EIGENVALUES OF THE LAPLACE OPERATOR ([4])

In this section we consider only reversible random walks (Γ,p) on a graph Γ with the properties:

$$\lambda_x p(x,y) = \lambda_y p(y,x) \qquad \text{(reversible)}$$

$$0 < \ell \leq \lambda_x p(x,y), \quad \lambda_x \leq L < \infty$$

for all vertices x and edges xy of Γ (e.g. the simple random walk on a graph with uniformely bounded degrees $d_x \leq D$ has these properties: $\ell = 1$, $L = D$).

The Laplace operator Δ is defined by

$$\Delta = I - P.$$

Then Δ is a selfadjoint operator with respect to the inner product $(\, , \,)$ (notation as in section 2) and Δ is positive. Therefore all eigenvalues of Δ are non negative; let

$$\lambda_o = \text{smallest eigenvalue of } \Delta.$$

There is a positive lower bound for λ_o if the graph Γ has a certain geometric property (discrete analogue of Cheeger's inequality), which we now introduce:

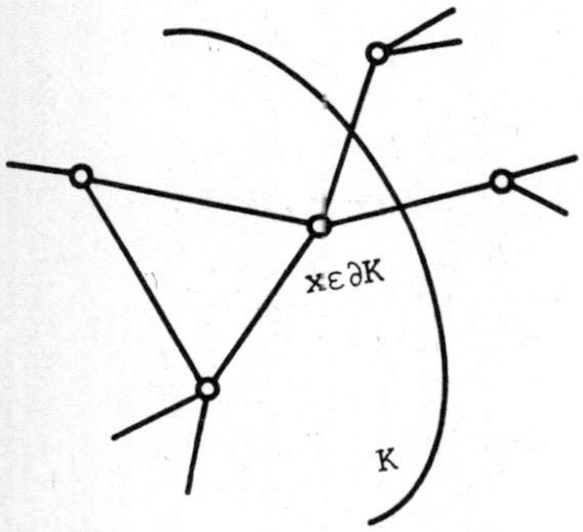

If K is a finite subgraph of Γ, we define ∂K, the boundary of K, to consist of all vertices x of K for which there is an edge (in Γ) joining x to a vertex not in K.

Let $|K|$ $(|\partial K|)$ denote the number of vertices of $K(\partial K)$. Then we have the following

<u>Theorem</u> ([4]): Suppose that (Γ,p) is a reversible random walk on the graph Γ and that

(i) $0 < \ell \leq \lambda_x p(x,y)$, $\lambda_x \leq L < \infty$

for all vertices x and edges xy,

(ii) there is a constant $\alpha > 0$ so that

$$\alpha\,|K| \leq |\partial K| \qquad \text{(IS)}$$

for all finite subgraphs K of Γ.

Then

$$\frac{(f,f)_D}{(f,f)} \geq \frac{\alpha^2 \ell^2}{2\,L^2}$$

for all $0 \neq f \in c_o(V)$.

Since by the Rayleigh principle

$$\lambda_o = \inf_{0 \neq f \in c_o(V)} \frac{(f,\Delta f)}{(f,f)} = \inf_{0 \neq f \in c_o(V)} \frac{(f,f)_D}{(f,f)}$$

we obtain immediately the

<u>Corollary</u>: Under the hypotheses of the theorem we have

$$\lambda_o \geq \frac{\alpha^2 \ell^2}{2\,L^2} .$$

<u>Consequences and remarks</u>:

1) The random walk (Γ,p) as in the theorem is transient ([4]).
2) A planar graph with $7 \leq d_x \leq D$ fulfills (IS) and is therefore transient ([4]). Of course the number 7 cannot be replaced by 6 since the usual graph of $\mathbb{Z}^2$ with diagonals in one direction through all vertices added has $d_x \equiv 6$ and is recurrent.

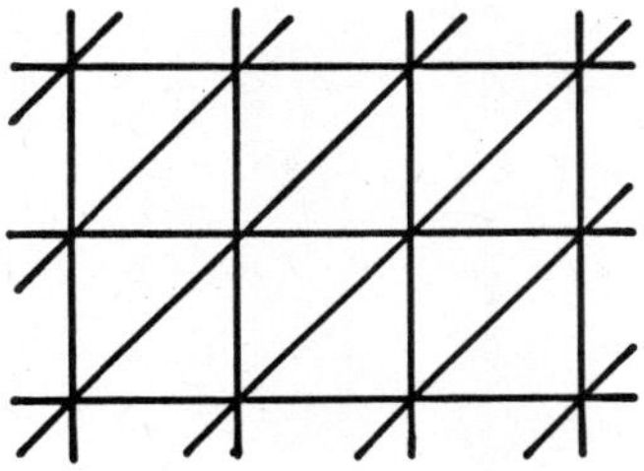

3) (IS) implies that Γ has <u>exponential growth</u>:

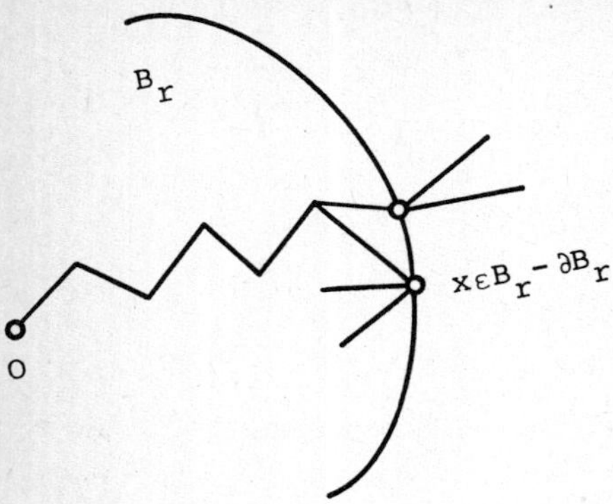

Fix a reference vertex O and let

B_r = subgraph of Γ consisting of all vertices that can be reached from O along $\leq r$ edges.

Then

$$\partial B_r \subseteq B_r \setminus B_{r-1}$$

and therefore

$$|B_r| \geq |B_{r-1}| + |\partial B_r| \overset{(IS)}{\geq} |B_{r-1}| + \alpha |B_r|,$$

which gives

$$(1-\alpha)\, |B_r| \geq |B_{r-1}|$$

and so

$$|B_r| \geq (1-\alpha)^{-r} \qquad (0 < \alpha < 1).$$

4) Since for $\Gamma = (V,E)$

$$\lambda_o = \inf_{0 \neq f \in c_o(V)} \frac{(f,\Delta f)}{(f,f)} \leq \frac{(f,\Delta f)}{(f,f)} = \frac{(f,f)_D}{(f,f)}$$

for all $0 \neq f \in c_o(V)$ the corollary (condition (IS)) implies:

There is a constant $c > 0$ such that

$$\sum_x \lambda_x f^2(x) \leq C \sum_{x \sim y} \lambda_x p(x,y)(f(x) - f(y))^2$$

for all $f \in c_o(V)$. This is discrete WIRTINGER (or SOBOLEV) inequality. The converse is also true: If we take in this inequality $f = 1_K$, the characteristic function of a finite subgraph K of Γ, then under the hypothesis (i) of the theorem we get

$$\ell |K| \leq \sum_x \lambda_x 1_K^2(x) \leq C \sum_{x \sim y} \lambda_x p(x,y)(1_K(x) - 1_K(y))^2 \leq C\, L\, |\partial K|.$$

So we obtain (under condition (i) of the theorem):

(IS) <=> WIRTINGER inequality.

5) For (Γ,p) an arbitrary random walk on a graph Γ we call

$$\sigma_p = \limsup_{n\to\infty} (p^n(x,y))^{1/n} \leqq 1$$

<u>the spectral radius of p</u>; it is independent of x,y. Since one has generally

$$p^n(x,y) = O(\sigma_p^n) \qquad (n \to \infty),$$

it is important to know conditions when $\sigma_p < 1$.

Now if the smallest eigenvalue λ_o of the Laplacian Δ is positive, $\lambda_o > 0$ (this holds by the corollary under condition (IS)) then there is a function $f \neq 0$ such that

$$\Delta f = \lambda_o f .$$

This implies

$$Pf = (1-\lambda_o)f$$

and therefore it follows (see [12]) that

$$\sigma_p \leqq 1 - \lambda_o < 1.$$

Unfortunately this estimate is in general rather poor but it is the only method I know to get nontrivial upper bounds for σ_p. Of course, $\sigma_p < 1$ implies the transience of (Γ,p).

It can be shown that the converse also holds ([9]) and so we obtain:

$$\text{(IS)} \iff \sigma_p < 1$$

under condition (i) of the theorem for (Γ,p).

6) (IS) is called a <u>(strong) isoperimetric inequality</u>. It would be interesting and important to have good characterizations of the class of graphs with property (IS).

Some examples are: (IS) holds if

- Γ is planar and $7 \leqq d_x \leqq d$ ([4])
- Γ is a tree (without dead ends), where

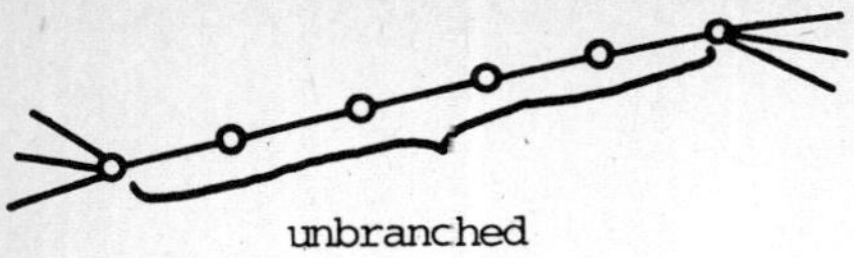

the length of the unbranched parts is uniformly bounded

- Γ is a homogeneous tree T_d ($d_x \equiv d \geq 3$); in this case (IS) holds with

$$\alpha = \frac{d-2}{d-1} \qquad ([2]).$$

4. ISOPERIMETRIC INEQUALITIES ([16])

We saw in section 3 (remark 5) that on a graph Γ where the strong isoperimetric inequality (IS) holds

$$p^n(x,y) \to 0 \quad \text{for} \quad n \to \infty$$

exponentially fast for certain random walks (Γ,p). Since for many irreducible random walks (Γ,p) on an infinite graph one has (see section 0)

$$p^n(x,y) \to 0 \quad \text{for} \quad n \to \infty$$

it seems interesting to consider other (weaker) isoperimetric inequalities for Γ and to try to relate these with the speed of convergence of $p^n(x,y) \to 0$; this is indeed possible but not easy.

We consider the following <u>isoperimetric inequalities</u> for a graph Γ:

There is a constant $C > 0$ such that

(IS_n) $$|K|^{n-1} \leq C\,|\partial K|^n$$

for all finite subgraphs K of Γ.

Then one has

$$(IS) \Rightarrow \ldots \Rightarrow (IS_n) \Rightarrow (IS_{n-1}) \Rightarrow \ldots \Rightarrow (IS_1)$$

and

$$(IS_1) \text{ holds} \iff \Gamma \text{ is infinite.}$$

Consider also the following conditions for a reversible random walk (Γ,p) (V = vertex set of Γ) and $n \in \mathbb{N}$:

There is a constant $C > 0$ such that

$$(S_n) \qquad \|f\|_{\frac{n}{n-1}} = \Big(\sum_{x\in V} \lambda_x |f(x)|^{\frac{n}{n-1}}\Big)^{\frac{n-1}{n}} \leq C \sum_{x\sim y} \lambda_x p(x,y)\,|f(x)-f(y)|$$

for all $f \in c_o(V)$.

There is a constant $C > 0$ such that

$$(D_n) \qquad \|f\|_{\frac{2n}{n-2}} = \Big(\sum_{x\in V} \lambda_x |f(x)|^{\frac{2n}{n-2}}\Big)^{\frac{n-2}{2n}} \leq C \Big(\sum_{x\sim y} \lambda_x p(x,y)(f(x)-f(y))^2\Big)^{1/2}$$

for all $f \in c_o(V)$.

The following results are shown in [16]:

$(IS_n) \iff (S_n)$ if Γ is the Cayley-graph of a group generated by a (symmetric) finite set A and p is a symmetric probability with support A

$(S_n) \implies (D_n)$ if $n \geq 2$ and p is reversible

$(D_n) \iff \sup_{x,y} \lambda_y^{-1} p^k(x,y) = O(k^{-n/2})$ if $n \geq 2$

$(S_n) \implies \sup_{x,y} \lambda_y^{-1} p^k(x,y) = O(k^{-n/2})$ if $n \geq 1$

Γ infinite $\implies (S_1)$ if $\inf \lambda_x p(x,y) > 0$, where the inf is extended over all edges of Γ.

The last two relations imply the following

Corollary: If Γ is an infinite group generated by the (finite) support of a symmetric probability p then

$$\sup_{x,y} p^k(x,y) = O(k^{-1/2}).$$

(see also [1]).

It is conjectured that the corollary also holds for arbitrary probabilities p (whose support generates Γ). For simple random walks on an infinite, locally finite, connected graph the same result is true but for other (even reversible) random walks on graphs the convergence of $p^k(x,y)$ to 0 (for $k \to \infty$) can be much slower ([6], Ex. 3.2, 3.3) or (Γ,p) can even be positive recurrent.

The above implications and equivalences show that

isoperimetric
(S_n) Sobolev
(D_n) Dirichlet

} inequalities give (in general a good)

information about $p^k(x,y)$ for large k for many random walks (Γ,p). Of course the above estimates are interesting only for random walks (Γ,p) with $\sigma_p = 1$ (because of $p^k(x,y) = O(\sigma_p^k)$); but if $\sigma_p < 1$ one can sometimes derive good results by transforming p with the help of a positive eigenfunction of p to the eigenvalue σ_p into a probability q and by considering (Γ,q) where now $\sigma_q = 1$ (see [8]).

REFERENCES

[1] Ph. BOUGEROL: Fonctions de concentration sur certains groupes localement compacts. Z. Wahrscheinlichkeitstheorie verw. Gebiete 45 (1978), 135-157

[2] R. BROOKS: The spectral geometry of the Appollonian packing. Comm. Pure Appl. Math. 38 (1985), 357-366

[3] K.L. CHUNG: Markov chains. Springer-Verlag

[4] J. DODZIUK: Difference equations, isoperimetric inequalities and transience of certain random walks. Trans. Amer. Math. Soc. 284 (1984), 787-794

[5] R.J. DUFFIN: Distributed and lumped networks. J. of Math. Mech.8 (1959), 793-826

[6] P. GERL: Continued fraction methods for random walks on $\mathbb{N}$ and on trees. Springer Lecture Notes 1064 (1984),131-146

[7] P. GERL: Rekurrente und transiente Bäume. IRMA (Strasbourg) 1984, 80-87

[8] P. GERL: Sobolev inequalities and random walks. Preprint (1986)

[9] P. GERL: Random walks on graphs with a strong isoperimetric property. Preprint (1986).

[10] P. GERL - W. WOESS: Simple random walks on trees. Europ. J. Comb. (in print)

[11] T. LYONS: A simple criterion for transience of a reversible Markov chain. Ann Probability 11 (1983), 393-402

[12] W.E. PRUITT: Eigenvalues of non-negative matrices. Ann. Math. Stat. 35 (1966), 1797-1800

[13] J.L. SNELL - P. DOYLE: Random walks and electric networks. Math. Assoc. America, 1985

[14] F. SPITZER: Principles of random walk. Van Nostrand 1964

[15] T. STEGER: Harmonic analysis for an anisotropic random walk on a homogeneous tree. Ph.D. thesis, Washington University in St. Louis (1985)

[16] N.Th. VAROPOULOS: Isoperimetric inequalities and Markov chains. J. Funct. Anal. 63 (1985), 215-239

[17] N.Th. VAROPOULOS: Brownian motion can see a knot. Math. Proc. Camb. Phil. Soc. 97 (1985), 299-309

Peter Gerl
Institut für Mathematik
Universität Salzburg
Petersbrunnstraße 19
A-5020 Salzburg / Austria

Stable probability measures on groups and on vector spaces

A survey

W. HAZOD

Universität Dortmund

The aim of this paper is to give a survey on recent developments in the theory of stability of probability measures on groups and on vector spaces. Especially we want to point out parallel features in the group and in the vector space case. Therefore the phenomenon of stability is studied from different points of view, parallel in the cases mentioned above.

In the introductory § 1 "motivation of a definition" we start with a definition in a quite general situation: Given a space of probability measures $\mathcal{M}$ endowed with a convolution structure $*$, and given a continuous group of automorphisms $T = (\tau_t : t > 0, \tau_t \tau_s = \tau_{ts}$ for $t,s > 0)$, a probability measure $\mu \in \mathcal{M}$ is called T - stable if $\{\mu_t := \tau_t(\mu),\ t \geq 0\}$ is a convolution semigroup, i.e. $\mu_t * \mu_s = \mu_{t+s}$, $t,s > 0$. This definition is motivated by the classical definition of (strictly) stable measures on the real line due to P. Lévy.

A sequence of examples shows that there are several attempts studying stability in more general situations, which are covered by the definition given above. In the sequel however we will restrict the considerations to two types of examples, measures on topological vector spaces and on locally compact groups. These examples are in some sense complementary, since they involve commutative respectively non-commutative convolution structures.

In order to obtain parallel descriptions we avoid the more general concept of stability in the wide sense: If the linear automorphisms τ_t of the vector space are replaced by affine transformations we obtain a larger class of stable measures. On the other hand it is possible, as pointed out in [22], to introduce a similar concept of stability in the group case, too. However, due to the lack of commutativity in the group case, stability in the wide sense has to be defined in terms of generating functionals instead of measures. In order to avoid this rather technical description and to lay stress on the parallel features, we will only consider strictly stable measures. Although the cited results in the literature are in most cases stated for stability in the wide sense. For a first reading one should pay attention to the fact that the differences between these two concepts of stability disappear if the measures are supposed to be symmetric.

In § 2 we consider the contractible part $C(T) := \{x : \tau_t x \underset{t\to o}{\to} o\}$ of the underlying vector space (resp. group). It is shown that in very general situations we may suppose that T acts contracting. Especially, if the vector space is finite dimensional, C(T) is a closed T - invariant subspace, on which every T - stable measure is concentrated. In the case of a locally compact group the contractible part is a Lie group and hence the problem of finding stable measures is reduced to the finite dimensional vector space case: Indeed, every stable semigroup of measures on the group is representable by a corresponding stable semigroup on the associated trangent space.

This is the motivation to study in § 3 und § 4 the case of finite dimensional vector spaces in more details. The vector space splits into a direct sum of T - invariant subspaces on which the Gaussian resp. the non - Gaussian part are concentrated. Considering the non - Gaussian part, it is possible to represent the Lévy-measure as a mixture of Lévy-measures concentrated on orbits $\{\tau_t x : t > 0\}$, where the mixing measure is concentrated on a suitable cross-section of the orbits of T. It is important for the limit theorems considered in § 6 that it is always possible to find a closed cross-section. The connection between stability on groups and on vector spaces pointed out in § 2 enables us to carry over the results mentioned above to the case of locally compact groups.

In § 4 we give an intrinsic description of stable measures in terms of the decomposability group and the invariance group of (a semigroup of) measures. Especially we describe the impact of the fullness - condition on the structure of the decomposability group.

In § 5 we study (operator-) stable laws on infinite dimensional (separable Banach -) spaces. Since there exist recent monographs and survey articles on stability with respect to homothetical transformations $x \mapsto t^{\alpha}x$ (see e.g. W. Linde [72], A. Weron [110]) we restrict our considerations here to the general situation: $T = (\tau_t)_{t>0}$ is considered as a strongly continuous group of bounded automorphisms (with additional assumptions on the behaviour for $t \to o$ and $t \to \infty$). The problem of desintegration of the Lévy - measure is studied and, especially for spaces of type p, satisfactory results are obtained.

In § 6 we consider the correspondence between the existence of a non-void domain of attraction and stability. Again the three cases are considered separately: finite dimensional vector spaces and separable Banach spaces resp. locally compact groups. The differences between the group and the vector space case are due to the fact that in general in contrast to the vector space case a convolution semigroup (μ_t) is not uniquely determined by μ_1. Satisfactory results are only known in special cases.

In § 7 we briefly sketch recent developments in stability not covered by the preceding paragraphs: Lévy - type representations, 0 - 1 - laws, densities of stable laws on special groups with respect to special automorphism groups, stability of Gaussian measures and central limit theorems, and the correspondence between stability of measures on (Lie-) groups and self - similar processes on the associated tangent spaces.

§ 1 Motivation of a definition

Let $(X_n)_{n \geq 1}$ be a sequence of identically distributed real random variables with common distribution $\nu \in M^1(\mathbb{R})$. Assume that there exist $a_n > 0$, $b_n \in \mathbb{R}$, such that the normalized (and centered) sums $\bar{S}_n := a_n \cdot \sum_1^n (X_i - b_n)$ converge weakly to a probability measure $\mu \in M^1(\mathbb{R})$. Then μ is called stable (in the wide sense).

By θ_c we denote the homothetical transformation $x \mapsto cx$ on $\mathbb{R}$ (for some $c > 0$). Then the distribution of $\bar{S}_n$ is

$$(1) \qquad \bar{S}_n(P) = \theta_{a_n}(\nu * \varepsilon_{-b_n})^n .$$

The limit measure μ is infinitely divisible, hence embeddable into a continuous convolution semigroup $(\mu_t,\ t \geq 0,\ \mu_0 = \varepsilon_0)$ such that $\mu = \mu_1$. And there exist some $\alpha > 0$, and a function $b : \mathbb{R}_+ \to \mathbb{R}$, such that the equation

$$(1') \qquad \theta_{t^\alpha}(\mu) = \mu_t * \varepsilon_{b(t)},\ t \geq 0$$

holds. If we put $\tau_t := \theta_{t^\alpha}$, $t > 0$, we obtain a continuous one-parameter group $T = (\tau_t)_{t > 0}$ of automorphisms of $\mathbb{R}$, such that $\tau_t \tau_s = \tau_{ts}$, $t,s > 0$. And equation (1') now reads

$$(1') \qquad \tau_t(\mu) = \mu_t * \varepsilon_{b(t)},\ t > 0.$$

μ is called strictly stable, if the limit is obtained without centering, i.e. if $b(t) \equiv 0$. (This is the original definition of P. Lévy). Hence in this case

$$(1'') \qquad \tau_t(\mu) = \mu_t,\ t > 0.$$

<u>Remark:</u> The convolution semigroup $(\mu_t)_{t \geq 0}$ is uniquely determined by $\mu_1 = \mu$ (this is true for any infinitely divisible measure) and we obtain

$$(2) \qquad [\theta_{a_n}(\nu * \varepsilon_{-b_n})]^{[nt]} \xrightarrow[n \to \infty]{} \mu_t,\ t \geq 0,$$

whence

(3) $\tau_t(\mu_s) = \mu_{st} * \varepsilon_{b(s,t)}, \quad s,t \geq 0$

(with $b(s,t) = b(st) - \tau_t(b(s))$) follows.

In the strictly stable case this reduces to

(2') $[\theta_{a_n}(\nu)]^{[nt]} \longrightarrow \mu_t$ resp.

(3') $\tau_t(\mu_s) = \mu_{st}, \quad t,s > 0.$

Now replace $\mathbb{R}$ by $\mathbb{C}$ and let in (1) the norming constants be elements of $\mathbb{C}$. We obtain two different possible types of limit laws, if we suppose in addition that the measures are full (i.e. not concentrated on a hyperplane):

(a) $a_n \in \mathbb{R}_+^*$, $b_n \in \mathbb{C} \cong \mathbb{R}^2$. Then the possible limiting measures are the usual stable measures on $\mathbb{R}^2$, characterized by the equation

(3a) $\tau_t(\mu) = \mu_t * \varepsilon_{b(t)}$, where

(4a) $\tau_t(x) = t^\alpha x$ for some $\alpha > 0$.

(b) $a_n \in \mathbb{C}^*$, $b_n \in \mathbb{C}$. In this case the limiting measures are "operator stable measures" in the sense of M. Sharpe [88], characterized by the equation

(3b) $\tau_t(\mu) = \mu_t * \varepsilon_{b(t)}$, where now

(4b) $\tau_t : \mathbb{C} \to \mathbb{C}$ is given by $\tau_t(x) = t^\alpha x, \; t > 0$

$\alpha \in \mathbb{C}$, $\mathrm{Re}\,\alpha > 0$, resp. $\alpha = a + ib$, $a > 0$, $b \in \mathbb{R}$

and $\tau_t(x) = t^{ib} \cdot t^a \cdot x$.

(c) Operator stable laws on $\mathbb{R}^n$ ($n \geq 2$) are defined as limit distributions of normed $\mathbb{R}^n$-valued random variables $A_n \sum_1^n (X_i - b_n)$, where $b_n \in \mathbb{R}^n$, $A_n \in \mathrm{Aut}(\mathbb{R}^n)$. If the measures are supposed to be full they are characterized by an equation of the type

(3c) $\tau_t(\mu) = \mu_t * \varepsilon_{b(t)}, \quad t > 0,$

where $\tau_t : \mathbb{R}^n \to \mathbb{R}^n$ is defined by

(4c) $\tau_t(x) = t^A x$, with $A \in \mathrm{Aut}(\mathbb{R}^n)$ such that

$\mathrm{Spec}(A) \subseteq \{z \in \mathbb{C} : \mathrm{Re}(z) \geq \frac{1}{2}\}$. (We put as usual $t^A := e^{(\log t)A}$).

Even in the case n = 2 there exist operator stable laws which are not of the form given in (a) and (b). (See S.V. Semovskii [86] § 5). (For the definition of operator stable laws see e.g. M. Sharpe [88], for the case n = 2 especially J. Michaliček [77,78]).

In order to simplify the formulas and to be able to point out the parallel development in the group case we will restrict the considerations later on to strictly stable measures. Then in the formulas (3a) - (3c) we have to put $b(t) \equiv 0$.

Now we are ready to give a general definition of stability motivated by the introductory examples above. The definition is general enough to cover a sequence of examples listed up at the end of this § 1. Although we will restrict our considerations in the following § 2 - § 7 only to two types of examples, it should be mentioned, that various results mentioned in § 2 - § 7 for measures on groups and on vector spaces could be obtained in the general framework.

Let Y be a completely regular topological Hausdorff space, let $M^1(Y)$ be the set of regular bounded measures on Y and let $\mathcal{M} \subseteq M^1(Y)$ be a weakly closed set. (The weak topology is the $\sigma(M^b(Y), C^b(Y))$ - topology).
Let $*$ be a convolution structure on $\mathcal{M}$, such that $\mathcal{M}$ endowed with the operation $*$ and with the weak topology becomes a topological semigroup.

Let $T = (\tau_t)_{t > 0}$ be a continuous one-parameter group of automorphisms of $\mathcal{M}$, i.e. $\tau_t : \mathcal{M} \to \mathcal{M}$ is a topological $*$ - automorphism of $\mathcal{M}$ for $t > 0$, $\tau_t \tau_s = \tau_{ts}$ for $t, s > 0$ and $t \mapsto \tau_t(\nu)$ is continuous for $\nu \in \mathcal{M}$.

1.1 <u>Definition</u> A subset $(\mu_t, t > 0) \subseteq \mathcal{M}$ is called a continuous convolution semigroup if

(i) $\mathbb{R}_+^* \ni t \mapsto \mu_t \in \mathcal{M}$ is weakly continuous and

$$\text{(ii)} \quad \mu_t * \mu_s = \mu_{t+s}, \quad t,s > 0.$$

In the sequel we will only consider examples of convolution semigroups for which $\mu_o := \lim_{t \to o} \mu_t$ exists. In this case $\mu_o^2 = \mu_o$ and

$$\mu_t * \mu_s = \mu_{t+s} \quad \text{for} \quad t,s \geq 0.$$

We start with the following

Definition $\mu \in \mathcal{M}$ is called T - stable if there exists a function

$r : \mathbb{R}_+^* \times \mathbb{R}_+^* \to \mathbb{R}_+^*, (t,s) \mapsto r = r(t,s)$, such that $\tau_t(\mu) * \tau_s(\mu) = \tau_r(\mu)$. (See the definition of stability used by K. Schmidt [85] resp. E. Kehrer [59].)

1.2 Lemma Assume that $\lim_{t \to o} \tau_t(\mu) =: \mu_o$ exists.

Then the following assertions are equivalent:

(i) μ is T - stable

(ii) There exists $\beta > o$ such that $(\mu_t := \tau_{t^\beta}(\mu))_{t > o}$ is a continuous convolution semigroup (with $\mu_o := \lim_{t \to o} \tau_t(\mu)$).
In this case we have

$\tau_{t^\beta}(\mu_s) = \mu_{st}$, $t,s \geq 0$. (Cf. (3').)

[The idea of the proof is the following:
Define $f : \mathbb{Q}_+^* \ni p \mapsto f(p) \in \mathbb{R}$ as follows:

For $n \in \mathbb{N}$ put $f(n) \in \mathbb{R}$, such that $\tau_{f(n)}(\mu) = \mu^n$, $f(1/n) = 1/f(n)$.

Now for $k,n \in \mathbb{N}$ $f(k/n) := f(k) / f(n)$. The behaviour of $\tau_t(\mu)$ at $t = o$ implies that $t \mapsto \tau_t(\mu)$ is injective.

Obviously f is welldefined then and $f(p \cdot q) = f(p) \cdot f(q)$ for $p,q \in \mathbb{Q}_+^*$.

Therefore for $p,q \in \mathbb{Q}_+^*$

$$\tau_{f(p)}(\mu) * \tau_{f(q)}(\mu) = \tau_{f(p+q)}(\mu)$$

as is easily shown.

$f : \mathbb{Q}_+ \to \mathbb{R}_+$ being a homomorphism with respect to multiplication, there exists $\beta \in \mathbb{R}$, such that $f(q) = q^\beta$, $q \in \mathbb{Q}_+^*$.

Define $\mu_t := \tau_{t^\beta}(\mu)$, $t \in \mathbb{R}_+^*$, then $\mu_t * \mu_s = \mu_{t+s}$, $t,s > o$ by the continuity of $t \mapsto \mu_t$.]

Replacing τ_{t^β} by τ_t $t > 0$, we are led to the following definition, which is equivalent to the definition above if $\lim_{t \to 0} \tau_t(\mu)$ exists:

1.3 Definition Let $T = (\tau_t)_{t>0}$ be a continuous one-parameter group of automorphisms of $\mathcal{M}$. Let $\mu \in \mathcal{M}$.
μ is called (strictly) T-stable if $(\mu_t := \tau_t(\mu))_{t>0}$ is a continuous convolution semigroup, i.e. if $\mu_t * \mu_s = \mu_{t+s}$, $t,s > 0$ and

(6) $\tau_t(\mu_s) = \mu_{st}$ for $t,s > 0$.

μ is called stable if there exists some group T, such that μ is T-stable. Similar, if (μ_t) is a given convolution semigroup, we call (μ_t) T-stable if (6) holds.

In most of the following examples automorphisms of $\mathcal{M}$ are induced by automorphisms of the state space Y, i.e $<\tau(\mu), f> = <\mu, f \circ \tau>$ for $\mu \in \mathcal{M}$, $f \in C^b(Y)$, where $\tau : Y \to Y$ is a homeomorphism.

Examples.

1. Stable measures on $\mathbb{R}$ and $\mathbb{R}^d$, usual convolution. As already mentioned in the introduction stable and operator stable measures on vector spaces form an important class of examples. The automorphisms under consideration are

a) homothetical transformations

$x \mapsto t^\alpha x$, $t > 0$

b) affine transformations

$x \mapsto t^\alpha x + b(t)$, $t > 0$

(with $b(ts) = t^\alpha b(s) + b(t)$, $t,s > 0$)

c) groups of linear automorphisms

$x \mapsto t^A x$, $t > 0$, for some $A \in \mathrm{Aut}(\mathbb{R}^d)$

d) affine transformations

$x \mapsto t^A x + b(t)$, $t > 0$,

(with $b(ts) = t^A b(s) + b(t)$).

In the sequel we will only consider the cases a) and c). A more general set-up is described in examples 6 - 8.

2. Let Y be a completely regular semigroup and let $T = (\tau_t)$ be a continuous group of (semigroup-) automorphisms. The convolution in $\mathcal{M} = M^1(Y)$ is defined as $< \mu * \nu, f > := \iint f(x \cdot y) d\mu(x) d\nu(y)$, $f \in C^b(Y)$. In this framework U. Grenander [18] discussed the so called extreme value distributions: Put $Y = \mathbb{R}$, $x \circ y := \max(x,y)$ and $\tau_t : x \mapsto t^\alpha x$. The stable laws in the sense of definition 1.3 are the so called maximum - stable laws.

For a recent treatment of max - stability on $\mathbb{R}$, $\mathbb{R}^d$ and $\mathbb{R}^\infty$ respectively see P. Vatan [108]. See also the literature cited there.

3. In a series of papers K. Urbanik ([104, 105, 106, 107]) studied convolution structures on $\mathbb{R}_+$ (i.e. $M^1(\mathbb{R}_+)$ endowed with a convolution $*$), the so - called generalized convolution algebras. Stable laws are studied in [104 - 107]. See also R. Jajte's generalization [40, 41] of this concept and N. Bingham [3].

4. Stable measures on the discrete space $\mathbb{Z}_+$ were studied by F.W. Steutel and K. van Harn [96]. Here we put $Y = \mathbb{Z}_+$, $\mathcal{M} := \{$infinitely divisible measures on $\mathbb{Z}_+\}$, and automorphisms $\tau_t \in \mathrm{Aut}(\mathcal{M})$ are defined via generating functions.

It is known that the stable measures in this case can be characterized in the following way: Let $(\Pi_\lambda = \sum_{k=0}^{\infty} e^{-\lambda} \frac{\lambda^k}{k!} \varepsilon_k)_{\lambda \geq 0}$ be the semigroup of Poisson measures on $\mathbb{Z}_+$. Define via subordination $\phi : M^1(\mathbb{R}_+) \to \mathcal{M}$ $\mu \mapsto \int_{\mathbb{R}_+} \Pi_\lambda \, d\mu(\lambda) =: \phi(\mu)$.

Then the stable measures on $\mathbb{Z}_+$ are just the ϕ - images of stable measures on $\mathbb{R}_+$. See G. Forst [12].

5. Z. Jurek [47] studied stability concepts which do not exactly fit into the framework of definition 1.3. Let $Y = \mathbb{R}$ (in [48] a separable Hilbert space) and define $\tau_t : x \mapsto (1 - \frac{t}{|x|}) \cdot 1_{\{|y| > t\}}(x)$. Then $T = (\tau_t)$

is a class of non-linear continuous non-injective transformations $\mathbb{R} \to \mathbb{R}$. Stability and limit behaviour of "shrunken variables" are studied. This example shows that the definition given in 1.3 is not the most general possible.

6. Let Y be a locally convex vector space, $\mathcal{M} := M^1(Y)$ the set of regular probabilities. Stability defined via the homothetical transformations $x \mapsto t^\alpha x$ has been extensively studied in the last years by many authors. At least for certain types of spaces there are satisfactory results characterizing stability. For a survey on recent results and on the literature on this field see W. Linde [72] resp. A. Weron [110]. We shall consider these examples only as a special case of the more general concept of operator stability:

7. Let $Y = E$ be a separable Banach space and put $\mathcal{M} := M^1(E)$. Let $T = (\tau_t)_{t>0}$ be a group of bounded (linear) automorphisms of E such that $t \mapsto \tau_t$ is continuous in the strong operator topology.
For $E = \mathbb{R}^d$ strong and uniform continuity coincide and we obtain just the operator stable laws on $\mathbb{R}^d$ introduced by M. Sharpe [88]. We will consider this example in § 3 and § 4. For dim $E = \infty$ see § 5.

8. Let $Y = G$ be a locally compact group and $T = (\tau_t)_{t>0} \subseteq \mathrm{Aut}(G)$ be a continuous group of automorphisms of G. We will show in the following § 2 - § 4 that there is a connection between operator stable measures in the sense of 7. and the stable measures on groups.

Due to the non-commutativity of convolution the definition of stable measures in the wide sense on G is more complicated than in the vector space case. In order to exhibit the common structure, we restrict the considerations to the case of strictly stable measures defined in 1.3.

In contrast to the vector space case for a locally compact group the measures $(\mu_t)_{t\geq 0}$ of a continuous convolution semigroup are in general not uniquely determined by μ_1. Therefore it is more precise to speak of stable convolution semigroups (μ_t). (For a general definition of stability on groups see e.g. [22 - 25]).

9. Stability on hypergroups. Probabilities on hypergroups became more and more popular in the past. There are a lot of features similar to

probabilities on locally compact groups. Although there seem to be no treatments of automorphisms and stability on general hypergroups, it is worthwhile to mention that the concept of stability leads in a natural way to hypergroups.

a) Let $(\mu_t)_{t \geq 0}$ be a T - stable continuous convolution semigroup on a locally compact group G. Let K be the compact subgroup, such that $\mu_o = \omega_K$. Then any measure μ_t is left and right K - invariant. Hence $(\mu_t)_{t \geq 0}$ may be regarded as a convolution semigroup on the double-coset space $K \backslash G/K$, which is a hypergroup.

b) We shall see later on that there is a representation $\mu_t = \nu_t \otimes \omega_K$, where $(\nu_t)_{t \geq 0}$ is a continuous convolution semigroup on a (simply connected nilpotent) Lie group N on which K acts as a group of automorphisms, and the measures ν_t are invariant under the action of K. Hence $(\nu_t)_{t \geq 0}$ can be considered as a convolution semigroup on the hypergroup of K - orbits N//K of N.
A recent survey on probabilities on hypergroups is given by H. Heyer [28].

10. A general investigation of stability in the context of dynamical systems and semiflows on affine semigroups (e.g. on spaces of probability measures) covering examples above can be found in [82].

§ 2 The contraction property.

As indicated before we restrict our considerations to the examples 7., 8., i.e. to probabilities on vector spaces and locally compact groups respectively.

A) Let E be a separable Banach space and $T = (\tau_t)_{t>0}$ be a group of automorphisms of E, such that $t \mapsto \tau_t$ is continuous in the strong operator topology. Define the T - contractible part:

2.1 Definition $C(T) := \{x \in E : \tau_t(x) \xrightarrow[t \to 0]{} 0\}$.

2.2 Lemma C(T) is a T - invariant linear subspace of E.

As is easily seen C(T) is in general not closed in E. But it is possible to find a suitable renorming:

2.3 Proposition Let $T = (\tau_t) \subseteq \mathrm{Aut}(E)$ be a continuous group. Then, endowed with the norm $\|x\|_1 := \sup_{0<t\leq 1} \|\tau_t(x)\|$.

$(C(T), \|\cdot\|_1)$ is a Banach space, the identity

id: $(C(T), \|\cdot\|_1) \longrightarrow (C(T), \|\cdot\|)$ is continuous and

$(\tau_t) \restriction_{(C(T), \|\cdot\|_1)}$ acts contracting.

[See Z. Jurek [57]].

2.4 Remark The geometry of the vector space may be changed by renorming. Especially if $(E, \|\cdot\|)$ is a Hilbert space (or a space type p or cotype q), $(C(T), \|\cdot\|_1)$ need not have this property.

The importance of the contractible part C(T) in connection with stability is shown by the following result:

2.5 Theorem Let $(\mu_t)_{t \geq 0}$ be T - stable on E. Then the measures μ_t are concentrated on the contractible part C(T), i.e. $\mu_t(\complement C(T)) = 0$, $t > 0$.

[See E. Siebert [92] for locally compact groups. The proof carries over to the vector space case.
The idea of the proof is the following: Let η be the Lévy-measure of (μ_t). Then for any neighbourhood $U \in \mathfrak{U}(0)$ we have

$\limsup_{t \to 0} \frac{1}{t} \mu_t(\complement U) \leq \eta(\complement U)$, hence there exist $1 > \varepsilon > 0, a > 0$, such that $\mu_t(\complement U) \leq a \cdot t$ for $0 < t \leq \varepsilon$.

Therefore for any fixed t_o, $0 < t_o \leq \varepsilon$, the relation $\mu_t = \tau_t(\mu_1)$ implies

$\int 1_{\complement U} \, d\mu_{t_o^n} = \int 1_{\complement U} \circ \tau_{t_o}^n \, d\mu_1 \leq t_o^n a$, hence $\sum_n \int |1_{\complement U} \circ \tau_{t_o}^n| \, d\mu_1 < \infty$.

This implies $1_{\complement U} \circ \tau_{t_o^n} \to 0 \ \mu_1$ - a.e.

If we consider a sequence of neighbourhoods $U_k \in \mathfrak{U}(o)$, $U_k \downarrow$, $\cap U_k = \{o\}$, we obtain the desired result.]

Thus in some sense we may always assume that $T = (\tau_t)$ acts contracting, i.e. $C(T) = E$. (But observe Remark 2.4). In the following we shall need a further property of $T = (\tau_t)$:

2.6 <u>Definition</u> The orbits of T are joining 0 and infinity if

(1) $\tau_t(x) \xrightarrow[t \to 0]{} 0$ for $x \in E$ and

(2) $\|\tau_t(x)\| \xrightarrow[t \to \infty]{} \infty$ for $x \in E \setminus \{o\}$ hold

((1) is just the contraction condition $E = C(T)$).

In the Ph. D. Thesis of E. Kehrer [59], which we shall follow to some extent, and in Z. Jurek [57] the conditions (1) and (2) are sometimes replaced by the stronger resp. weaker conditions

(1') $\|\tau_t\| \xrightarrow[t \to 0]{} 0$ resp.

(2') $\|\tau_t\| \xrightarrow[t \to \infty]{} \infty$.

From $\|x\| = \|\tau_t \tau_{1/t}(x)\| \leq \|\tau_t\| \cdot \|\tau_{1/t}(x)\|$ for $t \in \mathbb{R}_+^*$ it is evident that the following implications hold:

(1) => (2') and (1') => (2).

But in general the implication (1) => (2) does not hold. To see this consider the following example:

2.7 <u>Example</u> Define $\varphi : \mathbb{R} \to \mathbb{R}$, $\varphi(x) := \min(1, e^{-x})$.

Then $\varphi(x-y) \uparrow 1$, $y \to +\infty$, $\downarrow 0$, $y \to -\infty$. Put $\mu := \varphi \cdot \lambda^1$, $E := L^1(\mathbb{R}, \mu)$ and define $\tau_t f(x) := f(x + \log t)$, $x \in \mathbb{R}$, $t \in \mathbb{R}_+^*$. Then (1) holds, but $\|\tau_t f\|_E \uparrow \int |f| \, d\lambda^1$ for $t \to \infty$.

In the case of finite dimensional vector spaces the situation is simple:

2.8 <u>Proposition</u> Assume $\dim(E) < \infty$. Then $C(T)$ is a closed subspace of E. Furthermore $\|\tau_t(x)\| \xrightarrow[t \to \infty]{} \infty$ for $x \in C(T)$. Hence if $E = C(T)$ (1') and (2) hold.

Moreover, (if necessary we replace E by the complexification $E_{\mathbb{C}}$) put

$\Delta^+ := \{\lambda \in \operatorname{Spec}(\tau_{t_0}) : |\lambda| > 1\}$ for some $t_0 > 1$. Then

$C(T) = \{x : \tau_t(x) \xrightarrow[t \to 0]{} 0\} = \operatorname{Span}\{x : \exists\, n \in \mathbb{N},\ \lambda \in \Delta^+ : (\tau_{t_0} - \lambda \mathrm{id})^n x = 0\}$.

Assume $E = C(T)$. Then we obviously have:

$\|\tau_t\| \xrightarrow[t \to 0]{} 0$ and $\|\tau_t(x)\| \xrightarrow[t \to \infty]{} \infty$, $x \in E \setminus \{0\}$.

B) Let <u>G be a locally compact group</u>, K a compact subgroup. Let $T = (\tau_t)_{t > 0}$ be a continuous group in $\operatorname{Aut}(G)$. Assume K to be T-invariant.

Let $(\mu_t)_{t \geq 0}$ be a continuous convolution semigroup in $M^1(G)$ such that $\mu_0 = \omega_K$. Assume that (μ_t) is T-stable.

As in the vector space case it can be shown that μ_t is concentrated on the contractible part:

2.9 <u>Definition</u> $C_K(T) := \{x \in G : \tau_t(x) K \xrightarrow[t \to 0]{} K\}$ is called the K-contractible part.

$C_K(T)$ is a subgroup of G. If $K = \{e\}$ we write $C_{\{e\}}(T) =: C(T)$.

The structure of $C_K(T)$ is completely known:

2.10 <u>Proposition</u> a) $C_K(T)$ is a closed subgroup of G, isomorphic to the semidirect product of $C(T)$ and K.

b) $C(T)$ is isomorphic to a simply connected nilpotent Lie group, such that T acts contracting on $C(T)$.

[b) see E. Siebert [92], a) [26]].

Therefore one can prove:

2.11 <u>Theorem</u> Let $(\mu_t)_{t \geq 0}$, $\mu_0 = \omega_K$, be a T - stable continuous convolution semigroup on a locally compact group G.
Then $\mu_t(\complement C_K(T)) = 0$ for $t > 0$.

Furthermore, there exists a T - stable continuous convolution semigroup $(\nu_t)_{t \geq 0}$ on C(T), such that $\mu_t = \nu_t \otimes \omega_K$, $t > 0$, and such that ν_t is invariant under the action of K. On the other hand, any K - invariant T - stable semigroup (ν_t) on C(T) generates a T - stable semigroup $(\mu_t, t \geq 0, \mu_0 = \omega_K)$ on G.
[See E. Siebert and the author [26].]

Hence in view of 2.10 the description of stable convolution semigroups on locally compact groups is reduced to the case of simply connected nilpotent Lie groups.

Let G be a simply connected nilpotent Lie group with Lie algebra $\mathfrak{g}$. Then exp : $\mathfrak{g} \to$ G is a topological isomorphism. Let $\tau \in$ Aut(G) and let $d\tau$ be the differential of τ at e. Then for a continuous group $T = (\tau_t)_{t > 0} \subseteq$ Aut(G) the differentials $dT := (d\tau_t)_{t > 0}$ form a continuous group of automorphisms of the Lie algebra $\mathfrak{g}$ (and hence of a finite dimensional vector space).
A continuous convolution semigroup (μ_t) is characterized by its generating functional $A : \mathfrak{D}(G) \ni f \mapsto \langle A,f \rangle := \frac{d^+}{dt} \langle \mu_t, f \rangle |_{t=0}$.
$(\mu_t)_{t \geq 0}$ is T - stable in the sense of Definition 1.3

(3) iff $\tau_t(A) = tA$, $t > 0$.
Since via exp $\mathfrak{g}$ and G are isomorphic, the spaces of test functions $\mathfrak{D}(G)$ and $\mathfrak{D}(\mathfrak{g})$ are isomorphic. Therefore for the generating functional A of a semigroup $(\mu_t)_{t \geq 0}$ on G, there exists a (uniquely defined) generating functional $\mathring{A}$ (of a semigroup $(\mathring{\mu}_t)$ on the vector space $\mathfrak{g}$), such that $\langle A,f \rangle = \langle \mathring{A},\mathring{f} \rangle$ $(f \in \mathfrak{D}(G)$, $\mathring{f} = f \circ \exp \in \mathfrak{D}(\mathfrak{g}))$.
The stability condition (3) above and the definition of $d\tau_t$ imply:

2.12 <u>Theorem</u> A is the generating functional of a T - stable semigroup $(\mu_t)_{t \geq 0}$ on the group G iff $\mathring{A}$ is the generating functional of a dT - stable semigroup $(\mathring{\mu}_t)_{t \geq 0}$ on the vector space $\mathfrak{g}$.

[[21, 22, 23, 24]].

2.13 Remarks a) The correspondence between A and $\overset{\circ}{A}$ does not imply $\exp(\overset{\circ}{\mu}_t) = \mu_t$, $t \geq 0$. The correspondence is only infinitesimal :

$$< \overset{\circ}{A}, \overset{\circ}{f} > = \frac{d^+}{dt} \Big|_{t=0} < \overset{\circ}{\mu}_t, \overset{\circ}{f} > = \frac{d^+}{dt} \Big|_{t=0} < \mu_t, f > = < A, f >.$$

b) If K is a compact group of automorphisms acting on G, the differentials $d\tau(\tau \in K)$ act on $\mathfrak{g}$. The measures μ_t are invariant under the action of K iff the measures $\overset{\circ}{\mu}_t$ are invariant under the action of $d\tau$, $\tau \in K$. Hence the representation of μ_t obtained in 2.11 has a corresponding representation of $\overset{\circ}{\mu}_t$.

c) Automorphisms of the Lie algebra $\mathfrak{g}$ are vector space automorphisms. Therefore the measures $\overset{\circ}{\mu}_t$ in 2.12 are operator stable laws (cf. § 1).

d) Hence the problem of determining T-stable continuous convolution semigroups on a locally compact group G is completely reduced to the vector space case: 1. determine a T-invariant compact subgroup K and 2. a T-invariant closed subgroup N on which T acts contracting (then N is isomorphic to a nilpotent Lie group). 3. Let $dT = (d\tau_t)$. Determine the dT-stable convolution semigroups $(\overset{\circ}{\nu}_t)$ on the finite dimensional vector space $\mathfrak{g}$ (the Lie algebra of N) which are invariant under the action of K. 4. Define ν_t corresponding to $\overset{\circ}{\nu}_t$ according to a). Then $(\mu_t := \nu_t \otimes \omega_K)$ is a T-stable semigroup on G.

Therefore a more detailed study of the stable measures on finite dimensional vector spaces is in order.

§ 3 (Operator-) stable measures on finite dimensional vector spaces.

In [88] M. Sharpe introduced the concept of operator stable measures on a finite dimensional vector space V. A measure $\mu \in M^1(V)$ is called operator - stable, if it is the limit distribution of a normalized sum $A_n \sum_1^n \xi_i$, where $(\xi_i)_{i \geq 1}$ is a sequence of i.i.d. V - valued random variables and $A_n \in Aut(V)$. [Indeed, we suppress a centering term in order to obtain strict stability]. Assume the limit measure μ to be full, i.e. not concentrated on a hyperplane. Then there exists a continuous group $T = (\tau_t = t^A) \subseteq Aut(V)$, such that μ is T - stable in the sense of definition 1.3. (See M. Sharpe [88]).

On the other hand, let us start with a continuous convolution semigroup $(\mu_t)_{t \geq 0} \subseteq M^1(V)$. Assume (μ_t) to be T - stable for some group $T = (\tau_t) \subseteq Aut(V)$. Then μ_1 is stable in the sense of 1.3, hence T - stable in the sense of K. Schmidt [85].
Therefore ([85] Theorem 3.5) there exists a decomposition of V into a direct sum of T - invariant subspaces $V = V_1 \oplus V_2$, up to a shift μ_t is concentrated on V_1 and the restriction $(\mu_t \mid V_1)$ is a full and $T|_{V_1}$ - stable semigroup.

Hence in the secuel we may assume (μ_t) to be full.

The following results are well known:
Let $T = (\tau_t = t^A) \subseteq Aut(V)$ be a continuous group.

a) $C(T)$ is a closed T - invariant subspace (cf. Prop. 2.7).

b) $V = C(T)$ iff $Re\,\lambda > 0$ for $\lambda \in Spec(A)$.

Moreover the following representation holds:

3.1 <u>Theorem</u> Let $(\mu_t)_{t \geq 0}$ be full and T - stable. Then there exists a decomposition $V = V_g \oplus V_p$ into a direct sum of T - invariant subspaces. There exist semigroups $(\mu_t^g)_{t \geq 0}$ on V_g, $(\mu_t^p)_{t \geq 0}$ on V_p which are stable with respect to the restrictions $T|_{V_g}$ and $T|_{V_p}$ respectively, such that up to a shift $\mu_t = \mu_t^g \otimes \mu_t^p$. The measures μ_t^g are symmetric and Gaussian, the measures μ_t^p have no Gaussian component.

V_p corresponds to $\{\lambda \in \mathrm{Spec}(A) : \mathrm{Re}\,\lambda > \frac{1}{2}\}$,

V_g to $\{\lambda \in \mathrm{Spec}(A) : \mathrm{Re}\,\lambda = \frac{1}{2}\}$.

[See M. Sharpe [88], W.N. Hudson and J.D. Mason [33]; see also K. Schmidt [85].]

In order to obtain a decomposition of the non - Gaussian part $(\mu_t^p)_{t \geq 0}$ we need a polar decomposition with respect to T. Decompositions of this form were obtained by different authors (see e.g. M. Sharpe [88], W.N. Hudson and J.D. Mason [32], Z. Jurek [57]). Here we use a decomposition due to E. Siebert [91], which yields a closed cross - section:

3.2 <u>Propostion</u> Assume $T = (\tau_t)$ to act contracting on V. Then there exists a closed cross - section Q. Hence $Q \otimes \mathbb{R}_+^* \ni (x,t) \mapsto \tau_t x \in V \setminus \{0\}$ is a homeomorphism.

[E. Siebert [91], § 2, E. Kehrer [59] I. 43. See also Z. Jurek [57].]

(The importance of the closedness of Q will be clearified in § 6 in connection with limit laws.)

3.3 <u>Corollary</u> Let $T \subseteq \mathrm{Aut}(V)$ be a continuous group. Let $(\mu_t)_{t \geq 0}$ be T-stable and let η be the Lévy-measure of (μ_t) (representing the non - Gaussian part (μ_t^p)). Then there exists a desintegration

$$\eta = \int_Q \eta_x \, d\nu(x)$$

where η_x are Lévy-measures concentrated on the orbits $O_x = \{\tau_t x : t > 0\}$ such that $\tau_t \eta_x = t\eta_x$ and $\nu \in M^1(Q)$.

On the other hand, any Lévy-measure η of this form fulfills $\tau_t \eta = t\eta$. Hence, under a suitable integrability condition, η generates a (strictly) stable semigroup (μ_t) without Gaussian component.

[The integrability condition is necessary to obtain strict stability].

Gaussian T - stable measures are characterized in the following way:

3.4 <u>Theorem</u> Let $(\mu_t)_{t \geq 0}$ be Gaussian. Then (μ_t) is T - stable iff the covariance operator R fulfills the equations

$$\tau_t R \tau_t^* = t R \quad \text{resp.} \quad AR - RA^* = t R \quad (t > 0),$$

(where $T = (\tau_t = t^A)_{t>0}$).

[M. Sharpe [88], K. Schmidt [85] 5.1].

Now let, as in § 2, G be a simply connected nilpotent Lie group with Lie algebra $\mathfrak{g}$. Assume $T \subseteq \mathrm{Aut}(G)$ to act contracting on G. Let $(\mu_t)_{t\geq 0}$ be a T-stable semigroup on G.

Let $dT = (d\tau_t)$ be the group of differentials and $(\mathring{\mu}_t)_{t\geq 0}$ be the corresponding dT-stable semigroup on $\mathfrak{g}$.

Let A resp. $\mathring{A}$ be the corresponding generating functionals on G resp. $\mathfrak{g}$.

3.5 Theorem There exists a decomposition $\mathfrak{g} = \mathfrak{g}_g \oplus \mathfrak{g}_p$ into a direct sum of dT-invariant subspaces, and a decomposition $\mathring{\mu}_t = \mathring{\mu}_t^g \otimes \mathring{\mu}_t^p$ into the Gaussian and non-Gaussian part respectively. Hence $\mathring{A}$, and therefore A, admit decompositions $\mathring{A} = \mathring{\Gamma} + \mathring{P}$, $A = \Gamma + P$, where $\mathring{\Gamma}$ (resp. Γ) are the Gaussian generating functionals of dT-(resp. T-) stable semigroups, and $\mathring{P}$ (resp. P) are the functionals corresponding to the non Gaussian part.

$\mathfrak{g}_p$ is an ideal in $\mathfrak{g}$, hence $\exp(\mathfrak{g}_p) =: G_p$ is a (closed) normal subgroup of G, on which the semigroup generated by the non-Gaussian part μ_t^p is concentrated.

[See [22] § 3].

The decomposition of the Lévy-measure carries over to the group case via the exponential map:

3.6 Theorem There exists a compact cross-section $Q \subseteq G$ with respect to the orbits of T, such that

$$Q \otimes \mathbb{R}_+^* \ni (x,t) \mapsto \tau_t(x) \in G \setminus \{e\} \text{ is a homeomorphism.}$$

Hence the Lévy-measure η of a T-stable semigroup admits a decomposition

$$\eta = \int_Q \eta_x \, d\nu(x),$$

where η_x are T-stable Lévy-measures concentrated on the orbits

$O_x = \{\tau_t(x) : t > 0\}$, $x \in Q$.

More details about the general form of operator stable laws in finite dimensional vector spaces [30 - 34], [50 - 52], [59], [63], [74], [75], [78], [86], [91], [109].

An intrinsic description of stability, which clearifies the structure of the possible groups T is discussed in the following § 4. The results there could be used to obtain more information about suitable cross-sections and desintegrations of the Lévy measure.
The problem of finding an appropriate cross-section and a desintegration of the Lévy-measure is again treated in § 6 in connection with limit laws and domains of attraction.

§ 4 An intrinsic definition of stability.

In this § 4 we follow the ideas of A. Luczak [73] and E. Siebert [91]. See also W.N. Hudson and J.D. Mason [32]. (As before we restrict our considerations to the case of strictly stable measures.) Let V be a finite dimensional vector space.

4.1 Definition Let $(\mu_t)_{t \geq 0}$ be a continuous convolution semigroup. The decomposability group (in the strict sense) of (μ_t) is defined as

$\mathfrak{Z}((\mu_t)) := \{(\tau, c) \in \mathrm{Aut}(V) \otimes \mathbb{R}_+^* \;:\; \tau(\mu_t) = \mu_{ct},\ t > 0\}$.

The invariance group (in the strict sense) of (μ_t) is

$\mathfrak{I}((\mu_t)) := \{\tau \in \mathrm{Aut}(V) : (\tau, 1) \in \mathfrak{Z}((\mu_t))\}$ $(= \{\tau \in \mathrm{Aut}(V) : \tau(\mu_t) = \mu_t,\ t > 0\})$.

Analogously we define for a single measure μ

$$\mathfrak{I}(\mu) := \{\tau \in \mathrm{Aut}(V) : \tau(\mu) = \mu\}.$$

Obviously the following holds:

4.2 Proposition Let $(\mu_t) \subseteq M^1(V)$ be a continuous convolution semigroup resp. $\mu \in M^1(V)$. Then we have

(i) $\mathfrak{Z} = \mathfrak{Z}((\mu_t))$ is a closed subgroup of $\mathrm{Aut}(V) \otimes \mathbb{R}_+^*$.

(ii) $\mathfrak{I} = \mathfrak{I}((\mu_t)) = \mathfrak{I}(\mu_t)$ for any $t > 0$.

(iii) $\mathfrak{I}(\mu)$ is a closed subgroup of $\mathrm{Aut}(V)$.

(iv) $\phi : \mathfrak{Z} \ni (\tau, c) \mapsto c \in \mathbb{R}_+^*$ is a continuous homomorphism with kernel $\ker(\phi) = \mathfrak{I}$.

(v) $(\mu_t)_{t \geq 0}$ is stable (with respect to some continuous group $T \subseteq \mathrm{Aut}(V)$) iff there exists a continuous group $T = (\tau_t) \subseteq \mathrm{Aut}(V)$ such that $(\tau_t, t) \in \mathfrak{Z}((\mu_t))$ for all $t > 0$.

Hence we obtain the following intrinsic characterization of (operator) stability on finite dimensional vector spaces:

4.3 Theorem Let $(\mu_t)_{t \geq 0}$ be a continuous convolution semigroup on a finite dimensional vector space V. Then the assertions are equivalent:

(i) There exists a continuous group $T \subseteq \mathrm{Aut}(V)$ such that $(\mu_t)_{t \geq 0}$

is T - stable

(ii) $\phi(\mathfrak{Z}((\mu_t))) = \mathbb{R}_+^*$

(ϕ is the homomorphism defined in 4.2 (iv).)

In fact, (ii) can be replaced by the weaker condition

(ii') $\phi(\mathfrak{Z}((\mu_t)))$ is dense in $\mathbb{R}_+^*$.

[The equivalence of (i) and (ii) easily follows from the fact that ϕ is a continuous homomorphism from the Lie group $\mathfrak{Z}$ onto the Lie group $\mathbb{R}_+^*$. Hence to the continuous one - parameter group $\mathbb{R}_+^*$ in the image of ϕ there corresponds (at least) a one - parameter group in $\mathfrak{Z}$.
For the equivalence of (ii) and (ii') see [91] § 1 and [74] § 3.
In fact, the fullness condition imposed there, is superfluous: Assume that $\phi(\mathfrak{Z}) =: D$ is dense in $\mathbb{R}_+^*$. Then put
$\Gamma := \{\tau \in \mathrm{Aut}(V) : \exists\, c \in D,$ such that $(\tau, c) \in \mathfrak{Z}((\mu_t)) =: \mathfrak{Z}\}$.
For fixed $0 < c < 1$, $c \in D$, $\tau \in \Gamma$, such that

$$(\tau, c) \in \mathfrak{Z}, \quad \text{we have} \quad \tau(\mu_t) = \mu_{ct}, \quad t > 0.$$

Hence (μ_t) is operator semistable. For operator semistable semigroups there exists a decomposition $V = V_1 \oplus V_2$, such that (μ_t) is concentrated on V_1 and is full there. To see this, repeat the arguments of the proof in the stable case (K. Schmidt [85]) or apply the characterization of fullness given in A. Luczak [73] to the semistable symmetric semigroup $(\mu_t * \tilde{\mu}_t)$. Now (μ_t) is a full semigroup on V_1 and the equivalence of (ii) and (ii') is proved.]

The fullness condition imposed on $(\mu_t)_{t \geq 0}$ has a strong impact on the structure of the decomposability group:

4.4 <u>Lemma</u> A measure $\mu \in M^1(V)$ is full (i.e. not concentrated on a hyperplane) iff $\mathfrak{I}(\mu)$ is compact.

[Well known. E.g. simple consequence of M. Sharpe's compactness lemma].

4.5 <u>Proposition</u> Assume $(\mu_t)_{t \geq 0}$ to be a full T - stable continuous convolution semigroup. Then the decomposability group $\mathfrak{Z}((\mu_t))$ is isomorphic to the direct product $\mathfrak{I}(\mu_1) \otimes \mathbb{R}_+^*$.
With other words, there exists a continuous group
$\bar{T} = (\bar{\tau}_t)_{t > 0} \subseteq \mathrm{Aut}(V)$, such that $\bar{\tau}_t\, \rho = \rho\, \bar{\tau}_t$ for $\rho \in \mathfrak{I}(\mu_1)$ and

$$\mathfrak{Z}((\mu_t)) = \{(\rho\bar{\tau}_t, t) : \rho \in \mathfrak{I}(\mu_1),\ \bar{\tau}_t \in \bar{T},\ t > 0\}.$$

[Consequence of a splitting theorem of K.H. Hofmann, P. Mostert [29], since $\mathfrak{Z}$ is isomorphic to a compact semidirect extension $\mathbb{R}_+^* \circledS \mathfrak{I}$ of the real line ; see [25], 5.12 Remarque.]

Consequently we obtain the following result, due to W.N. Hudson and J.D. Mason [32, 34], see also [30]:

4.6 <u>Theorem</u> Let $(\mu_t)_{t \geq 0}$ be a full convolution semigroup stable with respect to some continuous group $T = (\tau_t = t^A,\ t > 0)$. Let $\mathfrak{B}$ be the set of "exponents", i.e. $\mathfrak{B} := \{B \in \mathrm{Aut}(V) : (\mu_t)_{t \geq 0}$ is $(t^B, t > 0)$ - stable$\}$.

Let further $\mathfrak{k}$ be the Lie algebra of the (compact) Lie group $\mathfrak{I}(\mu_1)$. Then $\mathfrak{B} = A + \mathfrak{k}$. Moreover, we can choose A in such a way that $[A, \mathfrak{k}] = 0$.

[Follows immediately from Prop. 4.5 since for every exponent $B \in \mathfrak{B}$ the group $(t^B, t)_{t > 0}$ is contained in the connected component $\mathfrak{Z}((\mu_t))_0$.]

Most of the results of § 4 carry over to the case of locally compact groups (see [25] § 5):

<u>Let G be a locally compact group,</u> let $(\mu_t)_{t \geq 0}$ be a continuous convolution semigroup, and let $\mu \in M^1(G)$.

4.7 <u>Definition</u> (i) $\mathfrak{I}(\mu) = \{\tau \in \mathrm{Aut}(G) : \tau(\mu) = \mu\}$.

(ii) $\mathfrak{Z}((\mu_t)) := \{(\tau, c) \in \mathrm{Aut}(G) \otimes \mathbb{R}_+^*$ such that $\tau(\mu_t) = \mu_{ct},\ t > 0\}$

(iii) $\mathfrak{I}((\mu_t)) := \{\tau \in \mathrm{Aut}(G) : \tau(\mu_t) = \mu_t,\ t > 0\}$.

Proposition 4.2 (i), (iii) - (v) holds also in this case.

4.8 <u>Theorem</u> Let G be a Lie group. Then the following assertions are equivalent:

(i) (μ_t) is T - stable with respect to some continuous group $T = (\tau_t) \subseteq \mathrm{Aut}(G)$.

(ii) $\phi(\mathfrak{Z}((\mu_t))) = \mathbb{R}_+^*$.

We obtain a version of Proposition 4.5.

4.9 Proposition Assume $(\mu_t)_{t \geq 0}$ be a T - stable continuous convolution group on G. Suppose that $\mathfrak{J}(\mu_1)$ is compact (hence $\mathfrak{J}(\mu_t)$ is compact for any $t > 0$). Then $\mathfrak{Z}((\mu_t))$ splits into a direct product $\mathfrak{Z}((\mu_t)) = \mathfrak{J}((\mu_t)) \otimes \bar{T}$.

[See [25]].

As already mentioned, in [81] K. Schmidt and K.R. Parthasarathy started the investigation of measures on a vector space which are stable in a more general sense (see especially K. Schmidt [85]): Let Γ be a group of automorphisms of V. $\mu \in M^1(V)$ is called Γ - stable if for $\gamma_1, \gamma_2 \in \Gamma$ there exists $\gamma_3 \in \Gamma$, such that $\gamma_1(\mu) * \gamma_2(\mu) = \gamma_3(\mu)$.

[If we put $\Gamma = T = (\tau_t)_{t > 0}$ this is just the definition of T - stability given in § 1; see also Kehrer [59]].

4.10 Proposition Let μ be Γ - stable. Then

(i) μ is embeddable into a continuous convolution semigroup $(\mu_t)_{t \geq 0}$ of Γ - stable measures.

(ii) Let $\bar{\Gamma}$ be the closure of Γ in Aut(V). The μ is $\bar{\Gamma}$-stable.

(ii) There exists a continuous group $T = (\tau_t)_{t > 0} \subseteq \bar{\Gamma}$, hence μ is T - stable

So Γ - stable measures are special types of (operator) stable measures. (See also [55], [80], [79]).

It would be possible to define Γ - stable measures on locally compact groups, but up to now there exists no investigation in this direction.

§ 5 (Operator) stable laws on infinite dimensional vector spaces

As already mentioned in § 1 there are different approaches to stability. Most of them start with a definition representing stable measures as certain limit distributions. We shall treat limit distributions and domains of attraction in the following § 6; therefore in 5.1 - 5.3 we sketch the different attempts using definition 1.3 instead of the original definitions.

Let E be an infinite dimensional locally convex vector space.

5.1 Let $T = (\tau_t := t^\alpha)_{t>0}$ be a group of homothetical transformations. There is a comprehensive literature concerning stable measures with respect to T. Especially we have a good knowledge of the canonical form of the characteristic functions, desintegration of the Lévy-measure and of the Gaussian measures (which are always stable with respect to $T = (t^{1/2})_{t>0}$). Since there exist recent surveys on this field we do not go into details here. See (for Banach spaces) e.g. W. Linde [72], A. Weron [110] and the literature cited there. For more general vector spaces see e.g. E. Dettweiler [6], A. Tortrat [100].

5.2 Now let E be a separable Banach space. K. Urbanik [103] studied for the first time (operator) stable laws in the infinite dimensional case, generalizing M. Sharpe's approach. It turns out that a full measure μ is stable in Urbanik's sense iff there exists a uniformly continuous group $T = (t^B)_{t>0}$ of automorphisms of E such that μ is T - stable. In fact, Urbanik studied self - decomposable measures, the class of stable measures is contained as a special case.

[The uniform continuity of $(\tau_t)_{t>0}$ is due to the definition: μ is strictly stable if there exist $\{A_n\} \subseteq \mathrm{Aut}(E)$ and $\nu \in M^1(E)$, such that $A_k(\nu^k) \to \mu$, and in addition the compactness condition

(C) the semigroup generated by $\{A_{k+1} \cdot A_k^{-1}\}$ is relatively compact with respect to the norm topology

holds.]

See also W. Krakowiak [62] . There the influence of the compactness condition is worked out clearly. For stability in the more general sense of K. Schmidt see e.g. [79]].
The structure of stable measures in this situation is quite similar to the finite - dimensional case.

5.3 Let again be E a separable Banach space. The first examples of stable measures with respect to only strongly continuous groups T were studied by R. Jajte [43]. See also Z. Jurek [52, 54]. The special groups T are of the form $\tau_t = t^{\alpha} V_t$, $t > 0$, where $\alpha > 0$, and $(V_t)_{t>0}$ is a strongly continuous group of isometries. Condition (C) is no longer valid. But we still have

(1') $\|\tau_t\| \xrightarrow[t \to 0]{} 0$ (Therefore especially $C(T) = E$)

and (hence)

(2) $\|\tau_t x\| \xrightarrow[t \to \infty]{} \infty$ for any $x \neq 0$.

(Cf. the discussion in § 2, especially 2.6).

5.4 We sketch the developments in a more general situation following the Ph.D. Thesis of E. Kehrer [59]. The conditions (1') and (2) are slightly weakened, but still strong enough to obtain results similar to the finite dimensional case. We repeat § 2. Definition 2.6:

5.5 Definition Let E be a separable Banach space and let $T = (\tau_t)_{t>0}$ be a strongly continuous group of automorphisms. T joins 0 and ∞ if

(1) $\tau_t(x) \xrightarrow[t \to 0]{} 0$ for $x \in E$ (i.e. if T acts contracting on E)

and

(2) hold. (Cf. Definition 2.6)

Under this hypothesis the existence of a cross-section and the corresponding polar decomposition is guaranteed:

5.6 Lemma Assume that T joins 0 and ∞. Then there exists a Borel-measurable cross-section Q, such that

$\Phi : Q \otimes \mathbb{R}_+^* \ni (x, t) \mapsto \tau_t(x) \in E \setminus \{0\}$ is continuous, bijective and and a Borel-isomorphism.

If in addition (1') holds we have in accordance with the finite dimensional case:

Q is closed iff Φ is a homeomorphism.

[E. Kehrer [59], Z. Jurek [57].]

In a more general situation we have the following result due to

Z. Jurek [54]:
Assume (1). Then (2) need not hold, but we have

(2') $\|\tau_t\| \xrightarrow[t \to \infty]{} \infty$. (Cf. § 2. 2.6, 2.7.)

Then for every Lévy-measure η fulfilling $\tau_t(\eta) = t\eta$, $t > 0$, there exists a measurable σ - compact T - invariant subset X_0 of E, on which η is concentrated, such that for the restriction $T|_{X_0}$ there exists a cross - section Q and a corresponding polar decomposition Φ, which is a Borel-isomorphism.

The existence of cross - sections with respect to T enables us to find a desintegration of the Lévy-measure of a T - stable convolution semigroup.

5.7 Theorem Let T be a continuous group in Aut(E) joining 0 and ∞. Let $(\mu_t)_{t \geq 0}$ be a T - stable continuous convolution semigroup (in the sense of Definition 1.3). Let η be the Lévy - measure and R the covariance operator. Let further Q be a cross - section with respect to T. Then we have:

(i) There exists a finite measure ν on Q such that
$\eta(B) = \int_Q \int_{\mathbb{R}_+^*} 1_B(\tau_t x) \cdot \frac{1}{t^2}\, dt d\nu(x)$ for every Borel set B, hence there exists a desintegration $\eta = \int_Q \eta_x \, d\nu(x)$, where the measures η_x are Borel - measures concentrated on the orbits $O_x = \{\tau_t x : t > 0\}$

(ii) $\tau_t R \tau_t^* = t R$, $t > 0$.

[[59], 1.3, 1.4 and the literature cited there.]

In the finite dimensional vector space case (and in the case of Lie groups) the orbital measures η_x are Lévy - measures (ν - almost every where) (cf. § 3). This is due to the fact that in these cases Lévy - measures are characterized by a simple integrability condition. This characterization does not longer hold for general infinite dimensional vector spaces. But for spaces of type p similar characterizations are available, hence we obtain:

5.8 Proposition Let E be a separable Banach space of type p. Let T be a strongly continuous group of automorphisms fulfilling (1') and therefore (2) (cf. 2.6.). Hence

$$\omega_0^- := \lim_{t\to 0} \log \|\tau_t\| / \log t \quad > \quad 0.$$

Assume $p\omega_0^- > 1$. Let $\eta = \int_Q \eta_x \, d\nu(x)$ be a desintegration of a T - stable Lêvy - measure η, obtained in 5.7. Then for ν - a.e. $x \in Q$ η_x is a Lêvy-measure.

[See [59] 1.5. The proof is based on the fact that the functions $y \mapsto \min(1, \|y\|^p)$ are η and η_x - integrable for ν - a.e. $x \in Q$. Then, since E is of type p, η_x is a Lêvy - measure] .

5.9 Remarks a. Let $(\tau_t = t^\alpha V_t)_{t>0}$ be the group considered by R. Jajte (see 5.3). Then $\|\tau_t\| = t^\alpha$, (1'), (2) hold and $\omega_0^- = \alpha$. Therefore Proposition 5.8 applies if E is of type $p > 1/\alpha$. Especially, if E is a Hilbert space (the situation in [43], [52]), then E is of type p for any $p \geq 2$. Therefore the proposition applies for any $\alpha > 0$.

b. In § 2 it was shown that a T - stable convolution semigroup is concentrated on the contractible part C(T). But C(T) need not be closed. By a renormining $\|\cdot\|_1$ it is possible to consider $(C(T), \|\cdot\|_1)$ as a Banach space. But, as already mentioned, by renormining the geometry of the space might be changed. So, if we start with a Banach-space $(E, \|\cdot\|)$ of type p, the renormed subspace $(C(T), \|\cdot\|_1)$ not necessarily inherits this property.

So the situation in the infinite dimensional case is more complicated then in the finite dimensional vector space case and in the case of a locally compact group.

5.10 Let T be a continuous group acting contracting on E. Then the following questions arise:

a. Do there exist full T - stable Gaussian semigroups,

b. full T-stable semigroups without Gaussian components,

c. a decomposition $E = E_g \oplus E_p$, where E_g supports the Gaussian part and E_p the non - Gaussian part?

There are partial answers, see e.g. [59]; the geometry of the space E is involved as well as the spectrum of the generator of T. We omit the details.

§ 6 Domains of attraction and limit theorems.

As pointed out in § 1 in most cases stability is defined via limit distributions of normed sums of identically distributed random variables, resp. of convolution powers of a probability distribution. Our approach starts with equation (6) of § 1 as a defined. Now we show that the definition via limit theorems is included. (Again we are only interested in strict stability).

A) Let V be a finite dimensional vector space and T a continuous group in Aut(V). Let $\mathfrak{N} \subseteq \mathrm{Aut}(V)$ be a subgroup.

6.1 Definition Let $\mu \in M^1(V)$.

The generalized domain of attraction of μ with respect to $\mathfrak{N}$ is defined as the set $GDA(\mu, \mathfrak{N}) := \{\nu \in M^1(V)$: There exists a sequence

$$\{ \sigma_n\} \subseteq \mathfrak{N} , \quad \text{such that} \quad \sigma_n(\nu^n) \to \mu\}.$$

If $\mathfrak{N} = \mathrm{Aut}(V)$ we simply write $GDA(\mu)$.
The domain of normal attraction of μ with respect to T is the set
$NDA(\mu, T) := \{\nu \in M^1(V) : \tau_{\frac{1}{n}} (\nu^n) \to \mu\}$.

6.2 Remarks

a) If μ is embeddable into a continuous convolution semigroup (μ_t) which is stable with respect to some group T, then obviously $\mu = \mu_1 \in NDA(\mu, T)$.
We always have $NDA(\mu,T) \subseteq GDA(\mu, \mathfrak{N})$ for any group $\mathfrak{N}$ containing T.

b) M. Sharpe's definition of operator stability now reads:
μ is operator stable iff $GDA(\mu) \neq \emptyset$. Hence the main results of Sharpe's paper [S8] may be formulated as follows:
The following assertions are equivalent for a full measure $\mu \in M^1(V)$:

(1) $GDA(\mu) \neq \emptyset$.

(2) There exists a continuous group $T \subseteq \mathrm{Aut}(V)$ such that $NDA(\mu, T) \neq \emptyset$.

(3) μ is embeddable into a continuous convolution semigroup $(\mu_t)_{t \geq 0}$, which is T-stable (with respect to some group T).

c) The "classical" situation, i.e. stability and domains of attraction with respect to homothetical transformations, is covered by the definition as follows:

Let Z be defined as $Z = \{u \cdot id_V : u \in \mathbb{R}\}$ and for fixed $\alpha > 0$ $H_\alpha := \{t^\alpha \cdot id_V =: \tau_t\}_{t>0}$; then $GDA(\mu, Z)$ resp. $NDA(\mu, H_\alpha)$ are the usual domain of attraction resp. domain of normal attraction of μ (without centering constants).

6.3 An interesting problem is the description of the domains of attraction of a given stable measure. Basic for the following investigations were the studies of E.L. Rvačeva [84], where the domain of attraction of homothetically stable measures was described. The generalized domain of attraction in the operator stable case is treated by M.G. Hahn, M.J. Klass [19], see especially [20]. For the domain of normal attraction see especially W.N. Hudson, J.D. Mason, J. A. Veeh [35], Z. Jurek [53, 54, 55], E. Kehrer [59].

6.4 Definition Let $\nu \in GDA(\mu)$ resp. $\nu \in NDA(\mu, T)$. Let $\{\sigma_n\}_{n\geq 1} \subseteq Aut(V)$, such that $\sigma_n(\nu^n) \to \mu$. Then the compound Poisson semigroups $\pi_n(t) := \exp(nt(\sigma_n(\nu) - \varepsilon_e))$, $t \geq 0$ resp. $\Pi_n(t) := \exp(nt(\tau_{\frac{1}{n}}(\nu) - \varepsilon_e))$ $t \geq 0$, are called the accompanying laws of μ (resp. of the semigroup $(\mu_t)_{t\geq 0}$ with $\mu = \mu_1$).

Convergence of the convolution powers $\sigma_n(\nu^n) \to \mu$ is compared with convergence of the accompanying laws to (μ_t), (see 6.9 ff). Note that the Lévy-measures of the semigroups $(\pi_n(t))_{t\geq 0}$ are $\eta_n = n\, \sigma_n(\nu)|_{V\setminus\{0\}}$ resp. $= n\,\tau_{1/n}(\nu)|_{V\setminus\{0\}}$.

In the following we try to describe the domains of attraction in terms of η_n and with aid of moment conditions.

But we have to pay attention to the fact, that we restricted our considerations to strictly stable measures: Most of the results in the papers mentioned above consider stable measures in the wide sense. Hence instead of $\sigma_n(\nu^n)$ resp. $\tau_{1/n}(\nu^n)$ centered products $\sigma_n((\nu * \varepsilon_{x_n})^n)$ resp. $\tau_{1/n}((\nu * \varepsilon_{x_n})^n)$ are involved where $(x_n)_{n\in\mathbb{N}}$ is a suitable sequence in V. But these concepts coincide for example if the measures are additionally assumed to be symmetric, or if the Lévy - measure fulfill some integrability conditions. So, sometimes in the sequel we assume symmetry in order to avoid these difficulties (though the symmetry condition is not supposed in the papers under consideration.)

But first we show once more the importance of the fullness condition:

6.5 Theorem Let $(\mu_t)_{t \geq 0}$ be a continuous convolution semigroup which is stable with respect to some continuous group T.
Assume that μ_1 (and hence any μ_t) is full. Then the domain of normal attraction $NDA(\mu_1, T)$ does not depend on T.

[The proof is based on the following idea:
The fullness condition implies the compactness of the invariance group $\mathfrak{I}(\mu_1) = \mathfrak{I}((\mu_t))$. Hence, as mentioned in § 4, 4.5 the decomposability group $\mathfrak{Z}((\mu_t))$ splits into a direct product $\mathfrak{Z}((\mu_t)) = \bar{T} \otimes \mathfrak{I}((\mu_t))$, where $\bar{T} = (\bar{\tau}_t)_{t>0}$ is a suitable continuous group of automorphisms.
Therefore $\tau_t = \bar{\tau}_t \cdot \rho_t$, $t > 0$, with $\rho_t = \tau_t \bar{\tau}_{1/t} \in \mathfrak{I}(\mu_1)$.
Now let $\nu \in NDA(\mu_1, \bar{T})$, hence $\bar{\tau}_{1/n}(\nu^n) \to \mu_1$.

Then $\tau_{1/n}(\nu^n) = \bar{\tau}_{1/n}\, \rho_{1/n}(\nu^n) \to \mu_1$, since the points of accumulation of $\{\rho_{1/n}\}$ lie in $\mathfrak{I}(\mu_1)$. See e.g. [35 , 30 , 51 , 59 ,].]

6.6 Corollary Let $(\mu_t)_{t \geq 0}$ be a full symmetric Gaussian semigroup. Then (μ_t) is stable with respect to the homothetical group $H_{1/2} = (t^{1/2}\, id_V)$. Therefore the domain of normal attraction $NDA(\mu_1, T)$ for any operator group $T \in \mathfrak{Z}((\mu_t))$ coincides with the "classical" domain of normal attraction $NDA(\mu_1, H_{1/2})$. With other words $\tau_{1/n}(\nu^n) \to \mu_1$ iff $\theta_{1/\sqrt{n}}(\nu^n) \to \mu_1$.

A measure μ is called elliptically symmetric if $\mathfrak{I}(\mu) = W\, O(\mathbb{R},d)\, W^{-1}$, where $d = \dim(V)$, $O(\mathbb{R},d)$ is the group of orthogonal transformations of V and W is some fixed operator in Aut(V).
Remarks. a) Any full Gaussian measure is elliptically symmetric.
b) μ is elliptically symmetric iff $\mathfrak{I}(\mu)$ is maximally compact in Aut(V).
Corollary 6.6 holds more generally:

6.7 <u>Corollary</u> Assume (μ_t) to be T-stable for some group T. Assume in addition the measures μ_t to be elliptically symmetric. Then for some $\alpha > 0$ (with $H_\alpha := (t^\alpha \, id_V)_{t>0}$) we have $NDA(\mu_1, T) = NDA(\mu_1, H_\alpha)$.

[Elliptical symmetry implies fullness. Hence
$\mathfrak{Z}((\mu_t))$ splits $\mathfrak{Z}((\mu_t)) \cong \bar{T} \otimes \mathfrak{I}(\mu_1)$ for some group $\bar{T} = (\bar{\tau}_t)$.
The operators $\bar{\tau}_t$ commute with every $\mathfrak{G} \in \mathfrak{I}(\mu_1) = W \; O(\mathbb{R},d) \; W^{-1}$, whence (see e.g. [26] proof of 3.5) $\bar{\tau}_t = t^\alpha \, id_V$ follows. Application of 6.5 yields the assertion.]

For the structure of elliptically symmetric measures see: J.P. Holmes, W.N. Hudson, J.D. Mason [30], A. Luczak [74], E. Siebert [91]]

As is well known the tail behaviour of a probability measure on $\mathbb{R}$ is closely related to the description of the domain of attraction of a stable measure, and there is a great difference between the Gaussian and the non-Gaussian case. In the multivariate case the situation is quite similar as the following result shows:

6.8 <u>Theorem</u> Let $T = (t^A)_{t>0}$ be a continuous group of automorphisms. Let $(\mu_t)_{t>0}$ be T-stable and full. Let further $\nu \in NDA(\mu_1, T)$.

a) Assume (μ_t) to have no Gaussian part.

(i) Put $\lambda_o := (\max\{Re\,\lambda : \lambda \in Spec(A)\})^{-1}$ and let $0 < r < \lambda_o$. Then for any $t > 0$ μ_t has an absolute moment of order r.

(ii) Put $\lambda_1 := (\min\{Re\,\lambda : \lambda \in Spec(A)\})^{-1}$ and let $r \geq \lambda_1$. Then the r-th absolute moment of ν is infinite.

b) Assume (μ_t) to be Gaussian. Then the 2nd absolute moment of ν is finite.

[a) (i) and b) see Z. Jurek [50].
For (ii) see E. Kehrer [59] II. 3.2].

For full Gaussian measures we already know (see 6.6) that $NDA(\mu_1, T) = NDA(\mu_1, H_{1/2})$, hence we have classical moment conditions describing the domain of normal attraction. (See e.g. [59], [84], [49]). For stable measures without Gaussian component the situation is different. If the measures are supposed to be elliptically symmetric we have a "classical" description via corollary 6.7. In the general situation the Lévy-measures of the accompanying laws are involved:

6.9 Theorem Let $(\mu_t)_{t\geq 0}$ be a full T - stable continuous convolution semigroup without Gaussian component. Let η be the Lévy - measure. Let $\nu \in M^1(V)$.
Then the following implications hold

(i) => (ii); and (ii) => (i) if in addition the measures are assumed to by symmetric.

(i) $\nu \in NDA(\mu_1, T)$ (which is independent of T, see 6.5).

(ii) For every $\delta > 0$ and every δ - neighbourhood U_δ of 0 with $\eta(\partial U_\delta)=0$

$$n\,\tau_{1/n}(\nu)\,|_{\complement U_\delta} \rightarrow \eta\,|_{\complement U_\delta} \quad \text{weakly.}$$

Let Q be a closed cross - section and $\eta = \int_Q \eta_x\, d\lambda(x)$ be a desintegration obtained in Corollary 3.3. Then (ii) is equivalent to

(iii) $t\nu\,\{\tau_s x : x \in B,\ s > t\} \xrightarrow[t\to\infty]{} \lambda(B)$

for every Borel set $B \subseteq Q$ with $\lambda(\partial B) = 0$.

[W.N. Hudson, J.D. Mason, J.A. Veeh [35], Z. Jurek [54], see especially E. Kehrer [59], II. 2.3, 2.4. The importance of the closedness condition in order to obtain the equivalence of (ii) and (iii) is pointed out in [59] 2.2.]

B) Let E be a separable Banach space.

Let T be a strongly continuous group in Aut(E) joining 0 and infinity. In addition we assume (1') $\|\tau_t\| \xrightarrow[t\to 0]{} 0$. Let $\mathfrak{N} \subseteq Aut(E)$ be a subgroup.
We define analogous to the finite dimensional case the domains of attraction $GDA(\mu,\mathfrak{N})$, $GDA(\mu)$, $NDA(\mu, T)$ for a given measure μ (see Definition 6.1) and the accompanying laws (see Definition 6.4).

For infinite dimensional spaces fullness does not imply compactness of $\mathfrak{I}(\mu_1)$. Hence Theorem 6.5 is replaced by a weaker version:

6.10 Theorem Let (μ_t) be T - stable. Let $\mathfrak{N} \subseteq Aut(E)$ be a subgroup such that $\mathfrak{N} \cap \mathfrak{I}(\mu_1)$ is compact and $T \subseteq \mathfrak{N}$. Then we have for any continuous group $\bar{T} = (\bar{\tau}_t) \subseteq \mathfrak{N}$ such that $(\bar{\tau}_t, t)_{t>0}$ and $(\tau_t, t)_{t>o} \subseteq \mathfrak{Z}((\mu_t))$:

$NDA(\mu_1, T) = NDA(\mu_1, \bar{T})$.

[The ideas of the proof of 6.5 carry over without changes.]

6.11 Remark Even for Gaussian measures fullness does not imply compactness of $\mathfrak{J}(\mu_1)$. Hence the corollaries 6.6 and 6.7 have no counterpart in the infinite dimensional case (except in a weak version according to 6.10).

It is possible to describe weak convergence of measures on E by convergence of finite dimensional marginals. By this way it is possible to find necessary and sufficient conditions for $\nu \in M^1(E)$ to belong to $NDA(\mu_1, T)$. (For sufficiency the geometry of the space is involved: E is supposed to be of type p). For domains of attraction with respect to $\mathcal{N} = \{u \cdot id_E, u \in \mathbb{R}\}$ resp. $T = \{t^\alpha\ id_E, t > 0\}$ see W. Linde [72], A. Weron [110], A. Araujo, E. Giné [1].

Due to the lack of an equivalent to 6.6 the description of $NDA(\mu_1, T)$ of a Gaussian semigroup (μ_t) depends on T: Let $(\mu_t)_{t>0}$ be a Gaussian semigroup with covariance operator R. Let $\nu \in M^1(E)$.

6.12 Let E be of type p. Then sufficient conditions are known for ν to belong to $NDA(\mu, T)$.

See e.g [59] III § 4. 4.1.

[The conditions are stated in terms of the tail behaviour of $\nu(\tau_s(x) : s > t, x \in B)$, the covariance operator R, and exhaustion by finite dimensional subspaces. We do not give the details here].

6.13 Let E be the dual of a space of type 2 and (μ_t), T, ν as before. Then we have necessary conditions for a measure ν to belong to $NDA(\mu, T)$.

See [59] III § 4 . 4.2.

[We don't give the details here.]

6.12 and 6.13 together give characterizations of $NDA(\mu_1, T)$ of symmetric Gaussian semigroups on Hilbert spaces.

For semigroups without Gaussian component we have the following:

6.14 There exist necessary conditions for $\nu \in M^1(E)$ to belong to $NDA(\mu_1, T)$ in terms of the Lêvy-measures of the accompanying laws, analogous to 6.9 (ii) and (iii), similar to the conditions obtained in [1] for stability with respect to homothetical transformations (see e.g. [59] III. § 3 . 3.2),
and considering the finite dimensional marginals and exhausting E by finite dimensional subspaces
(see [59] III § 3. 3.6).
[We omit the details.]

6.15 Sufficient conditions are obtained for spaces E of type p, see [59] III § 3. 3.8 ff, if the measures are assumed to be symmetric.

6.16 Assume (μ_t) to be symmetric without Gaussian component. Then 6.14 and 6.15 together yield a characterization (for symmetric measures) of $NDA(\mu_1, T)$ for spaces E of type p, where

$$p \cdot \sup_{t<1} \log \|\tau_t\| / \log t > 1.$$

See [59] III § 3. 3.11.
We omit the details.
For $T = (t^\alpha \, id_E)_{t>0}$ the characterization is just theorem 4.3 in A. Araujo, E. Giré [1].

C) Let G be a locally compact group.

Due to the fact that a convolution semigroup $(\mu_t)_{t \geq 0}$ is in general not determined by μ_1, the situation is more complicated in the group case. We have to distinguish between attraction to a single measure and attraction to a semigroup.

6.17 Definition Let $\mathcal{N} \subseteq Aut(G)$ be a subgroup. Let $T = (\tau_t)$ be a continuous group. Let $\mu \in M^1(G)$ and let $(\mu_t)_{t \geq 0}$ be a convolution semigroup.
$GDA(\mu, \mathcal{N}) := \{\nu \in M^1(G)$: There exist $a_n \in \mathcal{N}$, such that $a_n(\nu^n) \to \mu\}$.
$GDA((\mu_t)_{t \geq 0}, \mathcal{N}) := \{\nu \in M^1(G)$: There exist a_n such that $(a_n(\nu))^{[nt]} \to \mu_t$, $t > 0\}$. We suppress $\mathcal{N}$ if $\mathcal{N} = Aut(G)$.

Analogously we define the domain of normal attraction:

$$NDA(\mu, T) := \{\nu \in M^1(G) : \tau_{1/n}(\nu^n) \to \mu\}$$

$$NDA((\mu_t)_{t \geq 0}, T) := \{\nu \in M^1(G) : (\tau_{1/n}(\nu))^{[nt]} \to \mu_t, \; t > 0\}.$$

6.18 Remark a) It is well known that in the vector space case the notions of domains of attraction of a full measure μ_1 and of the corresponding convolution semigroup $(\mu_t)_{t \geq 0}$ coincide.

b) For special groups it is known that (for a suitable definition of fullness) the following assertions are equivalent:

(i) μ is embeddable into a continuous convolution semigroup (μ_t) which is T - stable with respect to some T (hence $NDA((\mu_t), T) \neq \emptyset$),

(ii) $GDA(\mu) \neq \emptyset$

[For motion groups see P. Baldi [2], Y.S. Khokhlov [60]; for Heisenberg groups and for the diamond group see T. Drisch, L. Gallardo [7 , 8]].

In general we have the following

6.19 Theorem Let G be a locally compact group, K a compact subgroup and let $T = (\tau_t)$ be a continuous automorphism group. Let $(\mu_t)_{t \geq 0}$ be a continuous convolution semigroup with idempotent $\mu_0 = \omega_K$. Finally let $\nu \in M^1(G)$, such that $\omega_K * \nu * \omega_K = \nu$. Then the following assertions are equivalent:

(i) $\nu \in NDA((\mu_t), T)$

(ii) $(\mu_t)_{t \geq 0}$ is T - stable.

[See [22] 2.7 for a proof if $K = \{e\}$.]

Remarks a) In the papers mentioned in 6.18 the assertions are more general (due to the restriction to special examples of groups): The measure μ is a priori not supposed to be embeddable, and the attracted measure ν is not supposed to be K - invariant. In addition ν is only supposed to lie in the generalized domain of attraction.

b) It is easily seen that the situation is less complicated if the measures (μ_t) and ν are supposed to be symmetric.

In the situation of 6.19 a reduction to nilpotent, simply connected

Lie groups is possible:

6.20 <u>Proposition</u> Suppose (μ_t), T, K, and ν to fulfill the conditions of 6.19. Then ν is concentrated on the contractible part $C_K(T)$. Hence, via 2.10, 2.11 the problem of describing $NDA((\mu_t), T) \cap \omega_K * M^1(G) * \omega_K$ is reduced to the case of nilpotent, simply connected Lie groups.

[Let η be the Lévy measure of $(\mu_t)_{t \geq 0}$. Then $\tau_{1/n}(\nu)^{[nt]} \to \mu_t$, $t \geq 0$, implies the convergence of the Lévy measures of the accompanying laws $n\, \tau_{\frac{1}{n}}(\nu) \to \eta$ vaguely on $G \setminus \{e\}$. Now we can repeat the ideas of the proof of 2.5, 2.11 in order to obtain $\nu(C_K(T)) = 1$.]

6.21 <u>Definition</u> A probability measure $\mu \in M^1(G)$ is called completely full if $\mathfrak{I}(\mu)$ is a compact subgroup of Aut (G).

[In the sense of this definition full measures on finite dimensional vector spaces and in the special examples mentioned above are completely full.]

Let $(\mu_t)_{t \geq 0}$ be a convolution semigroup which is T - stable with respect to some group T. Then obviously $\mathfrak{I}(\mu_t) = \tau_t\, \mathfrak{I}(\mu_1)\, \tau_{1/t}$ for $t > 0$. Hence if μ_1 is completely full then the same holds for any μ_t, $t > 0$. We call (μ_t) completely full in this case.

6.22 <u>Theorem</u> Let $(\mu_t)_{t \geq 0}$ be a completely full T - stable semigroup. Then for any continuous group $\bar{T}$, such that $(\bar{\tau}_t, t) \subseteq \mathfrak{Z}((\mu_t))$ we have $NDA((\mu_t), T) = NDA((\mu_t), \bar{T})$.

[Follows immediately from the splitting obtained in 4.9 and 6.5 $\mathfrak{Z}((\mu_t)) \cong \mathbb{R}_+^* \otimes \mathfrak{I}((\mu_t))$.]

The definition of elliptical symmetry depends on the structure of Aut(G) resp. of the decomposability group. Let $(\mu_t)_{t \geq 0}$ be a convolution semigroup and suppose without loss of generality G to be a nilpotent, simply connected Lie group.

(Hence $Aut(G) \cong Aut(\mathfrak{g}) \subseteq GL(\mathfrak{g})$.)

6.23 Definition A measure $\mu \in M^1(G)$ is called elliptically symmetric if $\mathfrak{J}(\mu)$ is a maximal compact subgroup of Aut(G).

6.24 Definition Let G be a (simply connected nilpotent) Lie group and let $T = (\tau_t)$ be a group of contractions. T is called a group of dilations if the generator A of the differentials $d\tau_t = t^A \in \mathrm{Aut}(\mathfrak{g})$ is representable as a diagonal matrix with positive entries. (See e.g. [94, 95], [37 - 39].)

Let G be a step two nilpotent group, (i.e. $[\mathfrak{g},\mathfrak{g}] \subseteq \mathfrak{z}(\mathfrak{g})$). T is called homothetical if $d\tau_t$ induces homothetical transformations on the abelian Lie algebras $[\mathfrak{g},\mathfrak{g}]$ and $\mathfrak{g}/[\mathfrak{g},\mathfrak{g}]$.

[E.g. if $\mathfrak{g} = \mathfrak{h}_n$ the 2n+1 - dimensional Heisenberg Lie algebra with canonical basis $X_1,\dots,X_n$, $Y_1,\dots,Y_n$, Z, then for $\alpha > 0$
$d\tau_t : X_i \mapsto t^\alpha X_i$, $Y_i \mapsto t^\alpha Y_i$, $Z \mapsto t^{2\alpha} Z$ is homothetical.].
Now we define by induction: Let $\mathfrak{g}$ be nilpotent of degree $n (> 2)$. Then T is homothetical if the induced transformations $d\bar{\tau}_t$ on $[\mathfrak{g},\mathfrak{g}]$ and on $\mathfrak{g}/[\mathfrak{g},\mathfrak{g}]$ are homothetical.

6.25 Definition A semigroup $(\mu_t)_{t \geq 0}$ is called completely elliptically symmetric if $\mathfrak{J}((\mu_t))$ is a maximal compact subgroup of Aut(G) and if any continuous group $T = (\tau_t)_{t > 0}$, such that $\tau_t \rho = \rho \tau_t$, $t \in \mathbb{R}_+^*$, $\rho \in \mathfrak{J}$ is of the form $\tau_t = \rho_t \bar{\tau}_t$, where $\rho_t \in \mathfrak{J}$ and $(\bar{\tau}_t)$ is homothetical.

6.26 Remarks a) There are profound investigations on stability with respect to homothetical automorphism groups, see e.g. P. Glowacki [13 , 14]. In connection with Gaussian measures and central limit theorems, see the comments in § 7. See also L. Gallardo [9, 10]. For the structure of groups admitting dilations or homothetical T's and harmonic analysis on these groups, see e.g. E. Stein et al. [94 ,95]. (The groups are called homogeneous groups, there).

b) In [26] § 3 E. Siebert and the author considered several examples of groups, for which elliptical symmetry implies complete elliptical symmetry.

6.27 Corollary (to Theorem 6.22)

Let (μ_t) be completely elliptically symmetric and T - stable. Then $NDA((\mu_t), T) = NDA((\mu_t), H)$, where H is a group of homothetical transformations.

6.28 Except in the case of Gaussian measures (see the comments in § 7) we have no explicit description of the domain of attraction in the group case. Nevertheless in concrete examples some results are possible (as pointed out in [22] 3.11.8): In a fundamental paper concerning limit theorems and non - commutative Fourier analysis [89] E. Siebert obtained conditions under which the limit behaviour of a triangular system and the corresponding accompanying laws coincide ([89] § 8, 8.1, 8.2). Consider for a given measure $\nu \in M^1(G)$ and a continuous group T the triangular system $\{\tau_{1/n}(\nu) =: \nu_{n,k}\}_{k=1}^{n}$, and the corresponding accompanying laws $\{\exp(\tau_{1/n}(\nu) - \varepsilon_e)) =: \Pi_{n,k}\}_{k=1}^{n}$. Then under the conditions of Siebert the limit behaviour of $\tau_{1/n}(\nu^n)$ and $\exp(n(\tau_{1/n}(\nu) - \varepsilon_e))$ coincide. Whence we obtain necessary conditions for ν to belong to $NDA((\mu_t), T)$ of some T - stable $(\mu_t)_{t \geq 0}$.

§ 7 Remarks and comments. Further developments in stability.

7.1 P. Lévy's representation of stable measures.

A stable measure on $\mathbb{R}$ is characterized by its Fourier transform. For non - Gaussian measures of arbitrary index $\alpha \in (0,2)$ the Fourier transform does not give a concrete representation of the measure. Therefore it is important that P. Lévy [71] obtained a representation of a stable distribution approximating the Lévy-measure by compound Poisson measures: Let $(N_t)_{t \geq 0}$ be a Poisson process, let $(\Gamma_j)_{j \geq 1}$ be the sequence of jumps in increasing order and let $(Y_j)_{j \geq 1}$ be a sequence of {-1, +1} - valued identically distributed random variables. (Notice that {-1, +1} is a cross - section with respect to an automorphism group $(\tau_t = \theta_{t^\alpha})_{t > 0}$). Suppose $\{\Gamma_j, Y_k\}$ to be independent. Then with appropriate centering constants $(a_j)_{j \geq 1}$ the series $\Sigma(\Gamma_j^{-1/\alpha} \cdot Y_j - a_j)$ converges a.e. to a random variable with the given stable distribution.
Recently similar representations were obtained for stable measures on vector spaces (M.B. Marcus, G. Pisier [76], see also W. Linde [72] 6.10; R. Le Page et al. [69]), for locally convex vector spaces (R. Sztencel [98]) and for operator stable laws in Urbanik's sense (R. Le Page [70]).
The variables (Y_j) are supposed to take values in an appropriate cross - section, then.
Recently H. Carnal showed the possibility of such a representation for locally compact groups, see these proceedings [4]. Indeed, Carnal's result is stated for nilpotent, simply connected Lie groups, but according to § 2, 2.9 ff it holds for arbitrary locally compact groups, then.

It seems worthwile to mention that similar representations are possible in other stability concepts: For max - stability (§ 1, ex. 2) see the paper of P. Vatan [108] and the literature cited there.

7.2 0 - 1 - laws and purity laws. It is well known that stable measures on a Banach space have the following property: For every measurable subspace $F \subseteq E$ we have $\mu(F) = 0$ or $= 1$. Indeed this is true for semi-stable laws μ and for subgroups $F \subseteq E$. For a recent survey on 0 - 1 - laws see A. Janssen [45] and the literature cited there. For a very general class of semi - stable laws E. Siebert ([90], Theorem 1) obtained purity laws and a general form of 0 - 1 - laws ([90], § 1, Corollary 1,2) which apply to (operator semi-) stable laws on vector spaces and on groups. See also the literature cited in [90]. A recent proof of a 0 - 1 - law is given in [99].

7.3 Absolute continuity, holomorphic semigroups and analytic densities.

Let G be a topological group (a vector space or a locally compact group). Let $(\mu_t)_{t \geq 0}$ be a continuous convolution semigroup. As already mentioned in 7.2 there exist general purity laws for semistable (hence for stable) semigroups. In [93] § 3 and § 4 E. Siebert studies connections between stability and holomorphicity, especially for absolutely continuous measures. This is applied to full (operator semi-) stable measures on $\mathbb{R}^n$. The case of stable holomorphic semigroups is treated in [93] § 5.

For special Lie groups, especially for the Heisenberg groups, P. Glowacki [13 - 17] considered homothetically - stable semigroups and smoothness properties. Conditions for absolute continuity are given in terms of non - commutative Fourier - transforms. See also the literature on stable Gaussian semigroups resp. their generators cited there.
For special examples see A. Hulanicki [38 - 39].
Special examples of stable semigroups on the Heisenberg group with non - smooth L^2 - densities are studied in [15].

7.4 Stability of Gaussian measures and central limit theorems.

Gaussian measures on groups and vector spaces are well known and there exists an immense literature on this field (see e.g [27] for locally compact groups). Stability of Gaussian measures arises in connection with central limit theorems. We only mention A. Araujo and E. Giné [1] for a recent treatment in the vector space case. In the case of Lie groups we refer to the fundamental contributions by Y. Guivarc'h, M. Keane, B. Roynette [11], P. Crépel [5], and A. Raugi [83]. For general locally compact groups see also the investigations of convergence of triangular systems to Gaussian measures in [89]. On the other hand stable Gaussian measures on Lie groups are characterized by the condition that the infinitesimal generator is a homogeneous differential operator of 2nd order. For the development in this field see e.g. E. Stein et. al. [94 , 95] and the literature cited there.

7.5 Self - similar processes.

J. Lamperti [68] studied a class of $\mathbb{R}$ - valued stochastic processes, originally called semi - stable, later on self - similar, which are characterized by the behaviour of the transition probabilities under homothetical transformations of the space and scaling of time. This concept is generalized to $\mathbb{R}^d$ (see Wah Kiu [61]), to Hilbert spaces E (R.G. Laha, V.K. Rohatgi [66]) and to operator stability on $\mathbb{R}^d = E$ (W. N. Hudson, J.D. Mason [36], R.G. Laha, V.K. Rohatgi [67]).
Let $(X_t)_{t \geq 0}$ be an E - valued stochastic process, let $T = (\tau_t)$ be a group of automorphisms of E and $\alpha > 0$. (X_t) is called self - decomposable with respect to T if for $t > 0$ $(X_{t^\alpha s})_{s \geq 0}$ is weakly equivalent to $(\tau_t(X_s))_{s > 0}$. (The increments are not supposed to be independent).
It is easily seen that there is a connection between stable semigroups on a Lie group and self-similar processes: Let (X_t) be a stochastic process with values in G according to the semigroup (μ_t), then $(\log X_t)$ is a self - similar stochastic process with values in $\mathfrak{g}$.

7.6. Final remarks.

On the real line generators of symmetric stable semigroups can be represented by fractional powers of Gaussian generators. Fractional powers of Gaussian semigroups and mixtures are considered in [21], see also [38, 39].
The behaviour of stable semigroups with respect to fractional powers is studied in [22, 24].

It is, as already stated, possible to describe measures on groups, which are stable in the wide sense in terms of generating functionals, see [22 - 25].

Generalizations of the concept of stability, e.g. semistability, self - decomposability etc. were recently studied by different authors, see e.g. [42, 102, 58]. We cannot go into details here. It should be noted that there is no reduction procedure as in § 2, hence the class of semistable laws on groups is much larger and the description is much more complicated than in the vector space case.

Finally we want to mention a recent parallel development in mathematical statistics. In the framework of statistical experiments the concept of a stable experiment was introduced. It can be shown that stable experiments correspond in an unique manner to stable measures on vector spaces. For details see e.g. [97, 46, 44].

References

[1] A. Araujo, E. Giné: The central limit theorem for real and Banach valued random variables. J. Wiley, New York (1980).

[2] P. Baldi: Lois stables sur les déplacements de $\mathbb{R}^d$. In: Probability measures on groups. Proceedings Oberwolfach (1978). Lecture Notes in Math. 706, 1 - 9. Springer (1979).

[3] N.H. Bingham: Factorization theory and domains of attraction for generalized convolution algebras. Proc. London Math. Soc. (3) 23, 16 - 30 (1971).

[4] H. Carnal: Les variables aléatoires de loi stable et leur représentation selon P. Lévy. In: Probability measures on groups VIII. Proceedings Oberwolfach (1985). Lecture Notes Math. Springer (1986).

[5] P. Crépel: Grenzwertsätze für abhängige Zufallsvariable und Irrfahrten auf Gruppen. In: Probability measures on groups. Proceedings Oberwolfach (1978). Lecture Notes Math. 706, 54 - 66. Springer (1979).

[6] E. Dettweiler: Stabile Maße auf Badrikianschen Räumen. Math. Z. 146, 149 - 166 (1976).

[7] T. Drisch, L. Gallardo: Stable laws on the Heisenberg group. In: Probability measures on groups. Proccedings Oberwolfach (1983). Lecture Notes Math. 1064, 56 - 79 (1984)

[8] T. Drisch, L. Gallardo: Stable laws on the diamond group. In preparation.

[9] L. Gallardo: Processuss subordonnés au mouvement brownien sur les groupes de Lie nilpotents. Compt. Rend. Acad. Sc. Paris 292, 413 - 416 (1981).

[10] L. Gallardo: Processus subordonnés et mouvement brownien sur les groupes de Lie nilpotents. In: Marches aléatoires et processus stochastiques sur les groupes de Lie, Nancy (1981) Inst. E. Cartan 40 - 52 (1983).

[11] Y. Guivarc'h, M. Keane, B. Roynette: Marches aléatoires sur les groupes de Lie. Lecture Notes Math. 624. Springer (1977).

[12] G. Forst: A characterization of self - decomposable probabilities on the half - line. Z. Wahrscheinlichkeitstheorie verw. Geb. 49, 349 - 352 (1979).

[13] P. Głowacki: A calculus of symbols and convolution semigroups on the Heisenberg group. Studia Math. 72, 291 - 321 (1982).

[14] P. Głowacki: Stable semigroups of measures on the Heisenberg group. Studia Math.

[15] P. Głowacki, A. Hulanicki: A semigroup of probability measures with non - smooth differentiable densities on a Lie group. Preprint (1984).

[16] P. Głowacki: On commutative approximate identities on non-graded homogeneous groups. Comm. Part. Differential Equ. 9, 979 - 1016 (1984).

[17] P. Głowacki: Stable semigroups of measures as commutative approximate identities on non-graded homogeneous groups. Preprint (1985).

[18] U. Grenander: Probabilities on algebraic structures. Almquist & Wiksell, Upsala (1963).

[19] M.G. Hahn, M.J. Klass: A survey of generalized domains of attraction and operator norming methods. In: Probability in Banach spaces III, Proceedings Medford (1980). Lecture Notes Math. 860, 187 - 218 (1981).

[20] M.G. Hahn, M.J. Klass: Affine normality of partial sums of i.i.d. random vectors; A characterization. Z. Wahrscheinlichkeitstheorie verw. Geb. 68, 479 - 505 (1985).

[21] W. Hazod: Subordination von Faltungs- und Operatorhalbgruppen. In: Probability measures on groups. Proceedings Oberwolfach (1978). Lecture Notes Math. 706, 144 - 202 (1979).

[22] W. Hazod: Stable probabilities on locally compact groups. In: Probability measures on groups. Proceedings Oberwolfach (1981). Lecture Notes Math. 928, 183 - 211 (1982).

[23] W. Hazod: Remarks on [semi-] stable probabilities. In: Probability measures on groups. Proceedings Oberwolfach (1983). Lecture Notes Math. 1064, 182 - 203 (1984).

[24] W. Hazod: Stable and semistable probabilities on groups and vector spaces. In: Probability theory on vector spaces III. Proceedings Lublin (1983). Lecture Notes Math. 1080, 69 - 89 (1984).

[25] W. Hazod: Semigroupes de convolution [demi-] stables et autodécomposables sur les groupes localement compacts. In: Probabilités sur les structures géometriques. Actes des Journees Toulouse (1984). Publ. du Lab. Stat. et Prob. Université de Toulouse, 57 - 85 (1985).

[26] W. Hazod, E. Siebert: Continuous automorphism groups on a locally compact group contracting modulo a compact subgroup and applications to stable convolution semigroups. Semigroup Forum (1986). To appear.

[27] H. Heyer: Probability measures on locally compact groups. Ergebnisse der Math. Berlin - Heidelberg - New York. Springer (1977).

[28] H. Heyer: Probability theory on hypergroups: A survey. In: Probability measures on Groups VII. Proceedings Oberwolfach (1983). Lecture Notes Math. 1064, 481 - 550 (1984).

[29] K.H. Hofmann, P. Mostert: Splitting in topological groups. Mem. Amer. Math. Soc. 43 (1963).

[30] J.P. Holmes, W.N. Hudson, J.D. Mason: Operator - stable laws: multiple exponents and elliptical symmetry. Ann. Probab. 10, 602 - 612 (1982).

[31] W.N. Hudson: Operator-stable distributions and stable marginals. J. Mult. Analysis 10, 26 - 37 (1980).

[32] W.N. Hudson, J.D. Mason: Exponents of operator stable laws. In: Probability in Banach spaces III. Proceedings Medford (1980). Lecture Notes in Math. 860, 291 - 298 (1981).

[33] W.N. Hudson, J.D. Mason: Operator stable laws. J. Mult. Analysis 11, 434 - 447 (1981).

[34] W.N. Hudson, J.D. Mason: Operator stable measures on $\mathbb{R}^2$ with multiple exponents. Ann. Prob. 9, 482 - 489 (1981).

[35] W.N. Hudson, J.D. Mason, J.A. Veeh: The domain of normal attraction of an operator-stable law. Ann. Prob. 11, 178 - 184 (1983).

[36] W.N. Hudson, J.D. Mason: Operator-self-similar processes in a finite dimensional space. Trans. Am. Math. Soc. 273, 281 - 297 (1982).

[37] A. Hulanicki: The distribution of energy in the Brownian motion in the gaussian field and analytic hypoellipticity of certain subelliptic operators on the Heisenberg group. Studia Math. 16, 165 - 173 (1976).

[38] A. Hulanicki: A Tauberian property of the convolution semigroup generated by $X^2 - |Y|^\gamma$ on the Heisenberg group. Proceedings Symposia Pure Math. (AMS) 35, 403 - 405 (1979).

[39] A. Hulanicki: A class of convolution semi-groups of measures on a Lie group. In: Probability theory on vector spaces II. Proceedings Błażejewko (1979). Lecture Notes Math. 828, 82 - 101 (1980).

[40] R. Jajte: Quasi stable measures in generalized convolution algebras. Bull. Acad.Sci.Pol. 24, 505 - 511 (1976).

[41] R. Jajte: Quasi stable measures in generalized convolution algebras II. Bull. Acad.Sci.Pol. 25, 67 - 72 (1977).

[42] R. Jajte: Semistable probability measures on $\mathbb{R}^N$. Studia Math. 61, 29 - 39 (1977).

[43] R. Jajte: V-decomposable measures on a Hilbert space. In: Probability theory on vector spaces II. Proceedings Błażejewko (1979). Lecture Notes in Math. 828, 108 - 127 (1980).

[44] A. Janssen, H. Milbrodt, H. Strasser: Infinitely divisible statistical experiments. Lecture Notes in Statistics 27 (1985).

[45] A. Janssen: A survey about zero-one-laws for probability measures on linear spaces and locally compact groups. In: Probability measures on groups VII. Proceedings Oberwolfach (1983). Lecture Notes Math. 1064, 551 - 563 (1984).

[46] A. Janssen: Unendlich teilbare statistische Experimente. Habilitationsschrift, Universität Dortmund (1982).

[47] Z.J. Jurek: A limit theorem for truncated random variables. Bull. Acad. Pol.Sci. 23, 911 - 913 (1975).

[48] Z.J. Jurek: Limit distributions for shrunken random variables. Diss. Math. 85, 1 - 46 (1981).

[49] Z.J. Jurek: On Gaussian measures on $\mathbb{R}^d$. In: Proceedings of the Sixth Conference on Probability Theory, Braşov (1979). Ed. Acad. Rep. S. Romania, Buckarest (1981).

[50] Z.J. Jurek: On stability of probability measures in Euclidean spaces. In: Probability theory on vector spaces II. Proceedings Błażejewko (1979). Lecture Notes Math. 828, 129 - 145 (1980).

[51] Z.J. Jurek: Central limit theorem in Euclidean spaces. Bull. Acad. Pol. Sci. Math. 28, 81 - 86 (1980).

[52] Z.J. Jurek: Limit distributions and one parameter groups of linear operators on Banach spaces. J. Multiv. Anal. 13, 578 - 604 (1983).

[53] Z. Jurek: Domains of normal attraction of operator - stable measures on Euclidean spaces. Bull. Acad. Pol. Sci., 28, 397 - 406 (1980).

[54] Z.J. Jurek: Remarks on V - decomposable measures. Bull. Acad. Pol. Sci. 30, 393 - 401 (1982).

[55] Z.J. Jurek: Domains of normal attraction for G - stable measures on $\mathbb{R}^d$. Theory Prob. Appl. 27, 396 - 400 (1982).

[56] Z.J. Jurek: Convergence of types, self - decomposibility and stability of measures on linear spaces. In: Probability in Banach spaces III. Proceedings Medford (1980). Lecture Notes in Math. 860, 257 - 284 (1981).

[57] Z.J. Jurek: Polar coordinates in Banach spaces. Bull. Acad. Polon. Sci. Math. 32, 61 - 66 (1984).

[58] Z.J. Jurek: Random integral representations for classes of limit distributions similar to Lévy class L_0^*. Preprint (1985).

[59] E. Kehrer: Stabilität von Wahrscheinlichkeitsmaßen unter Operatorgruppen auf Banachräumen. Dissertation, Universität Tübingen (1983).

[60] Yu.S. Khokhlow: On the convergence to a multi - dimensional stable law of the distribution of a shift parameter for the composition of random motions in Euclidean space. Theory Prob. Appl. 27, 363 - 365 (1982).

[61] S.Wah Kiu: Semistable Markov processes in $\mathbb{R}^n$. Stoch. Proc. Appl. 10, 183 - 191 (1980).

[62] W. Krakowiak: Operator stable probability measures on Banach spaces. Coll. Math. 41, 313 - 326 (1979).

[63] J. Kucharczak: On operator stable probability measures. Bull. Acad. Pol. Sci. 23, 571 - 576 (1975).

[64] J. Kucharczak: Remarks on operator stable measures. Coll. Math. 34, 109 - 119 (1976).

[65] J. Kucharczak, K. Urbanik: Operator stable probability measures on some Banach space. Bull. Acad. Pol. Sci. 25, 585 - 588 (1977).

[66] R.G. Laha, V.K. Rohatgi: Self - similar stochastic processes in a Hilbert space. Preprint (1982).

[67] R.G. Laha, V.K. Rohatgi: Operator self - similar stochastic processes in $\mathbb{R}_d$. Stoch. Proc. Appl. 12, 73 - 84 (1981).

[68] J. Lamperti: Semistable stochastic processes. Trans. Amer. Math. Soc. 104, 62 - 78 (1962).

[69] R. LePage, M. Woodruffe, J. Zinn: Convergence to a stable distribution via order statistics. Ann. Prob. 9, 624 - 632 (1981).

[70] R. LePage: Multidimensional infinitely divisible variables and processes. Part II. In: Probability in Banach spaces III. Proceedings Medford (1980). Lecture Notes Math. 860, 279 - 284 (1981).

[71] P. Lévy: Propriétés asymptotiques des sommes de variables aléatoires indépendantes ou enchaînées. Journal de Math. 14, fasc. IV. (1935).

[72] W. Linde: Infinitely divisible and stable measures on Banach spaces. Teubner Texte zur Math. vol.58, Leipzig (1983).

[73] A. Łuczak: Operator semi - stable probability measures on $\mathbb{R}^N$. Coll. Math. 45, 287 - 299 (1981).

[74] A. Łuczak: Elliptical symmetry and characterization of operator - stable and operator semi - stable measures. Ann. Probab. 12, 1217 - 1223 (1984).

[75] A. Łuczak: Independent marginals of a probability measure. Preprint (1983).

[76] M.B. Marcus, G. Pisier: Characterizations of almost surely continuous p - stable random Fourier series and strongly stationary processes. Acta Math. 152, 245 - 301 (1984).

[77] J. Michaliček: Der Anziehungsbereich von operatorstabilen Verteilungen im $\mathbb{R}_2$. Z. Wahrscheinlichkeitstheorie verw. Geb. 25, 57 - 70 (1972).

[78] J. Michaliček: Die Randverteilungen der operatorstabilen Maße im zweidimensionalen Raum. Z. Wahrscheinlichkeitstheorie verw. Geb. 21, 135 - 146 (1972).

[79] B. Mincer, K. Urbanik: Completely stable measures on Hilbert spaces. Coll. Math. 42, 301 - 307 (1979).

[80] K.R. Parthasarathy: Every completely stable distribution is normal. Sankhya 35, Ser. A, 35 - 38 (1973).

[81] K.R. Parthasarathy, K. Schmidt: Stable positive definite functions. Trans. Amer. Math. Soc. 203, 161 - 174 (1975).

[82] U. Pickartz: Semiflüsse auf Räumen von Wahrscheinlichkeitsmaßen. Dissertation, Universität Dortmund (1983).

[83] A. Raugi: Théorème de limite centrale pour un produit semi-direct d'un groupe de Lie résoluble simplement connexe de type rigide par un groupe compact. In: Probability measures on groups. Proceedings Oberwolfach (1978). Lecture Notes Math. 706, 257 - 324 (1979).

[84] E.L. Rvačeva: Domains of attraction of multidimensional distributions. Select. Transl. Math. Stat. Prob. 2, 183 - 205 (1962).

[85] K. Schmidt: Stable probability measures on $\mathbb{R}^{\nu}$. Z. Wahrscheinlichkeitstheorie verw. Geb. 33, 19 - 31 (1975).

[86] S.V. Semovskii: Operator stable laws of distributions. Sovjet Math. Doklady 20, 139 - 142 (1979).

[87] S.V. Semovskii: The central limit theorem for sums of random vectors normalized by linear operators. Sovjet Math. Doklady 20, 356 - 359 (1979).

[88] M. Sharpe: Operator stable probability measures on vector groups. Trans. Amer. Math. Soc. 136, 51 - 65 (1969).

[89] E. Siebert: Fourier analysis and limit theorems for convolution semigroups on a locally compact group. Adv. Math. 39, 111 - 154 (1981).

[90] E. Siebert: Semistable convolution semigroups on measurable and topological groups. Ann. Inst. H. Poincaré 20, 147 - 164 (1984).

[91] E. Siebert: Supplements to operator - stable and operator - semistable laws on Euclidean spaces. To appear in J. Multivariate Anal.

[92] E. Siebert: Contractive autmorphisms on locally compact groups. Math. Z. 191, 73 - 90 (1986).

[93] E. Siebert: Holomorphic convolution semigroups on topological groups. In: Probability measures on groups VII. Proceedings Oberwolfach (1983). Lecture Notes Math. 1064, 421 - 449 (1984).

[94] E. Stein, A. Nagel: Lectures on pseudodifferential - operators. Math. Notes, Princeton Univ. Press (1979).

[95] E. Stein, G.B. Folland: Hardy spaces on homogeneous groups. Math. Notes, Princeton Univ. Press (1982).

[96] F.W. Steutel, K. van Harn: Discrete analogues of self - decomposability and stability. Ann. Probability 7, 893 - 899 (1979).

[97] H. Strasser: Scale invariance of statistical experiments. Probability and Math. Stat. 5, 1 - 20 (1985).

[98] R. Sztencel: On the lower tail of stable seminorm. Bull. Pol. Acad. Sci. Math. 32, 11 - 12 (1984).

[99] R. Sztencel: Absolute continuity of the lower tail of stable seminorm. Preprint (1985).

[100] A. Tortrat: Lois stables dans un groupe. Ann. Inst. H. Poincaré, Sect. B. 17, 51 - 61 (1981).

[101] A. Tortrat: Lois zero - un et lois semi - stables dans un groupe. In: Probability measures on groups. Proceedings Oberwolfach (1981). Lecture Notes in Math. 928, 452 - 466 (1982).

[102] K. Urbanik: Lévy's probability measures on Euclidean space. Studia Math. 44, 119 - 148 (1972).

[103] K. Urbanik: Lévy's probability measures on Banach spaces. Studia Math. 63, 283 - 308 (1978).

[104] K. Urbanik: Generalized convolutions. Studia Math. 23, 217 - 245 (1964).

[105] K. Urbanik: Generalized convolutions II. Studia Math. 45, 57 - 70 (1973).

[106] K. Urbanik: Generalized convolutions III. Studia Math. 80, 167 - 189 (1984).

[107] K. Urbanik: Generalized convolutions IV. Preprint (1984).

[108] P. Vatan: Max - infinite divisibility and max - stability in infinite dimensions. In: Probability in Banach spaces V. Proceedings, Medford (1984). Lecture Notes Math. 1153, 400 - 425 (1985).

[109] J.A. Veeh: Infinitely divisible measures with independent marginals. Z. Wahrscheinlichkeitstheorie verw. Geb. 61, 303 - 308 (1982).

[110] A. Weron: Stable processes and measures: A survey. In: Probability theory on vector spaces III. Proceedings Lublin (1983). Lecture Notes Math. 1080, 306 - 364 (1984).

Wilfried Hazod

Fachbereich Mathematik
Universität Dortmund
Postfach 50 05 00

D - 4600 Dortmund

Towards a Duality Theory for Algebras*

MARTIN E. WALTER

0. Introduction.

In this article we discuss a duality theory for (not necessarily commutative) algebras that generalizes the Pontriagin-van Kampen duality for abelian locally compact groups. Though a more general theory undoubtedly exists we will restrict our attention to a large class of Banach algebras which includes matrix algebras, group algebras, Fourier algebras as well as others. Hopefully this duality theory will have many applications to such fields as the representation theory of groups and algebras, probability measures on groups and their duals, and mathematical physics. We hope eventually that this theory will shed additional light on the interrelationships of geometry and algebra.

1. Historical motivation.

In the mid-1930's Pontriagin and van Kampen developed a duality theory for abelian locally compact groups which was not only beautiful but useful mathematics. The basic construction, given a locally compact abelian group G, is to find the <u>dual</u> (or character) <u>group</u>, $\hat{G}$, of G.

Definition. The dual group, $\hat{G}$, of locally compact abelian group G, is the set of all continuous homomorphisms $\chi: G \to \mathbf{T}$, where $\mathbf{T}$ is the multiplicative group of complex numbers of length one, equipped with the pointwise product as the (dual) group multiplication and uniform convergence on compact subsets of G as the (dual) group topology.

Lemma. The dual group, $\hat{G}$, is a locally compact abelian group.

Proof. See [5] §23.

Some simple examples of this construction are

$$\hat{\mathbf{Z}} = \{n \mapsto e^{in\theta}: \ 0 \leq \theta < 2\pi\} \simeq \mathbf{T}$$

$$\hat{\mathbf{T}} = \{\theta \mapsto e^{i\theta n}: \ n \in \mathbf{Z}\} \simeq \mathbf{Z}$$

$$\hat{\mathbf{R}} = \{x \mapsto e^{ixy}: \ y \in \mathbf{R}\} \simeq \mathbf{R}$$

where $\mathbf{Z}$ and $\mathbf{R}$ are the additive group of integers and real numbers, respectively.

---

*This work was supported in part by a grant from the National Science Foundation.

We note in the above examples that if we construct the dual of the dual group of G we get back to where we started, namely G. More generally we have

Theorem [Pontriagin-van Kampen Duality Theorem]

If G is a locally compact abelian group, then $G \simeq \hat{\hat{G}}$, i.e., G is topologically isomorphic with the dual group of $\hat{G}$.

Proof: See [5] §24. Note that the isomorphism is carried out by a specific ("natural") map τ.

We now ask the fundamental and natural question: Does there exist such a duality theorem for locally compact non-abelian groups? We, of course, are not the first to ask this question. Very briefly the following is an incomplete list of some of the works dealing with the above question.

1) Tannaka-Krein duality for compact groups [1939-1949], see [6] §30.
2) Tatsuuma duality for locally compact groups [1967], see [12].
3) M. Takesaki, who gives a definitive treatment of the Hopf-von Neumann algebra approach to group duality in [10].
4) M. Walter, who gives the Fourier-Stieltjes approach to group duality in [13].
5) M. Enock and J. M. Schwartz develop what may be called the Kac algebra approach to duality in a series of papers, one of which is [4].
6) Others, including U. Haagerup, J. De Cannérie, Ocneanu, Stinespring, Eymard, Saito, to mention just a few, have all made significant constributions to the above problem of duality for nonabelian groups.

One lesson is apparent from the half century or so of work on the problem of formulating a duality theory for non-abelian groups. Namely, it appears to be impossible to formulate such a duality theory within the category of groups (in a non-trivial way). The dual of a non-abelian group G involves directly or indirectly an algebra.

This leads us to the following question: Can we formulate a duality theory for algebras that "includes" the Pontriagin-van Kampen duality as well as a duality theory for nonabelian groups? Although we will not demonstrate the following explicitly here we ideally want our duality theory to satisfy:

(1) Given an algebra A, there is a "dual algebra", D(A), obtained by applying D, a contravariant functor, to A.

(2) It is possible to obtain A from D(A) by applying D', a contravariant functor, to D(A).

Remark. If the above is satisfied, then no information is lost when applying D to A and D(A) is said to be a complete invariant for A. Also, it would be very nice if D' = D. In some sense the above two properties define an abstract duality theory for algebras.

2. Mathematical motivation from the real numbers, $\mathbf{R}$.

There are two notions of positivity applicable to complex-valued functions on $\mathbf{R}$.

Definitions. A function $f: \mathbf{R} \mapsto C$ is locally positive if $f(x) \geq 0$ for all x in $\mathbf{R}$.

A function $f: \mathbf{R} \mapsto C_n$ is globally positive if f is a positive definite function, i.e., $\sum_{i,j=1} \lambda_i \overline{\lambda}_j f(x_i - x_j) \geq 0$ for all $\lambda_1, \lambda_2, \ldots, \lambda_n \in \mathbf{C}$ and $x_1, \ldots, x_n \in \mathbf{R}$ and $n = 1, 2, \ldots$.

Remark. Note that these two notions of positivity are linked via the Fourier-Stieltjes transform by a famous theorem of Bochner, see [6] §33.

There are two notions of the product of functions on $\mathbf{R}$.

Definitions. The local product of two functions $f: R \mapsto \mathbf{C}$ and $g: R \mapsto \mathbf{C}$ is just the pointwise product, namely

$$fg(x) = f(x)g(x) \quad \text{for all} \quad x \in g.$$

The global product of f and g is given by convolution, namely

$$f*g(x) = \int_{-\infty}^{\infty} f(x)g(x-y)dy$$

Remark. Again the Fourier-Stieltjes transform links the two notions of product since the Fourier transform of a convolution of two functions is the pointwise product of the Fourier transforms (when they exist).

We contend that in a sense which can be made precise the global notions above are dual to the corresponding local notions. We note in passing that there are even dual notions of local and global continuity.

Now these notions of local and global products and positivity mesh together as follows:

Proposition 1. The local product of globally positive functions is globally positive.

Proof: See [6] Theorem (32.9).

Proposition 2. The global product of locally positive functions (when defined) is locally positive.

Proof: If $f(x) \geq 0$ and $g(z) \geq 0$ for all x, z in $\mathbf{R}$, then

$$0 \leqq \int_{-\infty}^{\infty} f(y)g(x-y)dy = f*g(x).$$

The reader may ask: Why aren't there 2 more propositions, one on the local products of locally positive functions the other on the global products of globally positive functions. The answer is that such propositions when looked at in the nonabelian case are either "self-dual" or false and hence have not been useful thus far. This comment may be further clarified for the reader by considering the next section of this article, in particular the remark following proposition 4.

A completely analogous situation exists, perhaps somewhat surprisingly, for M_n, the $n \times n$ matrices with complex number entries.

3. A duality for M_n.

We can define local and global positivity as in the case for $\mathbf{R}$.

Definitions. An $n \times n$ complex matrix $A = (a_{ij})$ is <u>locally</u> positive if $a_{ij} \geq 0$ $i,j = 1,\ldots,n$. Such an $A = (a_{ij})$ is <u>globally</u> positive if $\sum_{i,j=1}^{n} \lambda_i \bar{\lambda}_j \; a_{ij} \geq 0$ for all $\lambda_1,\ldots,\lambda_n \in \mathbf{C}$, i.e., $A = (a_{ij})$ is hermitian positive definite. We also have the following

Definitions. The <u>local</u> (or Schur-Hadamard) product of $A = (a_{ij})$, $B = (b_{ij})$ is

$$A \circ B = (a_{ij}b_{ij}).$$

The <u>global</u> product of A and B is $AB = \left(\sum_{k=1}^{n} a_{ik}b_{kj}\right)$, i.e., the usual matrix product.

Again as in the case of $\mathbf{R}$ we have

Proposition 3. The local product of globally positive matrices is globally positive.

Proof: See [6] (D.12) page 683.

Proposition 4. The global product of locally positive matrices is locally positive.

Proof: Clear.

Remark. Note that it is not true that the global product of globally positive matrices is globally positive. It is true if the matrices involved commute, however.

The one glaring omission thus far is the lack of a Fourier transform that links or makes "dual" the above local and global notions for M_n. It so happens that there is a transform whose existence can best be hinted at by the following calculation.

We can consider two Hilbert space structures on M_n which are analogous to $L^2(G,dx)$ and $L^2(\hat{G},d\hat{x})$ where G, $\hat{G}$ are dual locally compact abelian groups with (suitably normalized) Haar measures dx and $d\hat{x}$, respectively. Namely, given $A = (a_{ij})$ we can define

$$\|A\|_{L^2}^2 = \sum_{i,j=1}^{n} |a_{ij}|^2 \quad \text{and} \quad \|A\|_{\hat{L}^2}^2 = \mathrm{Tr}(A^*A)$$

where Tr is the usual trace, i.e., $\mathrm{Tr}(B) = \sum_{i=1}^{n} b_{ii}$, with $B = (b_{ij})$.

Just as in the case of the Fourier-Plancherel transform which establishes an isometry between $L^2(G,dx)$ and $L^2(\hat{G},d\hat{x})$ we have:

$$\|A\|_{\hat{L}^2}^2 = \mathrm{Tr}\left[A^*A\right]$$

$$= \sum_{i=1}^{n} \sum_{k=1}^{n} \overline{a}_{ki} a_{ki} = \|A\|_{L^2}^2$$

There is thus a strong analogy between the complex functions on **R** and M_n. This analogy is more clearly understood if one considers M_n as the complex valued functions on the (for the moment abstract) set $\{e_{ij}: 1 \leq i,j \leq n\}$ where the matrix $A = (a_{ij})$ corresponds to the function $A: e_{ij} \mapsto a_{ij}$, i.e., $A(e_{ij}) = a_{ij}$. If one equips the set $\{e_{ij}: 1 \leq i,j \leq n\}$ with an algebraic structure by declaring:

$$e_{i\ell}e_{kj} = e_{ij}, \quad \text{if} \quad k = \ell \quad \text{and}$$

$$e_{i\ell}e_{kj} \quad \text{is undefined if} \quad \ell \neq k,$$

then this abstract set becomes what some people call the "principal transitive groupoid on n elements", denoted here by G_n, cf. [9], [14].

The terminology may be considered formidable, but the idea is not. We can consider M_n to be the functions on G_n, where G_n is "like a group". We can then define local and global products of functions on G_n and we get, respectively, the Schur-Hadamard product and the ordinary matrix product. Thus the ordinary (global) product of matrices can be considered to be a convolution of functions on G_n, a "group-like" object.

Remark. This point of view, after a little reflection, makes the analogy above between functions on **R** and M_n, i.e., functions on G_n, more precise and almost "obvious". In fact the definition of convolution on a locally compact groupoid is well known, see [9]. The above analogy on **R** and G_n is a special case of a general theory of groupoid algebras, see [8].

The above "Fourier-Plancherel" transform for matrices can thus be considered to be an isometry between $L^2(G_n)$ and $L^2(\hat{G}_n)$ where $\hat{G}_n$ is at this moment undefined. We can, however, ask what $L^1(\hat{G}_n)$ might be, as well as $L^1(G_n)$. In analogy with the situation for $L^1(\mathbf{R})$ and $L^1(\hat{\mathbf{R}})$ or more generally, $L^1(G)$ and $L^1(\hat{G})$ where G is an abelian locally compact group we have.

Definitions. Put $L^1(G_n) = \{M_n, *, \|\cdot\|_{\hat{L}^1}\}$ where $A, B \in M_n$ give

$$A*B = \Big(\sum_{i=1}^{n} a_{ik}b_{kj}\Big) \quad\text{and}\quad \|A\|_{L^1(G_n)} = \sum_{i,j=1}^{n} |a_{ij}|.$$

Thus $L^1(G_n)$ is just M_n with its normal matrix product and the "pointwise" L^1-norm.

Put $L^1(G_n) = \{M_n, \circ, \|\cdot\|_{Tr}\}$ where $A, B \in M_n$ give $A \circ B = (a_{ij}b_{ij})$ and $\|A\|_{Tr} = Tr|A|$, with $|A| = (A^*A)^{1/2}$.

Remark. One could develop this example with $|A| = (AA^*)^{1/2}$.

Now in analogy with $L^1(G)$ and $L^1(\hat{G})$, G abelian, we might conjecture that $L^1(G_n)$ and $L^1(\hat{G}_n)$ are Banach algebras in "duality". Now we haven't defined what it might mean for algebras to be in duality (see the next section), but we have an even more fundamental problem. Are $L^1(G_n)$ and $L^1(\hat{G}_n)$ Banach algebras? The fact that $L^1(\hat{G}_n)$ <u>is</u> a Banach algebra is not only not obvious, we were the first (to our knowledge) to even ask the question, see [14]. We were led to ask the question as a natural consequence of the above analogy between functions on **R** and M_n, and by the belief that a general duality theory exists. That such a duality theory might exist then is hinted at by the following.

Proposition 5. $L^1(G_n)$ and $L^1(\hat{G}_n)$ are Banach algebras.

Proof: That $L^1(G_n)$ is a Banach algebra is a routine calculation, see [14]. We will prove that $L^1(\hat{G}_n)$ is a Banach algebra using a different proof that that given in [14]. Our proof is motivated by the "Hopf von Neumann algebra approach" of Takesaki.

Consider the map c: $e_{ij} \mapsto e_{ij} \otimes e_{ij}$ and its natural extension to M_n with its natural "operator norm", C*-algebra norm, C: $M_n \mapsto M_n \otimes M_n$, i.e. $c(A) = c(\sum a_{ij}e_{ij}) = \sum a_{ij}e_{ij} \otimes e_{ij}$, where $A = (a_{ij})$ and e_{ij} can be viewed as the matrix unit with a 1 in the ith row, j^{th} column, 0's everywhere else.

This map c is a co-multiplication in the sense that it is an isometric (cf. [3] 1.3.7 and 1.8.1.) matrix adjoint preserving isomorphism of M_n (as a C* algebra) into $M_n \otimes M_n$ (as a C* algebra), and its transpose defines an associative (and in this case commutative) product

$$^tc: \ M_n \otimes M_n \mapsto M_n .$$

Now M_n viewed as the Banach space dual of M_n with C* -norm is just M_n with the norm $\|\cdot\|_{Tr}$ defined above. Similarly $M_n \otimes M_n$ viewed as the Banach space dual of $M_n \otimes M_n$ as a C*-algebra is just $M_n \otimes M_n$ with the "product" trace norm, i.e.,

$$\|A \otimes B\|_{Tr\otimes Tr} = (Tr|A|)(Tr|B|) = \|A\|_{Tr}\|B\|_{Tr}$$

We also note that $\|{}^tc(A\otimes B)\|_{Tr} \leqq \|A\otimes B\|_{Tr\otimes Tr}$, i.e., ^tc is norm decreasing and that $^tc(A\otimes B) = A\circ B \in M_n$, the local product of A and B. This completes the proof.

Thus there is hope that there is a general duality theory that includes group algebras and matrix algebras (at least).

4. Duality for algebras.

We begin this section with a question: When are two Banach algebras A and B dual to each other? The rough answer to this question is this. Algebras A and B are dual to each other if A is an algebra of completely bounded (linear) maps of C*(B) into C*(B) and vice-versa, where C*(B) is a faithful C*- completion of B.

We will now use the remainder of our alloted space to make this notion of duality more precise, give examples of where this duality actually occurs, and then give some hints of possible applications. We begin with the notion of the universal C*-algebra completion of a Banach algebra (with involution) A, cf. [3] 2.7.

Definition. Suppose A is a Banach algebra with involution $*A$ (and bounded approximate unit). Define, for $a \in A$,

$$\|a\|_{C^*(A)} = \sup\{\|\pi(a)\|_{\mathcal{L}(H_\pi)} : \pi \text{ is a *-homomorphism of } A \text{ into } \mathcal{L}(H_\pi)\}$$

Note: $\mathcal{L}(H_\pi)$ denotes the C*-algebra of all bounded linear operators on a Hilbert space H_π.

We will assume that $\|\cdot\|_{C^*(A)}$ is a norm on A, i.e., $\|a\|_{C^*(A)} = 0$ if and only if $a = 0$. This is the case for convolution group algebras of absolutely left Haar integrable functions, viz., $L^1(G)$, G a locally compact group. It is also the case for $A(G)$ the Fourier algebra of G, see [14], [3].

Definition. $C^*(A)$ is the abstract C*-algebra obtained by taking the completion of A with respect to the $\|\cdot\|_{C^*(A)}$ norm.

Examples. $C^*(L^1(G))$ is the universal group C*-algebra, $C^*(G)$, cf. [3], 13.9.

$C^*(A(G))$ is just $C_0(G)$, the C*-algebra with sup norm of all continuous complex functions on G which vanish at infinity.

$C^*(L^1(G_n))$ is just M_n with its usual C*-operator norm and product.

$C^*(L^1(\hat{G}_n))$ is just M_n with Schur-Hadamard product and $\|A\|_\infty = \max|a_{ij}|$ for the norm.

We now will construct the "universe" in which we will search for the dual algebra of A, $D(A)$. That universe is:

Definition. If $C^*(A)$ is a C*-algebra, then $\mathcal{B}(C^*(A))$ is the (Banach) algebra of completely bounded linear maps T, of $C^*(A)$ into $C^*(A)$, with $\|T\|_{\mathcal{B}}$ the completely bounded norm of T.

We need to explain a few of the terms we just used. First of all if X is a Banach space $\mathcal{L}(X)$ is the Banach algebra of all continuous (bounded) linear maps T of X into X with $\|T\|_{\mathcal{L}(T)} = \sup\{\|Tz\|_X : z \in X, \|z\|_X \leqq 1\}$. Next, if $C^*(A)$ is a C*-algebra, then so is $C^*(A) \otimes M_n$, cf. [11] p. 192. In fact, as an algebra, $C^*(A) \otimes M_n$ is just the $n\times n$ matrices with entries from $C^*(A)$. Now for each n we have a map $T \otimes I_n : C^*(A) \otimes M_n \mapsto C^*(A) \otimes M_n$, where $T \in \mathcal{L}(C((A))$ and I_n is the identity map of M_n to M_n. For example, if $(z_{ij}) \in C^*(A) \otimes M_n$, then $T \otimes I_n(z_{ij}) = (Tz_{ij})$, i.e., just let T act on each matrix entry. Now $\|T \otimes I_n\|_{\mathcal{L}(C^*(A) \otimes M_n)}$ is a number. If $\sup_n \|T \otimes I_n\|_{\mathcal{L}(C^*(A) \otimes M_n)} \leqq K < \infty$, we say T is completely bounded with completely bounded norm, $\|T\|_{c.b.} = \sup_n \|T \otimes I_n\|_{\mathcal{L}(C^*(A) \otimes M_n)}$. Thus as an algebra $\mathcal{B}(C^*(A)) \subset \mathcal{L}(C^*(A))$. But $\|T\|_{\mathcal{B}(C^*(A))} \geq \|T\|_{\mathcal{L}(C^*(A))}$ with strict inequality possible.

Now $\mathcal{B}(C^*(A))$ has an additional, simple but important piece of structure; namely, it has a conjugation.

Definition: given an algebra $\mathfrak{A}$ over $\mathbf{C}$, $\overline{}$ is a conjugation on $\mathfrak{A}$ if

(i) $\overline{\lambda x} = \overline{\lambda}\,\overline{x}$, for $x \in \mathfrak{A}$, $\lambda \in \mathbf{C}$,

(ii) $\overline{x+y} = \overline{x} + \overline{y}$, for $x,y \in \mathfrak{A}$

(iii) $\overline{xy} = \overline{x}\,\overline{y}$, for $x,y \in \mathfrak{A}$

(iv) $\overline{\overline{x}} = x$, for $x \in \mathfrak{A}$.

Remark. A conjugation is just like an involution, cf. [3] 1.1, except that it does not reverse the products.

Examples. $L^1(G)$ for G a locally compact group has an involution, cf. [3] 13.2, $f \in L^1(G) \mapsto f^\# = \Delta^{-1} f^b \in L^1(G)$, where $f^b(x) = \overline{f}(x^{-1})$, and a conjugation $f \in L^1(G) \mapsto \overline{f} \in L^1(G)$, where $\overline{}$ is just complex conjugation.

$A(G)$ has a $\in A(G) \mapsto a^b \in A(G)$ as a conjugation and a $\in A(G) \mapsto \overline{a} \in A(G)$, i.e., ordinary complex-conjugation, as involution. [No we have not made a mistake, that is exactly what we mean! See the second remark following Theorem 2.

The conjugation on $\mathcal{B}(C^*(A))$ is defined by the following formula for $T \in \mathcal{B}(C^*(A))$, $\overline{T}(x) = (Tx^*)^*$ where $*$ is the involution in $C^*(A)$. We say that $\overline{}$ is dual to $*$ in this case.

Proposition 6. $\mathcal{B}(C^*(A))$ is a Banach algebra with conjugation $\overline{}$, composition of operators as product, and completely bounded norm.

We can now give a more precise definition of duality for algebras A and B.

Definition. Banach algebras A and B with involutions $^*A$ and $^*B$, respectively, and conjugations ^-A and ^-B, respectively, are <u>dual</u> to each other if there are faithful <u>involution preserving</u> monomorphisms ω_A and ω_B onto dense *-subalgebras of C*-algebras, $C^*(A)$, $C^*(B)$, respectively, i.e.,

$$\omega_A: A \mapsto C^*(A) \quad \text{and}$$

$$\omega_B: B \mapsto C^*(B)$$

such that there are conjugation preserving monomorphisms τ_A and τ_B of A and B into $\mathcal{D}(C^*(B))$ and $\mathcal{D}(C^*(A))$, respectively.

Diagrammatically we can express this in at least two ways

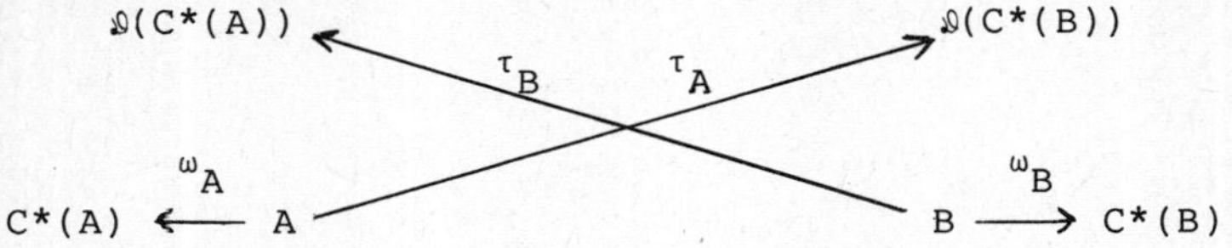

or

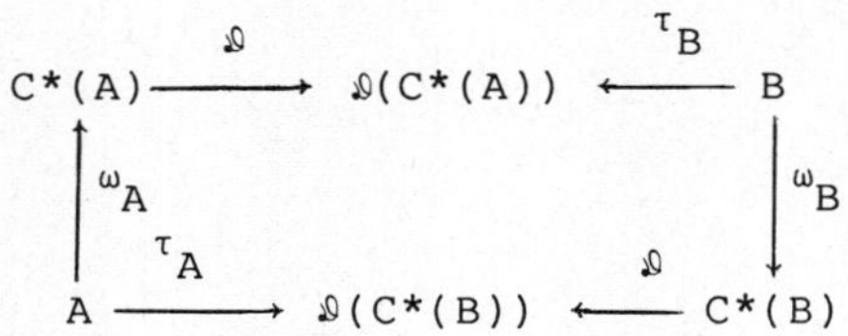

As examples we quote the following theorems from [14].

Theorem 1. If G is a locally compact group, and $L^1(G)$ is the L^1 group algebra and $A(G)$ is the Fourier algebra of G, then $L^1(G)$ and $A(G)$ are dual to one another.

Theorem 2. If G_n is the principal transitive groupoid on n elements and $L^1(G_n)$ and $L^1(\hat{G}_n)$ are as above, then $L^1(G_n)$ and $L^1(\hat{G}_n)$ are dual to each other.

Remark. The ω maps in the definition of dual algebras are not isometric in any of our examples, but the τ maps are isometric for the algebras of Theorem 1, but not isometric for the algebras in Theorem 2.

Remark. Consider $A(G)$, the Fourier algebra. It has two involutive type maps, namely, $a \in A(G) \mapsto a^b \in A(G)$ and $a \in A(G) \mapsto \bar{a} \in A(G)$. Which is the involution, which is the conjugation? It is not immediately obvious, since $A(G)$ is a commutative algebra. However, the complex conjugation map $a \mapsto \bar{a}$ is the <u>involution</u> in this case since this is the map which becomes the involution in the enveloping C^*-algebra, $C_0(G)$. The map $a \mapsto a^b$ is thus the <u>conjugation</u>.

We still do not have an answer to the following: How do we find $D(A)$, (the) dual algebra of A in $\mathcal{D}(C^*(A))$? A partial answer is the following.

Theorem [Landstad and Walter]. If A is a Banach algebra with C^*-completion $C^*(A)$, then $D(A) \in \mathcal{A}$ where $\mathcal{A} = \{s \in \mathcal{D}(C^*(A)):$ $sC^*(A) \subset \overline{\mathcal{F}}$ and $C^*(A)s \subset \overline{\mathcal{F}}\}$. Here $C^*(A)$ is viewed as a subalgebra of $\mathcal{B}(C^*(A)$ via left multiplication and $\mathcal{F}$ is the closure in $\mathcal{B}(C^*(A))$ of the operators of finite rank on $C^*(A)$.

Another question answered affirmatively by Tomiyama and Walter is: Does there always exist a "nontrivial" dual algebra $D(A)$? See [7].

5. Applications.

In the case where this duality theory reduces to a duality theorem about groups (and their algebras), the Fourier Stieltjes algebra, $B(G)$, cf. [13], is "the unitary representation theory of G". Thus this theory has applications to group representations.

In the case of G_n, a calculation of the completely bounded norm of the action of $L(G_n)$ on $C_o(G_n)$ gives a solid theoretical significance to a matrix norm considered in numerical analysis, cf. [2], p. 44.

In the October 1984 issue of the Bulletin of the American Math. Society, Doplicher and Roberts discuss a computation of the gauge group of a quantum field by means of a duality theory which they must create in order to overcome shortcomings in the classical Tannaka-Krein duality for compact groups. We claim that our duality theory is applicable in this situation, in particular, because it is "coordinate" free.

References

[1] Akemann, C., Walter, M.E., Nonabelian Pontriagin duality, Duke Math. J., 39 (1972), 451-463.

[2] Dennis, J. E. Jr., Schnabel, Robert B., Numerical Methods for Unconstrained Optimization and Nonlinear Equations, Prentice-Hall, 1983.

[3] Dixmier, J., Les C*-Algèbres et leur Représentations, Cahiers Scientifiques, Fasc. 29, Gauthier-Villars, Paris, 1964.

[4] Enock, M., Schwartz, J.M., Une dualité dans les algebres de von Neumann, Bull. Soc. Math. France, Suppl. Mém. 44 (1975), 1-144.

[5] Hewitt, Edwin and Ross, Kenneth A., Abstract Harmonic Analysis, Vol. 1, Springer-Verlag, 1963.

[6] Hewitt, Edwin and Ross, Kenneth A., Abstract Harmonic Analysis, Vol. 2, Springer-Verlag, 1970.

[7] Landstad, M., Walter, M.E., A Duality Theory for Algebras, in preparation.

[8] Ramsay, A. and Walter, M.E., Groupoid Algebras, in preparation.

[9] Renault, J., A Groupoid Approach to C*-algebras, Springer-Verlag, Lecture Notes in Math., Vol. 793, 1980.

[10] Takesaki, M., A characterization of group algebras as a converse of Tannaka-Stinespring-Tatsumma duality theorem, Amer. J. Math., 91 (1969), 529-564.

[11] Takesaki, M., Theory of Operator Algebras I, Springer-Verlag, 1979.

[12] Tatsumma, N., A duality for locally compact group, J. Math. Kyoto University 6, (1967), 187-293.

[13] Walter, M.E., A duality between locally compact groups and certain Banach algebras, J. Functional Analysis 17 (1974), 131-160.

[14] Walter, M.E., Dual algebras, to appear in Math. Scand.

Martin E. Walter
Department of Mathematics
University of Colorado
Campus Box 426
Boulder, CO 80309 U.S.A.

RANDOM FIELDS ON NONCOMMUTATIVE LOCALLY COMPACT GROUPS

Kari Ylinen
Department of Mathematics, University of Turku
SF-20500 Turku, Finland

1. Introduction

The expression "random field" has several distinct definitions in the literature of probability theory and mathematical physics, see e.g. [2, 22, 42 - 46, 49 - 58]; for us it signifies a special type of random function, namely, a family of (equivalence classes of) random variables indexed by a locally compact group G. The cases of G = **Z** and G = **R** of course yield the vast theories of random sequences and random processes; these we barely touch. Neither do we have much to say about more general abelian groups. Passing from **Z** and **R** to a general abelian G one loses the order structure which plays a basic role e.g. in prediction theory and in questions related to the Markov property; passing from an abelian G to the noncommutative case one must also dispense with the smoothly functioning apparatus of abelian harmonic analysis. We survey some aspects of what remains.

The bulk of the literature on random fields on noncommutative locally compact groups may be roughly classified under one of the following two headings: ergodic theorems, generalizations of the spectral representation theory. Interest in questions related to the first aspect goes back at least to the work of Alaoglu and Birkhoff [3, 4] on ergodic theory in the late thirties. We discuss in Section 5 the development of this problem. More space will be devoted to questions under the second heading: Sections 2, 3 and 4 mainly deal with these. The concluding Section 6 briefly mentions extensions to e.g. homogeneous spaces and generalized random fields.

The great classic in the spectral type representations of random fields in the noncommutative setting is the paper [58] by Yaglom. It is inherent in the nature of the matter that one deals with second order random fields, i.e., functions defined on a locally compact group and with values in the (complex) space $L^2(\Omega,\mathcal{A},P)$ for some probability space $(\Omega,\mathcal{A},P)$. Indeed, the existence of finite second moments for the random variables is necessary and sufficient for the existence of a workable covariance structure of the random field. It is essentially the covariance structure that allows the use of Fourier analytic techniques in the abelian case, and these methods one attempts to extend to the general situation. Traditionally one considers random functions which are centered at expectations; the discussion by Loève in [36, pp. 465-466] shows that this does not essentially restrict the generality. We assume the same convention, and so our random functions are families in $L_o^2(\Omega,A,P) = \{f \in L^2(\Omega,A,P) \mid \int f dP = 0\}$. In most (though not all) situations in the sequel it is only the covariance structure defined in terms of the ordinary inner product of $L_o^2(\Omega,A,P)$ that counts. We therefore choose a somewhat abstract point of view. Unless otherwise specified, a random field is a function ϕ: G $\rightarrow$ H where G is a locally compact group and H is an arbitrary complex Hilbert space with inner product $(\cdot|\cdot)$. This convention is merely a matter of convenience motivated by our functional analytic approach and the kind of problems we deal with. The emphasis in this survey is on those features which are peculiar to the noncommutativity of the "time domain". The probabilistic structure of a random function in the usual sense (involving the finite-dimensional distributions, trajectories etc.) is here relevant only exceptionally; when it is, we revert to $L_o^2(\Omega,A,P)$ in place of H.

In the work of Yaglom referred to above, almost exclusively (left or right) homogeneous random fields occur. Such random fields have great practical importance as the examples studied in [58] indicate, and under fairly general conditions they allow a detailed analysis in terms of direct integral representations. A different type of representation of a more general class of random fields as "Fourier transforms" of bounded linear H-valued operators on the group C*-algebra C*(G) was described in [59]; see Section 2. The random fields in question generalize the weakly harmonizable stochastic processes on **R**. These have a close connection with the continuous stationary processes as shown by Niemi [37]: the former are precisely those obtainable by orthogonal projection from the latter. In Section 4 we consider the problem of extending this result to the noncommutative situation. There is a satisfactory extension based on Pisier's generalization of the famous Grothendieck inequality, but this entails the introduction of a weak notion of homogeneity which we call hemihomogeneity. The same concept plays a key role in Section 3 which treats noncommutative generalizations of the fact that in the abelian case the continuous stationary processes are precisely those expressible as Fourier transforms of orthogonally scattered stochastic measures. We refer to Rao [47] for a background in these matters in the classical commutative situation.

2. Fourier type representations of random fields

Unless otherwise explicitely stated, G will always be an arbitrary locally compact (Hausdorff) group with a fixed left Haar measure λ, and the notation ds is used in integration with respect to λ. For any locally compact Hausdorff space X, M(X) will denote the Banach space of the (automatically bounded) regular complex Borel measures on X, identified with the dual $C_o(X)^*$ of the C*-algebra of the continuous complex functions on X vanishing at infinity. The subspace of M(X) consisting of all finite linear combinations of the Dirac measures δ_s, $s \in X$, is denoted by $M_{dd}(X)$. As usual, the Banach space $L^1(G)$ of (equivalence classes of) λ-integrable complex Borel functions is regarded as a subspace of M(G). For $\mu \in M(G)$, $\|\mu\|'$ denotes the supremum of the operator norms $\|\pi(\mu)\|$ where $\pi: G \to L(H_\pi)$ ranges over the continuous unitary representations of G (and π also denotes the natural extension of π to the convolution Banach-*-algebra M(G) so that $(\pi(\mu)\xi|\eta) = \int(\pi(s)\xi|\eta)d\mu(s)$ for $\mu \in M(G)$, $\xi,\eta \in H_\pi$). The completion of $L^1(G)$ with respect to the norm $\|\cdot\|'$ is the group C*-algebra C*(G) of G. We let $\omega: C^*(G) \to L(H_\omega)$ denote the universal representation. If $\tau: L^1(G) \to C^*(G)$ is the inclusion map, $\omega\circ\tau$ is a nondegenerate *-representation, and so it corresponds canonically to a continuous unitary representation also denoted by $\omega: G \to L(H_\omega)$; again the same notation is used for its natural extension to M(G). The von Neumann algebra generated in $L(H_\omega)$ by $\omega(C^*(G))$, i.e., the enveloping von Neumann algebra of C*(G), will be denoted by W*(G) and identified as usual with the bidual C*(G)** of C*(G). We have $\omega(M(G)) \subset W^*(G)$. The set of linear combinations of all continuous positive-definite complex functions on G will be denoted by B(G). There is a unique bijection $T: B(G) \to C^*(G)^*$ satisfying $\langle Tf,\mu\rangle = \int f d\mu$ for all $f \in B(G)$ and $\mu \in L^1(G) \subset C^*(G)$. Moreover, $\langle\omega(\mu),Tf\rangle = \int f d\mu$ for all $f \in B(G)$, $\mu \in M(G)$. For the background and further details (also summarized in [59, pp. 357-360]) related to this discussion we refer to Eymard [11] and Dixmier [7] ; [7] is also our general reference on the terminology of operator algebras and group representations.

The following definition was introduced in [59, p. 360].

2.1. DEFINITION. Let E be a (complex) Banach space and $\Phi: C^*(G) \to E$ a weakly compact linear operator (so that its second adjoint Φ^{**} maps W*(G) into E, [9 , p. 482]). The function $\hat{\Phi}: G \to E$ defined by the formula

$$\hat{\Phi}(s) = \Phi^{**}(\omega(s)),$$

$s \in G$, is called the Fourier transform of Φ.

The Fourier transform $\hat{\Phi}$ is strongly continuous, bounded in norm by $\|\Phi\|$, and it determines Φ uniquely [59, p. 360]. We are mainly interested in the case where E = H. Every bounded linear operator $\Phi: C^*(G) \to H$ is weakly compact (since the Hilbert space H is reflexive) and thus has a Fourier transform in the above sense. A random field $\phi: G \to H$ will be called <u>weakly harmonizable</u> if $\phi = \hat{\Phi}$ for some bounded linear operator $\Phi: C^*(G) \to H$.

2.2. EXAMPLE. The motivation for the above terminology of course comes from the theory of second order stochastic processes, the relevant part of which readily generalizes to any locally compact abelian group G in place of the time domain. Indeed, denoting by Γ the dual group of G we can identify $C^*(G)$ with $C_o(\Gamma)$, and so bounded linear maps from $C^*(G)$ to H are in a bijective correspondence with regular Borel vector measures on Γ. The weakly harmonizable random fields $\phi: G \to H$ in the above sense are then precisely the Fourier(-Stieltjes) transforms of regular Borel vector measures on Γ in the usual sense (see [59, §5] for details). Historically, the word "weakly" is added to make a distinction with the older and more restrictive notion of (strong) harmonizability. We refer to Rao [47] for a discussion of the interrelationship of these concepts. Suffice it to say here that in contrast to the latter case, (complex) measures on $\Gamma \times \Gamma$ are in general not enough for the Fourier type representation of the covariance function of a weakly harmonizable process; one has to resort to bimeasures.

In the noncommutative case the natural counterpart of the bimeasures mentioned above are the bounded complex bilinear forms on $C^*(G) \times C^*(G)$. Using the fact that any bounded linear map from $C^*(G_1)$ to $C^*(G_2)^*$ for any locally compact groups G_1 and G_2 is weakly compact one can show that any bounded bilinear form $B: C^*(G_1) \times C^*(G_2) \to \mathbf{C}$ has a unique separately weak* continuous extension $\tilde{B}: W^*(G_1) \times W^*(G_2) \to \mathbf{C}$ (called the <u>canonical extension</u>) when $C^*(G_i)$ is embedded in $W^*(G_i)$ as usual (see [29, pp. 75-77] or [59, pp. 365-366]). The <u>Fourier transform</u> of B is the function $\hat{B}: G_1 \times G_2 \to \mathbf{C}$ defined by the formula $\hat{B}(s,t) = \tilde{B}(\omega_1(s), \omega_2(t))$ for $s \in G_1$, $t \in G_2$ where ω_j has the obvious meaning. The function $\hat{B}$ is jointly continuous, bounded by $\|B\|$, and it determines B uniquely [59, p. 366]. In the commutative situation $\hat{B}$ has a natural interpretation in terms of integration with respct to a bimeasure [59, §5]. We mention in passing that in recent years there has been some interest in the harmonic analysis of bimeasures, see e.g. [13, 17, 18] and their references. In these works the canonical extension of a bimeasure and its Fourier transform are constructed with the help of Grothendieck's inequality (to be discussed in Section 4).

The covariance function $R: G \times G \to \mathbf{C}$ of a random field $\phi: G \to H$ is defined by the formula $R(s,t) = (\phi(s)|\phi(t))$. If ϕ is weakly continuous, then R is separately continuous, and the converse is easily seen to hold if ϕ is e.g. taken to be locally bounded. We collect in the following theorem (remembering that H is reflexive) several characterizations of weakly harmonizable random fields from [59, pp. 361, 363, and 379].

2.3. THEOREM. Let $\phi: G \to H$ be a weakly continuous random field and R its covariance function. The following fifteen conditions are equivalent:

(i) ϕ is weakly harmonizable;

(ii) for every $f \in H^*$ we have $f \circ \phi \in B(G)$, and $\sup \{ \|T(f\circ\phi)\| \mid f \in H^*, \|f\| \leq 1 \} < \infty$;

(iii) ϕ is bounded, and $\sup \{ \|\int \phi d\mu \| \mid \mu \in L^1(G), \|\mu\|' \leq 1 \} < \infty$;

(iv) (resp. (v)) the same as (iii) but $L^1(G)$ replaced by $M(G)$ (resp. $M_{dd}(G)$);

(vi) there is a bounded bilinear form $B: C^*(G) \times C^*(G) \to \mathbf{C}$ whose Fourier transform $\hat{B}$ satisfies $R(s,t) = \hat{B}(s,t^{-1})$ for all $s, t \in G$;

(vii) R is bounded, and $\{\int\int R(s,t)d\mu(s)d\nu(t) \mid \mu \in L^1(G), \nu \in M_{dd}(G), \|\mu\|' \leq 1, \|\nu\|' \leq 1\}$ is a bounded set of complex numbers;

(viii) to (xv) the same as (vii) but $L^1(G)$ replaced by $M(G)$ or $M_{dd}(G)$, and/or $M_{dd}(G)$ replaced by $L^1(G)$ or $M(G)$.

In the original "concrete" setting where we have $L_o(\Omega,A,P)$ in place of H, condition (v) may be reformulated in terms of the finite-dimensional distributions of the random function as follows (see [59, p. 379]):

(v)' there is a constant $C > 0$ such that for every finite set $\{s_1,\ldots,s_n\} \subset G$ and every vector $z = (z_1,\ldots,z_n) \in \mathbf{C}^n$ for which $\| \sum_{j=1}^{n} z_j \delta_{s_j} \|' \leq 1$, we have $E|z\cdot\underline{u}|^2 \leq C$ where $z\cdot\underline{u}$ is the scalar product of z with the random vector $u = (\phi(s_1),\ldots,\phi(s_n))$, and E denotes expectation.

In condition (vi), B is called the covariance bilinear form of ϕ. It satisfies $B(x,y) = (\Phi x|\Phi y^*)$ for all $x, y \in C^*(G)$ where $\Phi: C^*(G) \to H$ is the bounded linear map having ϕ as its Fourier transform [59, p. 379]. A weakly continuous random field $\phi: G \to H$ satisfying (iii) is sometimes said to be V-bounded (following Bochner [5]); this notion is thus equivalent to weak harmonizability.

Conditions (iii) and (v) generalize the conditions appearing in the well-known Eberlein-Schoenberg theorem [26, p. 304] characterizing the Fourier-Stieltjes transforms of scalar measures. The history of the problem in the abelian but vector valued case is explained in [47].

In the absence of commutativity the most complete theory of harmonic analysis is available for compact groups. Yaglom [58] makes efficient use of this theory in the study of continuous random fields on compact groups. We now describe a Fourier series representation (to be used in Section 3) in a slightly more general setting.

In the rest of Section 2 G is compact, and $\lambda(G) = 1$. We mainly follow the notation of Hewitt-Ross [26]. In particular, Σ is the set of equivalence classes of irreducible continuous (automatically finite-dimensional) unitary representations of G. For each $\sigma \in \Sigma$ we choose a fixed member $U^{(\sigma)}$ of σ with representation space H_σ whose dimension is d_σ. We also select a fixed orthonormal basis $\{\zeta_1^{(\sigma)}, \zeta_2^{(\sigma)}, \ldots, \zeta_{d_\sigma}^{(\sigma)}\}$ in H_σ, and write $u_{jk}^{(\sigma)}(s) = (U_s^{(\sigma)}\zeta_k^{(\sigma)} | \zeta_j^{(\sigma)})$ for $j, k \in \{1,\ldots,d_\sigma\}$, $s \in G$ (so that $(u_{jk}^{(\sigma)}(s))_{j,k=1}^{d_\sigma}$, denoted in the sequel by $[U_s^{(\sigma)}]$, is the matrix of $U_s^{(\sigma)}$ with respect to this basis).

In the following definition E is assumed to be an arbitrary complex Banach space, and we write indifferently $\alpha\xi$ or $\xi\alpha$ for the product of $\alpha \in \mathbf{C}$ and $\xi \in E$.

2.4. DEFINITION. Let ϕ be an E-valued Bochner λ-integrable function on the compact group G. For all $\sigma \in \Sigma$ and $j, k \in \{1,\ldots,d_\sigma\}$ we denote

$$a_{jk}^{(\sigma)} = \int \overline{u_{kj}^{(\sigma)}(s)}\phi(s)ds,$$

and by A_σ the matrix $(a_{jk}^{(\sigma)})_{j,k=1}^{d_\sigma}$ (with entries from E). In analogy with the scalar case we denote the sum of the entries in the main diagonal of the product matrix $A_\sigma[U_s^{(\sigma)}]$ by $tr(A_\sigma[U_s^{(\sigma)}])$, i.e,

$$tr(A_\sigma[U_s^{(\sigma)}]) = \sum_{j,k=1}^{d_\sigma} a_{jk}^{(\sigma)} u_{kj}^{(\sigma)}(s).$$

We call the formal expression $\sum_{\sigma \in \Sigma} d_\sigma tr(A_\sigma[U_s^{(\sigma)}])$ the <u>Fourier series</u> of ϕ.

In [58, p. 596] Yaglom considers a Fourier series representation of continuous random fields on a compact group. In the following more general result we denote by $L^2(G,\lambda,H)$ the Hilbert space of (equivalence classes of) λ-measurable H-valued functions for which the square of the norm is λ-integrable. For the proof we refer to [64].

2.5. PROPOSITION. If $\phi \in L^2(G,\lambda,H)$, then the Fourier series of ϕ converges (independently of order) to ϕ with respect to the norm of $L^2(G,\lambda,H)$.

Besides its tractability in direct analysis, the case of a compact group may be used as an illustration of the general method outlined in the first part of this section. (A more detailed presentation is given in [64].) We denote by $C_o(\Sigma)$ the C*-algebra of the families $S = (S_\sigma)_{\sigma \in \Sigma}$ where $S_\sigma \in L(H_\sigma)$ and for each $\varepsilon > 0$ the set of those $\sigma \in \Sigma$ for which $\|S_\sigma\| \geq \varepsilon$ is finite [26, p. 72]. For $\mu \in L^1(G) \subset M(G)$ we define $F_o(\mu) = (\hat{\mu}(\sigma))_{\sigma \in \Sigma}$ where

$$(\hat{\mu}(\sigma)\xi|\eta) = \int (U_s^{(\sigma)}\xi|\eta)d\mu(s), \quad \xi, \eta \in H_\sigma.$$

Then F_o is an injective *-homomorphism from $L^1(G)$ onto a dense *-subalgebra of $C_o(\Sigma)$; moreover, its inverse extends uniquely to an isometric *-isomorphism θ: $C_o(\Sigma) \to C^*(G)$. Now the dual of $C_o(\Sigma)$ is naturally identified with $L^1(\Sigma)$, the space of families $R = (R_\sigma)_{\sigma \in \Sigma}$ with

$\|R\|_1 = \sum_{\sigma \in \Sigma} d_\sigma tr((R^*R)^{1/2}) < \infty$, the duality being implemented by

$$\langle S,R \rangle = \sum_{\sigma \in \Sigma} d_\sigma tr(S_\sigma R_\sigma).$$

It can be shown that if for each $f \in B(G)$ the Fourier series is defined as a special case of Definition 2.4 (as in [26, §34]), and R_σ is the operator corresponding to the matrix A_σ (with $\phi = f$), then the family $R = (R_\sigma)_{\sigma \in \Sigma} \in L^1(\Sigma)$ satisfies $\theta^* o T(f) = R$. We have the following commutative diagrams

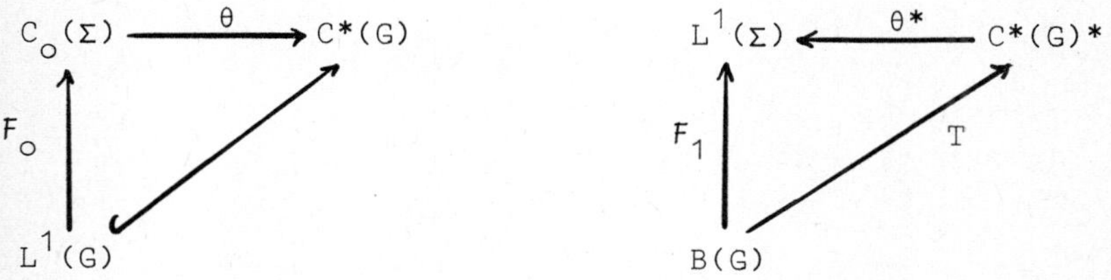

where F_1 denotes the mapping $f \to R$ described above. These identifications make possible a more concrete interpretation e.g. of Theorem 2.3; we again refer to [64] for details. We return to this set-up for compact groups in the next section.

3. The problem of homogeneity and orthogonally scattered vector Gleason measures

One of the basic features of a continuous stationary stochastic process on an abelian locally compact group is its spectral representation in terms of a regular orthogonally scattered vector measure on the Borel σ-algebra of the dual group, and it is natural to ask to what extent a similar result obtains in the noncommutative situation. The present section is devoted to this problem.

Again, G is a general locally compact group. We fix a random field ϕ: G → H and let R denote its covariance function as in Section 2, i.e., $R(s,t) = (\phi(s)|\phi(t))$. In the absence of commutativity several notions related to stationarity suggest themselves. In agreement with Yaglom [58] we call ϕ left (resp. right) homogeneous if $R(s,t) = R(us,ut)$ (resp. $R(s,t) = R(su,tu)$) for all s, t, u ∈ G. Yaglom calls ϕ two-way homogeneous if both of these conditions hold, and homogeneous if one of them holds. Obviously left (resp. right) homogeneity means that there is a (clearly positive-definite) function ρ: G → **C** such that $R(s,t) = \rho(t^{-1}s)$ (resp. $R(s,t) = \rho(st^{-1})$) for all s, t ∈ G. Then the continuity of ϕ, R and ρ, and the weak continuity of ϕ are easily seen to be equivalent. We now introduce still another related concept (compare [16, p. 255]); in fact much of what follows propagates the view that this is in many ways the "proper" generalization of weak stationarity to the noncommutative context.

3.1. DEFINITION. We say that a random field ϕ: G → H is hemihomogeneous if there are two continuous positive-definite functions ρ_1, ρ_2: G → **C** such that $R(s,t) = \rho_1(t^{-1}s) + \rho_2(st^{-1})$ for all s, t ∈ G.

In particular, the continuity of ϕ is built into this definition. Every hemihomogeneous random field ϕ is known to be weakly harmonizable [61], and so there arises the problem of characterizing hemihomogeneity in terms of properties of the bounded linear map Φ: C*(G) → H having a weakly harmonizable random field ϕ as its Fourier transform. If G is abelian, hemihomogeneity simply amounts to homogeneity (i.e., stationarity), so it has a well-known characterization in terms of the regular Borel vector measure μ_Φ on the the dual group Γ of G corresponding to Φ (see Example 2.2 and [59, §5]): μ_Φ should be orthogonally scattered in the sense that $(\mu_\Phi(B_1)|\mu_\Phi(B_2)) = 0$ for any disjoint Borel sets B_1 and B_2 in Γ (see e.g. [59, p. 381] for references, and [32]). The fact that an analogous characterization is possible for a large class of noncommutative groups is one important reason for studying hemihomogeneity. Before elaborating this statement we consider the case of compact groups where the hemihomogeneity of a weakly harmonizable random field has a concrete characterization in terms of its Fourier series.

3.2. EXAMPLE. In this example G is compact, and we resume the notational conventions used at the end of Section 2. For details we refer to [64]. Suppose ϕ: G → H is a weakly harmonizable random field and

Φ: $C^*(G) \to H$ is the bounded linear map having ϕ as its Fourier transform. For any $\sigma \in \Sigma$ we denote by ι_σ the natural injection of $L(H_\sigma)$ into $C_0(\Sigma)$ (making $S \in L(H_\sigma)$ correspond to the family having the value S at σ and zero elsewhere), and we write $\Phi_\sigma = \Phi \circ \theta \circ \iota_\sigma$. We identify $L(H_\sigma)$ by means of the basis $\{\zeta_1^{(\sigma)}, \ldots, \zeta_{d_\sigma}^{(\sigma)}\}$ with the space $M^{(\sigma)}$ of complex $d_\sigma \times d_\sigma$-matrices, and so we obtain a family $(\Phi_\sigma)_{\sigma \in \Sigma}$ of linear maps $\Phi_\sigma: M^{(\sigma)} \to H$. Then $\Phi_\sigma(R) = d_\sigma \mathrm{tr}(A_\sigma R)$ in the notation of Definition 2.4, and for any $B_\sigma = (b_{jk}^{(\sigma)})_{j,k=1}^{d_\sigma}$, $C_\sigma = (c_{jk}^{(\sigma)})_{j,k=1}^{d_\sigma} \in M^{(\sigma)}$ the following conditions are equivalent (with the notation of Definition 2.4):

(i) $(\Phi_\sigma R | \Phi_\sigma S) = \mathrm{tr}(B_\sigma S^* R) + \mathrm{tr}(C_\sigma R S^*)$

for all $R, S \in M^{(\sigma)}$;

(ii) $(d_\sigma a_{jk}^{(\sigma)} \mid d_\sigma a_{mn}^{(\sigma)}) = b_{jm}\delta_{kn} + c_{nk}\delta_{jm}$

for all $j, k, m, n \in \{1, \ldots, d_\sigma\}$ (where δ_{kn} is the Kronecker symbol). As a consequence one obtains the result that ϕ is hemihomogeneous if, and only if, there are two families $B = (B_\sigma)_{\sigma \in \Sigma}$ and $C_\sigma = (C_\sigma)_{\sigma \in \Sigma}$ of positive-definite matrices such that $\sum_{\sigma \in \Sigma} \mathrm{tr}(B_\sigma) < \infty$, $\sum_{\sigma \in \Sigma} \mathrm{tr}(C_\sigma) < \infty$, and the equation

$$(d_{\sigma_1} a_{jk}^{(\sigma_1)} | d_{\sigma_2} a_{mn}^{(\sigma_2)}) = b_{jm}^{(\sigma_1)} \delta_{kn} \delta_{\sigma_1 \sigma_2} + c_{nk}^{(\sigma_1)} \delta_{jm} \delta_{\sigma_1 \sigma_2}$$

holds for all $\sigma_1, \sigma_2 \in \Sigma$, $j, k \in \{1, \ldots, d_{\sigma_1}\}$, $m, n \in \{1, \ldots, d_{\sigma_2}\}$. In [58, pp. 596-597] Yaglom has analogous characterizations of left, right and two-way homogeneity. Condition (i) above of course implies that $(\Phi_\sigma R | \Phi_\sigma S) = 0$ whenever R and S are mutually orthogonal projections (i.e., self-adjoint idempotents) in $M^{(\sigma)}$. It is a remarkable fact that, conversely, the latter condition implies the existence of B_σ, $C_\sigma \in M^{(\sigma)}$ satisfying (i). This was proved (along with more general results) by Goldstein [15], extending a theorem of Jajte and Paszkiewics [28]. In fact, B_σ and C_σ may be taken to be positive-definite, and a fairly straightforward argument shows that ϕ is hemihomogeneous if, and only if, $(\Phi R | \Phi S) = 0$ whenever R and S are mutually orthogonal projections in $C^*(G)$.

The extension of the last statement in this example to more general groups is an important problem. Since in general there need not be a sufficient supply of projections in $C^*(G)$ for a locally compact group G, we rather consider the generalization of an equivalent statement. To this end we state a definition.

3.3. DEFINITION. A bounded linear map Φ: $C^*(G) \to H$ is said to be <u>orthogonally scattered</u> if $(\Phi^{**}p \mid \Phi^{**}q) = 0$ whenever p and q are mutually orthogonal projections in $W^*(G)$.

This notion generalizes the concept of orthogonal scatteredness in the abelian case referred to earlier in this section [59, p. 382]. In [59 , Theorem 6.14] it was proved that a bounded linear map Φ: $C^*(G) \to H$ is orthogonally scattered if, and only if, there is $f \in C^*(G)^*$ such that the canonical extension $\tilde{B}$ of the covariance bilinear form of ϕ satisfies $\tilde{B}(u,v) = \langle uv,f \rangle$ whenever u and v are commuting normal elements of $W^*(G)$, and in this case f is a uniquely determined positive linear functional, and the continuous positive-definite function $T^{-1}f = \rho$ on G satisfies $(\phi(s) \mid \phi(t)) = \rho(st^{-1})$ whenever s and t are commuting elements of G. The commutation condition is an obvious drawback, but for certain groups the theorem of Goldstein referred to in Example 3.2 is valid for the von Neumann algebra $W^*(G)$ thus making it possible to obtain a more satisfactory result where the commutation condition is absent but hemihomogeneity enters. For instance, if G is of type I so that G is a type I von Neumann algebra, Corollary 4.2 in [15] may be combined with the techniques of [59] to show that a bounded linear map Φ: $C^*(G) \to H$ is orthogonally scattered if, and only if, its Fourier transform is a hemihomogeneous random field.

The orthogonal scatteredness of a bounded linear map Φ: $C^*(G) \to H$ was defined above in terms of the restriction of its second adjoint to the projection lattice of $W^*(G)$. The restriction of Φ^{**} to the projection lattice of $W^*(G)$ is easily seen to be completely additive in the sense that $\Phi^{**}(\sum_{e \in \Lambda} e) = \sum_{e \in \Lambda} \Phi^{**}(e)$ for any orhogonal set Λ of projections in $W^*(G)$. Such functions on the projection lattice are sometimes called <u>vector Gleason measures</u> in honor of the famous theorem of Gleason [14] characterizing positive real valued completely additive functions on the lattice of projections on a separable Hilbert space of dimension at least three. The Fourier analysis of vector Gleason measures and their connection to random fields was briefly considered in [62]. H-valued vector Gleason measures on the projection lattice of an arbitrary von Neumann algebra and their orthogonal scatteredness are defined in the obvious way. It may be noted that one important reason for studying orthogonal scatteredness in the noncommutative case - besides the fact that it is a natural extension of the corresponding abelian notion - is its close connection with noncommutative anlogues of projection valued measures (see e.g. Theorem 3.4 in [60] and its generalization in [34]).

In [57] and [58] Yaglom develops for certain random fields a representation theory in terms of direct integrals over the dual space $\hat{G}$ of G. Specifically, G is taken to be separable and of type I, and ϕ is assumed continuous and left or right homogeneous. This type of representation has the advantage of certain concreteness, although the scope is somewhat limited. We shall not go into technicalities here; a succinct account is given in the review of [57] by Fell in Mathematical Reviews 23 A 2936.

4. Dilations and the Grothendieck-Pisier-Haagerup inequality

In the case of an abelian locally compact group G (in particular, $\mathbf{R}$) the weakly harmonizable or, equivalently, weakly continuous V-bounded processes - being the Fourier transforms of general regular Borel vector measures on the dual group instead of just the orthogonally scattered ones - appear to be the most natural generalization of the continuous wide sense stationary processes. Between these two classes of processes there is an important connection exhibited by Niemi [37], generalizing a special case treated by Abreu [1]. Namely, a process ϕ: G → H (G abelian) is weakly harmonizable if, and only if, ϕ can be dilated to a continuous stationary process ψ: G → K where K is a Hilbert space containing H as a subspace; in other words, $\phi(s) = P_H\psi(s)$, $s \in G$, where P_H is the orthogonal projection of K onto H.

In Niemi's proof a key role was played by a consequence of the famous Grothendieck inequality or "the fundamental theorem of the metric theory of tensor products", which in this particular application implies the 2-majorizability of Hilbert space valued vector measures. The Grothendieck inequality in its original formulation [21] reads as follows: There is a universal constant $K > 0$ such that whenever X_1 and X_2 are compact Hausdorff spaces and u is a bounded complex bilinear form on $C(X_1) \times C(X_2)$, there exist probability measures μ_i on X_i, i = 1, 2, such that

$$|u(f,g)| \leq K\|u\|\left(\int |f|^2 d\mu_1\right)^{1/2}\left(\int |g|^2 d\mu_2\right)^{1/2}$$

for all $f \in C(X_1)$, $g \in C(X_2)$. Here we are only concerned with the complex case; the smallest possible K is called the (complex) Grothendieck constant. A matrix version of Grothendieck's inequality was presented in [35], and in [39] Niemi showed that his dilation theorem implies (as well as it is implied by) the special case of this matrix version where attention is restricted to positive-definite marices. We let these remarks on the commutative background suffice, and turn to the noncommutative situation.

The gate to the proper treatment of the general case was opened by Pisier's [40] important generalization of the Grothendieck inequality. Again, a special instance of the result is enough for the application to random fields, but in view of its central role we give a brief account of the the history of the general problem. Pisier proved the following results:

(P1) [40, p. 400] Let A be a C*-algebra and E a Banach space of cotype 2 (e.g., E could be a Hilbert space). If u is a continuous finite rank linear map from A into E, then there is a positive linear form f on A with $\|f\| \leq 1$ such that

$$(1) \quad \|u(x)\| \leq 6^{1/2}(C_E)^2\|u\|[f((x^*x + xx^*)/2)]^{1/2}$$

for all $x \in A$. (Here C_E is a constant depending only on E; in the Hilbert space case it may be taken to be one.)

(P2) [40, p. 400] Let A and E be as in (P1). Assume that $u\colon A \to E$ is a continuous linear map such that for some net $(u_i)_{i \in I}$ of continuous finite rank linear maps $u_i\colon A \to E$ with $\sup_{i \in I}\|u_i\| = M < \infty$ we have $\liminf_i \|u_i(x)\| \geq \|u(x)\|$ for all $x \in A$. Then the conclusion of (P1) holds for u when the right hand side of (1) is multiplied by M.

A consequence of (P2) is the following:

(P3) [40, p. 408] Let A and B be two C*-algebras and $u\colon A \times B \to \mathbf{C}$ a bounded bilinear form. Suppose that there is a net $(u_i)_{i \in I}$ of finite rank bounded bilinear forms on $A \times B$ such that $\sup\|u_i\| = M < \infty$ and $\liminf_i |u_i(x,y)| \geq |u(x,y)|$ for all $(x,y) \in A \times B$. Then there are positive linear forms $f\colon A \to \mathbf{C}$, $g\colon B \to \mathbf{C}$, $\|f\| \leq 1$, $\|g\| \leq 1$, such that

$$|u(x,y)| \leq KM[f((x^*x + xx^*)/2)g((y^*y + yy^*)/2)]^{1/2}$$

for all $(x,y) \in A \times B$, where K is a universal constant.

For example, if E = H, the approximation condition in (P2) holds with M = 1, and this result is sufficient for the basic application to random fields we are about to discuss. To get a feeling of what is going on, we mention, however, some later developments. In [24] Haagerup proved the following:

(H1) [24, p. 94] Let A and B be C*-algebras and $u\colon A \times B \to \mathbf{C}$ a bounded bilinear form. There exist two states f_1, $f_2\colon A \to \mathbf{C}$ and two states g_1, $g_2\colon B \to \mathbf{C}$ such that

$$|u(x,y)| \leq \|u\|(f_1(x^*x) + f_2(xx^*))^{1/2}(g_1(y^*y) + g_2(yy^*))^{1/2}$$

for all $x \in A$, $y \in B$. This result improves (P3) by removing the approximation condition and also by lowering the constant K; in this formulation the inequality is the best possible in the sense that if one could exchange $\|u\|$ with $C\|u\|$ for all A, B and u, then $C \geq 1$ [24, p. 94]. The following is an immediate consequence of (H1):

(H2) [24, p. 94] If u is a bounded linear map from a C*-algebra A into a Hilbert space H, then there are two states $f_1, f_2: A \to \mathbf{C}$ such that

$$\|ux\|^2 \leq \|u\|^2(f_1(x^*x) + f_2(xx^*))$$

for all $x \in A$.

This is the most convenient version for the applications below. A direct proof of this version already appears in [23, p. 237]. The general (bilinear form) case is also dealt with in [30] and [31].

The first noncommutative dilation theorem inspired by Niemi's work in the commutative case and based on (P2) was proved by Goldstein and Jajte:

4.1. THEOREM [16, pp. 256-257]. Let A be a W*-algebra with predual A_*, H a hilbert space and $\Phi: A \to H$ a $\sigma(A,A_*)$-$\sigma(H,H^*)$-continuous (and hence bounded) linear map.
(a) There is a normal positive linear functional $f: A \to \mathbf{C}$ (i.e. $f \in A_*^+$) such that $\|f\| \leq 3\|\Phi\|^2$, and $\|\Phi x\|^2 \leq f(x^*x + xx^*)$ for all $x \in A$.
(b) There is a Hilbert space K containing H as a subspace and a linear map $\Psi: A \to K$ (which is automatically $\sigma(A,A_*)$-$\sigma(H,H^*)$-continuous) such that $\|\Psi\|^2 \leq 6\|\Phi\|^2$, for some $f \in A_*^+$ $(\Psi x|\Psi y) = f(xy^* + y^*x)$ for all $x, y \in A$, and $\Phi = P_H \circ \Psi$, where P_H is the orthogonal projection from K onto H.

Analogously with the commutative situation treated by Niemi [39] (in the matrix version) it may be observed that in very general circumstances a dilation property of the above sort is equivalent to a Grothendieck type inequality for <u>positive-definite</u> bilinear forms. (For an involutive algebra A, a bilinear form $B: A \times A \to \mathbf{C}$ is said to be positive-definite, if $B(x^*,x) \geq 0$ for all $x \in A$.)

4.2 PROPOSITION. (a) Let A be an involutive algebra, $\Phi: A \to H$ a linear map and $f_1, f_2: A \to \mathbf{C}$ positive linear forms (i.e., $f_j(x^*x) \geq 0$ for all $x \in A$). The following two conditions are equivalent:
(i) $\|\Phi x\|^2 \leq f_1(x^*x) + f_2(xx^*)$ for all $x \in A$;

(ii) there is a Hilbert space K such that for some linear map $\Psi: A \to K$ and some isometric linear map $V: H \to K$, $\Phi = V^*\Psi$ and $(\Psi x|\Psi y) = f_1(y^*x) + f_2(xy^*)$ whenever $x, y \in A$.

(b) Let A be an involutive Banach algebra. The following three conditions are equivalent:

(i) for every bounded positive-definite bilinear form $B: A \times A \to \mathbb{C}$ there are two continuous positive linear forms $f_1, f_2: A \to \mathbb{C}$ such that $|B(x,y)|^2 \leq f_1(x^*x + xx^*)f_2(y^*y + yy^*)$ for all $x, y \in A$;

(ii) for every Hilbert space H and every bounded linear operator $\Phi: A \to H$ there are two continuous positive linear forms $f_1, f_2: A \to \mathbb{C}$ such that $\|\Phi x\|^2 \leq f_1(x^*x) + f_2(xx^*)$ for all $x \in A$;

(iii) for every Hilbert space H and every bounded linear operator $\Phi: A \to H$ there is a continuous positive linear form $f: A \to \mathbb{C}$ such that $\|\Phi x\|^2 \leq f(x^*x + xx^*)$ for all $x \in A$.

Proof. (a) [61, p. 378].

(b) (i)$\Rightarrow$(ii): As in [24, p. 94], choose $B(x,y) = (\Phi x|\Phi y^*)$. Then for $f = f_1 + f_2$ where f_1 and f_2 are as in (i) we get

$$\|\Phi x\|^2 = B(x,x^*) \leq f_1(x^*x + xx^*)^{1/2} f_2(xx^* + x^*x)^{1/2}$$

$$\leq f(x^*x + xx^*).$$

Obviously (iii) implies (ii) and conversely (take $f = f_1 + f_2$).

(iii)$\Rightarrow$(i): Let $B: A \times A \to \mathbb{C}$ be a bounded positive-definite bilinear form. There is a Hilbert space H along with a linear map $\Phi: A \to H$ such that $B(x,y) = (\Phi x|\Phi y^*)$ for all $x, y \in A$ (for the well-known simple argument see e.g. [59, pp. 370-371]). Then Φ is bounded, in fact $\|\Phi\|^2 = \|B\|$, and

$$|B(x,y)|^2 \leq \|\Phi(x)\|^2 \|\Phi(y^*)\|^2 \leq f(x^*x + xx^*)f(yy^* + y^*y). \quad \square$$

We leave aside the question of the best possible norms of the functionals appearing in (b) (or of the best possible constants in 4.1), since such questions are not of great importance for the application to random fields to which we now turn.

4.3 THEOREM [61, p. 379]. For a random field $\phi: G \to H$ the following two conditions are equivalent:

(i) ϕ is weakly harmonizable;

(ii) there is a Hilbert space K along with a hemihomogeneous random field $\psi: G \to K$ such that for some isometric linear map $V: H \to K$ the equation $\phi = V^* \circ \psi$ obtains.

The proof is based on (P2) (or on (H2)). There is nothing unique about ψ. In fact ψ can be so chosen that for it $\rho_1 = \rho_2$ in Definition 3.1.

5. Ergodic theorems

More than half a century ago Khintchine [33] recognized the close connection between von Neumann's mean ergodic theorem and a weak law of large numbers for weakly stationary stochastic processes. Since then a wide variety of ergodic theorems for stochastic processes and random fields have been published. We give a brief account of the history of the problem mainly in the case of a "noncommutative time domain".

For a continuous wide sense stationary process the mean ergodic theorem is an easy consequence of the spectral representation; a standard reference is Doob [8] . Such a process ϕ on $\mathbf{R}$ may be expressed in the form $t \to U(t)\phi(0)$ where U is a continuous unitary representation of $\mathbf{R}$, so that the result is an instance of more general operator function versions of the ergodic theorem. As early as 1939 the work of Alaoglu and Birkhoff [3, 4] contains ergodic theorems involving a norm bounded group or semigroup G of bounded linear operators on a Banach space E such that G has a "nearly invariant" sequence (μ_n) of measures in the sense that $\mu_n(G) = 1$ and for any fixed $s \in G$

$$\lim_{n\to\infty} (|\mu_n(As) - \mu_n(A)| + |\mu_n(sA) - \mu_n(A)|) = 0$$

uniformly for A in the common domain of the μ_n. A key result (Theorem 7 in [4, p. 305] in somewhat different notation) states that if G has such a sequence of measures and if for some $x \in E$ the integrals $x_n = \int_G sx d\mu_n(s)$ exist, then $\lim_{n\to\infty} x_n = a \in E$ if, and only if, x is "ergodic" with limit fixed point a, i.e., the convex combinations $\sum_{i=1}^{n} \lambda_i s_i x$, $\lambda_i \geq 0$, $\sum_{i=1}^{n} \lambda_i = 1$, converge to a in a generalized sense. The elegant proof (related to ideas from F. Riesz's [48] phenomenally short proof of a von Neumann type mean ergodic theorem) is especially to be noted, since it has served as a model for many later variations of the theme.

Another major step relevant to our topic was taken by Calderón [6]. He considers a locally compact group G and a family $(N_t)_{t>0}$ of compact symmetric neighborhoods of the identity in G satisfying $N_t N_s \subset N_{t+s}$ and $|N_{2t}| \leq \alpha |N_t|$ for a constant α independent of t (here $|N_t| = \lambda(N_t)$ in the notation of Section 2). Calderón's mean ergodic theorem [6, p. 184] deals with the convergence of the averages

$$p_t(u) = \frac{1}{|N_t|} \int_{N_t} T_s(u)\,d\mu(s)$$

where $s \to T_s$ is a weakly continuous norm bounded homomorphism of G into the space L(E) of bounded linear operators on a Banach space E such that for some weakly compact convex set $C \subset E$, $T_s C \subset C$ for all $s \in G$. The conclusion is that for any $u \in C$, $p_t(u)$ converges to some $p(u) \in C$ as $t \to \infty$, and $T_s p(u) = pT_s(u) = p(u)$ for all $s \in G$. This result of course directly bears upon the abstract (second order) random field case, but it is to be noted that [6] also contains much more difficult dominated and individual ergodic theorems in the case where G acts as one-to-one measure preserving transformations on a finite measure space. One main drawback in Calderón's approach is the rather restrictive character of the averaging family $(N_t)_{t>0}$. In fact, if G is generated by the sets N_t, it must be unimodular [6, p. 183].

After Calderón's work the importance of a certain class of locally compact groups as a natural setting for ergodic theorems has become increasingly evident: the amenable groups. These have a host of characterizations (see e.g. [19, 12]), but from our point of view the key property is the fact that for an amenable locally compact group G there is a net $(\mu_i)_{i \in I}$ of probability measures in M(G) strongly convergent to left invariance [19, p. 33] (or asymptotically equidistributed in the terminology of [27, p. 131]) in the sense that for all $s \in G$ $\lim_i \|\delta_s * \mu_i - \mu_i\| = 0$. In fact one can take for μ_i the normalized characteristic functions (so that they are elements of norm one in $L^1(G) \subset M(G)$) of suitable compact sets in G [19, pp. 64-65]). In the case of a σ-compact locally compact group G the amenability of G is known to be equivalent to the existence of a <u>sequence</u> of compact sets $S_n \subset G$ of positive measure such that

(i) $S_n \subset S_{n+1}$ for all $n \in \mathbf{N}$,

(ii) $G = \bigcup_{n \in \mathbf{N}} S_n$,

(iii) $\lim_{n \to \infty} \lambda(S_n)^{-1} \lambda(sS_n \,\Delta\, S_n) = 0$ for all $s \in G$

(see [10]). Such sequences have been used to prove generalizations of mean and individual ergodic theorems close in spirit to the work of Calderón referred to above. One considers averages over the sets S_n of certain functions on G with values in an L^p-space over a measure space. The functions arise from an action of G on the measure space, and one tries to prove the convergence of the averages in the L^p-norm or almost everywhere. We shall not go into the technicalities of this field; the delicate questions especially of almost everywhere convergence are relevant to random fields with a rather special structure. Fairly definitive results appear e.g. in [20], where one may also find references to related work. We mention, however, some works dealing with ergodic theorems for random fields of the type we are mainly studying in this survey.

Tempelman [52] has an ergodic theorem for left or right homogeneous second order random fields on a locally compact group G. The result deals with averages with respect to certain functions on G forming "ergodic nets". Left and right hand versions are considered in [52], but we choose one for definiteness. A net $(f_i)_{i \in I}$ of functions from G to $\mathbf{C}$ is said to be <u>left ergodic</u>, if each f_i belongs to $L^1(G)$,

$$\sup_{i \in I} \int |f_i(s)|\,ds < \infty, \quad \lim_i \int f_i(s)\,ds = 1, \text{ and } \lim_i \int |f_i(ts) - f_i(s)|\,ds = 0$$

for all $t \in G$. For example, the condition (iii) above means that the sequence of normalized characteristic functions $\lambda(S_n)^{-1}\chi_{S_n}$ of the sets S_n is left ergodic. While amenability is not explicitely mentioned in [52], it is not difficult to show that the existence of a left ergodic net of functions is equivalent to the amenability of G. We quote the following result:

5.1. THEOREM [52, p. 203]. Let $(f_i)_{i \in I}$ be a left ergodic net of functions on G. If $\phi: G \to L_0(\Omega,A,P)$ is a right homogeneous random field which is measurable in the sense that $(x,s) \to [\phi(s)](x)$ is a measurable function on $\Omega \times G$, then the limit

$$\lim_i \int f_i(s)\phi(s)\,ds$$

exists with respect to the norm of $L_0^2(\Omega,A,P)$. Moreover, the limit is independent of the choice of the net $(f_i)_{i \in I}$ and invariant under left and right translations of ϕ.

The ergodic nets of Tempelman were picked up by Ponomarenko [41] who considered mean values over G of certain positive-definite operator valued functions and the representation of these mean values as limits of averages in terms of the ergodic nets. In the abelian case Ponomarenko proved an ergodic theorem for the Fourier transform of a Banach space valued Borel vector measure on the dual group [41, Theorem 3]. In particular, this result specializes to a mean ergodic theorem for weakly harmonizable random fields on locally compact abelian groups.

We conclude this section by announcing a result along analogous lines, but involving a more general averaging procedure, a noncommutative group and the noncommutative Fourier transform. The notation is as in Section 2. We say that a net $(\mu_i)_{i \in I}$ in M(G) is ω-<u>strongly convergent to right invariance</u>, if

$$\lim_i \|\omega(\mu_i * \delta_s - \mu_i)\xi\| = 0$$

for all $s \in G$ and $\xi \in H_\omega$.

5.2. THEOREM [63]. Let $(\mu_i)_{i \in I}$ be a $\|\cdot\|'$-bounded net in M(G) with $\lim_i \mu_i(G) = 1$. Suppose that both $(\mu_i)_{i \in I}$ and $(\mu_i^*)_{i \in I}$ are ω-strongly convergent to right invariance. If E is a Banach space and Φ: C*(G) → E is a weakly compact linear operator with Fourier transform ϕ: G → E, then

$$\lim_i \| \int \phi d\mu_i - \Phi^{**}(p_\omega) \| = 0,$$

where p_ω is the orthogonal projection of H_ω onto the space of the common fixed points of all $\omega(s)$, $s \in G$.

6. <u>Some generalizations</u>

The purpose of this concluding section is to be a brief guide to the literature on some topics closely related to but not directly touched by the preceding discussion. Again, the paper [58] by Yaglom is basic. More than one half of it is concerned with (homogeneous) random fields on homogeneous spaces. It turns out that much of the analysis of the group case carries over. We also refer to [25, §13.2] for a short exposition. The study of important concrete examples in [58] is especially noteworthy. Moreover, multidimensional random fields as well as generalized homogeneous random fields and fields with random homogeneous increments are considered. The generalized random fields are defined in terms of the structure of a finite-dimensional differentiable manifold which may be a Lie group itself or a homogeneous space G/K where G is a Lie group and K a compact subgroup. A generalized random field is then a random linear functional on the space of infinitely differentiable complex functions with compact support.

The paper [43] by Ponomarenko contains a representation theorem for so-called pseudohomogeneous random fields on homogeneous spaces much in the spirit of [58]. Ponomarenko has also studied generalized random fields on locally compact groups, see [42] for the commutative and [44] for the noncommutative case. The basis is the theory of distributions on locally compact groups developed by F. Bruhat, G. I. Kac and K. Maurin in the early sixties. If $\mathcal{D}(G)$ is the space of test functions - generalizing the Schwartz space of infinitely differentiable functions with compact support - a generalized random field is a continuous linear map from $\mathcal{D}(G)$ to $L^2(\Omega,A,P)$. In the abelian case considered in [42] a spectral representation is obtained for generalized homogeneous random fields, and an ergodic theorem is proved for them as well as for what are called generalized harmonizable fields. In [44] along with left or right homogeneous generalized random fields their various generalizations are studied, and [45] lists results about certain types generalized random fields on homogeneous spaces.

Finally, we mention Savichev-Tempelman [51] and Tempelman [53, 54] (and some of their references) as sources for results in the ergodic theory of random fields.

REFERENCES

[1] Abreu, J. L.: A note on harmonizable and stationary sequences. Bol. Soc. Mat. Mex. 15, 48-51 (1970).

[2] Adler, R. J.: The geometry of random fields. Chichester-New York-Brisbane-Toronto: John Wiley & Sons 1981.

[3] Alaoglu, L., Birkhoff, G.: General ergodic theorems. Proc. Natl. Acad. Sci. USA 25, 628-630 (1939).

[4] Alaoglu, L., Birkhoff, G.: General ergodic theorems. Ann. Math., II. Ser. 41, 293-309 (1940).

[5] Bochner, S.: Stationarity, boundedness, almost periodicity of random valued functions. Proc. Third Berkeley Symp. Math. Statist. and Prob. 2, 7-27 (1956).

[6] Calderón, A. P.: A general ergodic theorem. Ann. Math., II. Ser. 58, 182-191 (1953).

[7] Dixmier, J.: C*-algebras. North-Holland Mathematical Library Vol. 15, Amsterdam-New York-Oxford: North Holland Publishing Co. 1977.

[8] Doob, J. L.: Stochastic processes. New York: John Wiley & Sons 1953.

[9] Dunford, N., Schwartz, J. T.: Linear operators, Vol. I.: General theory. Pure and Appl. Math. Vol. 7, New York: Interscience 1958.

[10] Emerson, W. R., Greenleaf, F. P.: Covering properties and Følner conditions for locally compact groups. Math. Z. 102, 370-384 (1967).

[11] Eymard, P.: L'algèbre de Fourier d'un groupe localement compact. Bull. Soc. Math. Fr. 92, 181-236 (1964).

[12] Eymard, P.: Moyennes invariantes et représentations unitaires. Lecture Notes in Math. Vol. 300. Berlin-Heidelberg-New York: Springer-Verlag.

[13] Gilbert, J. E., Ito, T., Schreiber, B. M.: Bimeasure algebras on locally compact groups. J. Funct. Anal. (to appear).

[14] Gleason, A.: Measures on the closed subspaces of a Hilbert space. J. Math. Mech. 6, 885-893 (1957).

[15] Goldstein, S.: Orthogonal scalar products on von Neumann algebras. Stud. Math. 80, 1-15 (1984).

[16] Goldstein, S., Jajte, R.: Second order fields over W*-algebras. Bull. Acad. Pol. Sci., Ser. Sci. Math. 30, 255-259 (1982).

[17] Graham, C. C., Schreiber, B. M.: Bimeasure algebras on LCA groups. Pac. J. Math. 115, 91-127 (1984).

[18] Graham, C. C., Schreiber, B. M.: Sets of interpolation for Fourier transforms of bimeasures. Colloq. Math. (to appear).

[19] Greenleaf, F. P.: Invariant means on topological groups and their applications. Van Nostrand Math. Studies Series No. 16, New York: Van Nostrand-Reinhold 1969.

[20] Greenleaf, F. P.:, Emerson, W. R.: Group structure and the pointwise ergodic theorem for connected amenable groups. Adv. Math. 14, 153-172 (1974).

[21] Grothendieck, A.: Résumé de la théorie métrique des produits tensoriels topologiques. Bol. Soc. Mat. São Paulo 8, 1-79 (1956).

[22] Gudder, S. P.: Stochastic methods in quantum mechanics. North Holland Series in Probability and Applied Mathematics, New York-Oxford: North Holland 1979.

[23] Haagerup, U.: Solution of the similarity problem for cyclic representations of C*-algebras. Ann. Math., II. Ser. 118, 215-240 (1983).

[24] Haagerup, U.: The Grothendieck inequality for bilinear forms on C*-algebras. Adv. Math. 56, 93-116 (1985).

[25] Hannan, E. J.: Group representations and applied probability. Methuen's Review Series in Applied Probability Vol. 3, London: Methuen & Co. 1965.

[26] Hewitt, E., Ross, K. A.: Abstract harmonic analysis, Vol. II: Structure and analysis for compact groups, analysis on locally compact Abelian groups. Die Grundlehren der math. Wissenschaften, Band 152, Berlin-Heidelberg-New York: Springer-Verlag 1970.

[27] Heyer, H.: Probability measures on locally compact groups. Ergebnisse der Mathematik und ihrer Grenzgebiete 94, Berlin-Heidelberg-New York: Springer-Verlag 1977.

[28] Jajte, R., Paszkiewicz, A.: Vector measures on the closed subspaces of a Hilbert space. Stud. Math. 80, 229-251 (1978).

[29] Johnson, B. E., Kadison, R. V., Ringrose, J. R.: Cohomology of operator algebras. III. Reduction to normal cohomology. Bull. Soc. Math. Fr. 100, 73-96 (1972).

[30] Kaijser, S.: A simple-minded proof of the Pisier-Grothendieck inequality. In: Banach spaces, harmonic analysis, and probability theory, pp. 33-55. Lecture Notes in Math. Vol. 995, Berlin-Heidelberg-New York: Springer-Verlag 1983.

[31] Kaijser, S., Sinclair, A. M.: Projective tensor products of C*-algebras. Math. Scand. (to appear).

[32] Kampé de Fériet, J.: Analyse harmonique des fonctions aléatoires stationnaires d'ordre 2 définies sur un groupe abélien localement compact. C. R. Acad. Sci., Paris 226, 868-870 (1948).

[33] Khintchine, A.: Korrelationstheorie der stationären stochastischen Prozesse. Math. Ann. 109, 604-615 (1934).

[34] Kruszyński, P.: Orthogonally scattered measures on relatively orthocomplemented lattices. (Preprint).

[35] Lindenstrauss, J., Pelczyński, A.: Absolutely summing operators in L_p-spaces and their applications. Stud. Math. 29, 275-326 (1968).

[36] Loève, M.: Probability theory. 3rd ed. Princeton, N.J.,- Toronto-London-Melbourne: D. van Nostrand 1963.

[37] Niemi, H.: On orthogonally scattered dilations and the linear prediction of certain stochastic processes. Comment. Phys.-Math. 45, 111-130 (1975).

[38] Niemi, H.: On orthogonally scattered dilations of bounded vector measures. Ann. Acad. Sci. Fenn., Ser. A I 3, 43-52 (1977).

[39] Niemi, H.: Grothendieck's inequality and minimal orthogonally scattered dilations. In: Probability Theory on Vector Spaces III, pp. 175-187. Lecture Notes in Math. Vol. 1080, Berlin-Heidelberg-New York: Springer-Verlag 1984.

[40] Pisier, G.: Grothendieck's theorem for non-commutative C*-algebras with an appendix on Grothendieck's constants. J. Funct. Anal. 29, 397-415 (1978).

[41] Ponomarenko, A. I.: On the mean value of a positive definite operator-valued function on a group. Theory Probab. Math. Stat. 1, 155-161 (1974). (Translated from Teor. Veroyatn. Mat. Stat. 1, 159-165 (1970).)

[42] Ponomarenko, A. I.: Harmonic analysis of generalized wide-sense homogeneous random fields on a locally compact commutative group. Theory Probab. Math. Stat. 3, 119-137 (1974). (Translated from Teor. Veroyatn. Mat. Stat. 3, 117-134 (1970).)

[43] Ponomarenko, A. I.: Pseudohomogeneous random fields on groups and homogeneous spaces. Theory Probab. Math. Stat. 4, 108-114 (1974). (Translated from Teor. Veroyatn. Mat. Stat. 4, 117-122 (1971).)

[44] Ponomarenko, A. I.: Generalized second-order random fields on locally compact groups. Theory Probab. Math. Stat. 29, 125-133 (1984). (Translated from Teor. Veroyatn. Mat. Stat. 29, 100-109 (1983).)

[45] Ponomarenko, A. I.: Generalized random fields of second order on homogeneous spaces. (Russian). Dokl. Akad. Nauk Ukr. SSR, Ser. A 1984, No. 3, 12-15 (1984).

[46] Preston, C.: Random fields. Lecture Notes in Math. Vol. 534, Berlin-Heidelberg-New York: Springer-Verlag 1976.

[47] Rao, M. M.: Harmonizable processes: structure theory. Enseign. Math., II. Ser. 28, 295-351 (1982).

[48] Riesz, F.: Some mean ergodic theorems. J. London Math. Soc. 13, 274-278 (1938).

[49] Rosenblatt, M.: Stationary sequences and random fields. Basel-Boston-Stuttgart: Birkhäuser Verlag 1985.

[50] Rozanov, Yu. A.: Markov random fields. Berlin-Heidelberg-New York: Springer-Verlag 1982.

[51] Savichev, A., Tempelman, A. A.: Ergodic theorems on mixing homogeneous spaces. (Russian). Litov. Mat. Sb. 24, 167-175 (1984).

[52] Tempelman, A. A.: Ergodic theorems for homogeneous generalized stochastic fields and homogeneous stochastic fields on groups. (Russian). Litov. Mat. Sb. 2, 195-213 (1962).

[53] Tempelman, A. A.: The ergodicity of homogeneous Gaussian random fields on homogeneous spaces. Theory Probab. Appl. 18, 173-175 (1973). (Translated from Teor. Veroyatn. Primen. 18, 177-180 (1973).

[54] Tempelman, A. A.: Specific characteristics and variational principle for homogeneous random fields. Z. Wahrscheinlichkeitstheor. Verw. Geb. 65, 341-365 (1984).

[55] Vanmarcke, E.: Random fields: analysis and synthesis. Cambridge, Mass.: MIT Press 1983.

[56] Yadrenko, M. I.: Spectral theory of random fields. New York: Optimization Software, Inc. 1983.

[57] Yaglom, A. M.: Positive-definite functions and homogeneous random fields on groups and homogeneous spaces. Sov. Math. Dokl. 1, 1402-1405 (1961). (Translated from Dokl. Akad. Nauk SSSR 135, 1342-1345 (1960).)

[58] Yaglom, A. M.: Second order homogeneous random fields. Proc. Fourth Berkeley Symp. Math. Statist. and Prob. 2, 593-622 (1961).

[59] Ylinen, K.: Fourier transforms of noncommutative analogues of vector measures and bimeasures with applications to stochastic processes. Ann. Acad. Sci. Fenn., Ser. A I 1, 355-385 (1975).

[60] Ylinen, K.: Vector measures on the projections of a W*-algebra. Ann. Univ. Turku., Ser. A I 186, 129-135 (1984).

[61] Ylinen, K.: Dilations of V-bounded stochastic processes indexed by a locally compact group. Proc. Am. Math. Soc. 90, 378-380 (1984).

[62] Ylinen, K.: Vector Gleason measures and their Fourier transforms. In: Proc. of the OATE Conf., Buşteni, Romania 1983, pp. 589-594. Lecture Notes in Math. Vol. 1132, Berlin-Heidelberg-New York-Tokyo: Springer-Verlag 1985.

[63] Ylinen, K.: Ergodic theorems for Fourier transforms of noncommutative analogues of vector measures. (Preprint).

[64] Ylinen, K.: Random fields on compact groups. (In preparation).

Vol. 1062: J. Jost, Harmonic Maps Between Surfaces. X, 133 pages. 1984.

Vol. 1063: Orienting Polymers. Proceedings, 1983. Edited by J. L. Ericksen. VII, 166 pages. 1984.

Vol. 1064: Probability Measures on Groups VII. Proceedings, 1983. Edited by H. Heyer. X, 588 pages. 1984.

Vol. 1065: A. Cuyt, Padé Approximants for Operators: Theory and Applications. IX, 138 pages. 1984.

Vol. 1066: Numerical Analysis. Proceedings, 1983. Edited by D. F. Griffiths. XI, 275 pages. 1984.

Vol. 1067: Yasuo Okuyama, Absolute Summability of Fourier Series and Orthogonal Series. VI, 118 pages. 1984.

Vol. 1068: Number Theory, Noordwijkerhout 1983. Proceedings. Edited by H. Jager. V, 296 pages. 1984.

Vol. 1069: M. Kreck, Bordism of Diffeomorphisms and Related Topics. III, 144 pages. 1984.

Vol. 1070: Interpolation Spaces and Allied Topics in Analysis. Proceedings, 1983. Edited by M. Cwikel and J. Peetre. III, 239 pages. 1984.

Vol. 1071: Padé Approximation and its Applications, Bad Honnef 1983. Prodeedings. Edited by H. Werner and H. J. Bünger. VI, 264 pages. 1984.

Vol. 1072: F. Rothe, Global Solutions of Reaction-Diffusion Systems. V, 216 pages. 1984.

Vol. 1073: Graph Theory, Singapore 1983. Proceedings. Edited by K. M. Koh and H. P. Yap. XIII, 335 pages. 1984.

Vol. 1074: E. W. Stredulinsky, Weighted Inequalities and Degenerate Elliptic Partial Differential Equations. III, 143 pages. 1984.

Vol. 1075: H. Majima, Asymptotic Analysis for Integrable Connections with Irregular Singular Points. IX, 159 pages. 1984.

Vol. 1076: Infinite-Dimensional Systems. Proceedings, 1983. Edited by F. Kappel and W. Schappacher. VII, 278 pages. 1984.

Vol. 1077: Lie Group Representations III. Proceedings, 1982–1983. Edited by R. Herb, R. Johnson, R. Lipsman, J. Rosenberg. XI, 454 pages. 1984.

Vol. 1078: A. J. E. M. Janssen, P. van der Steen, Integration Theory. V, 224 pages. 1984.

Vol. 1079: W. Ruppert. Compact Semitopological Semigroups: An Intrinsic Theory. V, 260 pages. 1984

Vol. 1080: Probability Theory on Vector Spaces III. Proceedings, 1983. Edited by D. Szynal and A. Weron. V, 373 pages. 1984.

Vol. 1081: D. Benson, Modular Representation Theory: New Trends and Methods. XI, 231 pages. 1984.

Vol. 1082: C.-G. Schmidt, Arithmetik Abelscher Varietäten mit komplexer Multiplikation. X, 96 Seiten. 1984.

Vol. 1083: D. Bump, Automorphic Forms on GL (3,IR). XI, 184 pages. 1984.

Vol. 1084: D. Kletzing, Structure and Representations of Q-Groups. VI, 290 pages. 1984.

Vol. 1085: G. K. Immink, Asymptotics of Analytic Difference Equations. V, 134 pages. 1984.

Vol. 1086: Sensitivity of Functionals with Applications to Engineering Sciences. Proceedings, 1983. Edited by V. Komkov. V, 130 pages. 1984

Vol. 1087: W. Narkiewicz, Uniform Distribution of Sequences of Integers in Residue Classes. VIII, 125 pages. 1984.

Vol. 1088: A. V. Kakosyan, L. B. Klebanov, J. A. Melamed, Characterization of Distributions by the Method of Intensively Monotone Operators. X, 175 pages. 1984.

Vol. 1089: Measure Theory, Oberwolfach 1983. Proceedings. Edited by D. Kölzow and D. Maharam-Stone. XIII, 327 pages. 1984.

Vol. 1090: Differential Geometry of Submanifolds. Proceedings, 1984. Edited by K. Kenmotsu. VI, 132 pages. 1984.

Vol. 1091: Multifunctions and Integrands. Proceedings, 1983. Edited by G. Salinetti. V, 234 pages. 1984.

Vol. 1092: Complete Intersections. Seminar, 1983. Edited by S. Greco and R. Strano. VII, 299 pages. 1984.

Vol. 1093: A. Prestel, Lectures on Formally Real Fields. XI, 125 pages. 1984.

Vol. 1094: Analyse Complexe. Proceedings, 1983. Edité par E. Amar, R. Gay et Nguyen Thanh Van. IX, 184 pages. 1984.

Vol. 1095: Stochastic Analysis and Applications. Proceedings, 1983. Edited by A. Truman and D. Williams. V, 199 pages. 1984.

Vol. 1096: Théorie du Potentiel. Proceedings, 1983. Edité par G. Mokobodzki et D. Pinchon. IX, 601 pages. 1984.

Vol. 1097: R. M. Dudley, H. Kunita, F. Ledrappier, École d'Éte de Probabilités de Saint-Flour XII – 1982. Edité par P. L. Hennequin. X, 396 pages. 1984.

Vol. 1098: Groups – Korea 1983. Proceedings. Edited by A. C. Kim and B. H. Neumann. VII, 183 pages. 1984.

Vol. 1099: C. M. Ringel, Tame Algebras and Integral Quadratic Forms. XIII, 376 pages. 1984.

Vol. 1100: V. Ivrii, Precise Spectral Asymptotics for Elliptic Operators Acting in Fiberings over Manifolds with Boundary. V, 237 pages. 1984.

Vol. 1101: V. Cossart, J. Giraud, U. Orbanz, Resolution of Surface Singularities. Seminar. VII, 132 pages. 1984.

Vol. 1102: A. Verona, Stratified Mappings – Structure and Triangulability. IX, 160 pages. 1984.

Vol. 1103: Models and Sets. Proceedings, Logic Colloquium, 1983, Part I. Edited by G. H. Müller and M. M. Richter. VIII, 484 pages. 1984.

Vol. 1104: Computation and Proof Theory. Proceedings, Logic Colloquium, 1983, Part II. Edited by M. M. Richter, E. Börger, W. Oberschelp, B. Schinzel and W. Thomas. VIII, 475 pages. 1984.

Vol. 1105: Rational Approximation and Interpolation. Proceedings, 1983. Edited by P. R. Graves-Morris, E. B. Saff and R. S. Varga. XII, 528 pages. 1984.

Vol. 1106: C. T. Chong, Techniques of Admissible Recursion Theory. IX, 214 pages. 1984.

Vol. 1107: Nonlinear Analysis and Optimization. Proceedings, 1982. Edited by C. Vinti. V, 224 pages. 1984.

Vol. 1108: Global Analysis – Studies and Applications I. Edited by Yu. G. Borisovich and Yu. E. Gliklikh. V, 301 pages. 1984.

Vol. 1109: Stochastic Aspects of Classical and Quantum Systems. Proceedings, 1983. Edited by S. Albeverio, P. Combe and M. Sirugue-Collin. IX, 227 pages. 1985.

Vol. 1110: R. Jajte, Strong Limit Theorems in Non-Commutative Probability. VI, 152 pages. 1985.

Vol. 1111: Arbeitstagung Bonn 1984. Proceedings. Edited by F. Hirzebruch, J. Schwermer and S. Suter. V, 481 pages. 1985.

Vol. 1112: Products of Conjugacy Classes in Groups. Edited by Z. Arad and M. Herzog. V, 244 pages. 1985.

Vol. 1113: P. Antosik, C. Swartz, Matrix Methods in Analysis. IV, 114 pages. 1985.

Vol. 1114: Zahlentheoretische Analysis. Seminar. Herausgegeben von E. Hlawka. V, 157 Seiten. 1985.

Vol. 1115: J. Moulin Ollagnier, Ergodic Theory and Statistical Mechanics. VI, 147 pages. 1985.

Vol. 1116: S. Stolz, Hochzusammenhängende Mannigfaltigkeiten und ihre Ränder. XXIII, 134 Seiten. 1985.

Vol. 1117: D.J. Aldous, J.A. Ibr... Probabilités de Saint-Flour XIII – 1983. Edité par P.L. Hennequin. IX, 409 pages. 1985.

Vol. 1118: Grossissements de filtrations: exemples et applications. Seminaire, 1982/83. Edité par Th. Jeulin et M. Yor. V, 315 pages. 1985.

Vol. 1119: Recent Mathematical Methods in Dynamic Programming. Proceedings, 1984. Edited by I. Capuzzo Dolcetta, W.H. Fleming and T. Zolezzi. VI, 202 pages. 1985.

Vol. 1120: K. Jarosz, Perturbations of Banach Algebras. V, 118 pages. 1985.

Vol. 1121: Singularities and Constructive Methods for Their Treatment. Proceedings, 1983. Edited by P. Grisvard, W. Wendland and J.R. Whiteman. IX, 346 pages. 1985.

Vol. 1122: Number Theory. Proceedings, 1984. Edited by K. Alladi. VII, 217 pages. 1985.

Vol. 1123: Séminaire de Probabilités XIX 1983/84. Proceedings. Edité par J. Azéma et M. Yor. IV, 504 pages. 1985.

Vol. 1124: Algebraic Geometry, Sitges (Barcelona) 1983. Proceedings. Edited by E. Casas-Alvero, G.E. Welters and S. Xambó-Descamps. XI, 416 pages. 1985.

Vol. 1125: Dynamical Systems and Bifurcations. Proceedings, 1984. Edited by B.L.J. Braaksma, H.W. Broer and F. Takens. V, 129 pages. 1985.

Vol. 1126: Algebraic and Geometric Topology. Proceedings, 1983. Edited by A. Ranicki, N. Levitt and F. Quinn. V, 523 pages. 1985.

Vol. 1127: Numerical Methods in Fluid Dynamics. Seminar. Edited by F. Brezzi, VII, 333 pages. 1985.

Vol. 1128: J. Elschner, Singular Ordinary Differential Operators and Pseudodifferential Equations. 200 pages. 1985.

Vol. 1129: Numerical Analysis, Lancaster 1984. Proceedings. Edited by P.R. Turner. XIV, 179 pages. 1985.

Vol. 1130: Methods in Mathematical Logic. Proceedings, 1983. Edited by C.A. Di Prisco. VII, 407 pages. 1985.

Vol. 1131: K. Sundaresan, S. Swaminathan, Geometry and Nonlinear Analysis in Banach Spaces. III, 116 pages. 1985.

Vol. 1132: Operator Algebras and their Connections with Topology and Ergodic Theory. Proceedings, 1983. Edited by H. Araki, C.C. Moore, Ş. Strătilă and C. Voiculescu. VI, 594 pages. 1985.

Vol. 1133: K.C. Kiwiel, Methods of Descent for Nondifferentiable Optimization. VI, 362 pages. 1985.

Vol. 1134: G.P. Galdi, S. Rionero, Weighted Energy Methods in Fluid Dynamics and Elasticity. VII, 126 pages. 1985.

Vol. 1135: Number Theory, New York 1983–84. Seminar. Edited by D.V. Chudnovsky, G.V. Chudnovsky, H. Cohn and M.B. Nathanson. V, 283 pages. 1985.

Vol. 1136: Quantum Probability and Applications II. Proceedings, 1984. Edited by L. Accardi and W. von Waldenfels. VI, 534 pages. 1985.

Vol. 1137: Xiao G., Surfaces fibrées en courbes de genre deux. IX, 103 pages. 1985.

Vol. 1138: A. Ocneanu, Actions of Discrete Amenable Groups on von Neumann Algebras. V, 115 pages. 1985.

Vol. 1139: Differential Geometric Methods in Mathematical Physics. Proceedings, 1983. Edited by H. D. Doebner and J. D. Hennig. VI, 337 pages. 1985.

Vol. 1140: S. Donkin, Rational Representations of Algebraic Groups. VII, 254 pages. 1985.

Vol. 1141: Recursion Theory Week. Proceedings, 1984. Edited by H.-D. Ebbinghaus, G.H. Müller and G.E. Sacks. IX, 418 pages. 1985.

Vol. 1142: Orders and their Applications. Proceedings, 1984. Edited by I. Reiner and K. W. Roggenkamp. X, 306 pages. 1985.

Vol. 1143: A. Krieg, Modular Forms on Half-Spaces of Quaternions. XIII, 203 pages. 1985.

Vol. 1144: Knot Theory and Manifolds. Proceedings, 1983. Edited by D. Rolfsen. V, 163 pages. 1985.

...quet Order and Simplices. VI, 143 pages. 1985.

Vol. 1146: Séminaire d'Algèbre Paul Dubreil et Marie-Paule Malliavin. Proceedings, 1983–1984. Edité par M.-P. Malliavin. IV, 420 pages. 1985.

Vol. 1147: M. Wschebor, Surfaces Aléatoires. VII, 111 pages. 1985.

Vol. 1148: Mark A. Kon, Probability Distributions in Quantum Statistical Mechanics. V, 121 pages. 1985.

Vol. 1149: Universal Algebra and Lattice Theory. Proceedings, 1984. Edited by S. D. Comer. VI, 282 pages. 1985.

Vol. 1150: B. Kawohl, Rearrangements and Convexity of Level Sets in PDE. V, 136 pages. 1985.

Vol 1151: Ordinary and Partial Differential Equations. Proceedings, 1984. Edited by B.D. Sleeman and R.J. Jarvis. XIV, 357 pages. 1985.

Vol. 1152: H. Widom, Asymptotic Expansions for Pseudodifferential Operators on Bounded Domains. V, 150 pages. 1985.

Vol. 1153: Probability in Banach Spaces V. Proceedings, 1984. Edited by A. Beck, R. Dudley, M. Hahn, J. Kuelbs and M. Marcus. VI, 457 pages. 1985.

Vol. 1154: D.S. Naidu, A.K. Rao, Singular Pertubation Analysis of Discrete Control Systems. IX, 195 pages. 1985.

Vol. 1155: Stability Problems for Stochastic Models. Proceedings, 1984. Edited by V.V. Kalashnikov and V.M. Zolotarev. VI, 447 pages. 1985.

Vol. 1156: Global Differential Geometry and Global Analysis 1984. Proceedings, 1984. Edited by D. Ferus, R.B. Gardner, S. Helgason and U. Simon. V, 339 pages. 1985.

Vol. 1157: H. Levine, Classifying Immersions into $\mathbb{R}^4$ over Stable Maps of 3-Manifolds into $\mathbb{R}^2$. V, 163 pages. 1985.

Vol. 1158: Stochastic Processes – Mathematics and Physics. Proceedings, 1984. Edited by S. Albeverio, Ph. Blanchard and L. Streit. VI, 230 pages. 1986.

Vol. 1159: Schrödinger Operators, Como 1984. Seminar. Edited by S. Graffi. VIII, 272 pages. 1986.

Vol. 1160: J.-C. van der Meer, The Hamiltonian Hopf Bifurcation. VI, 115 pages. 1985.

Vol. 1161: Harmonic Mappings and Minimal Immersions, Montecatini 1984. Seminar. Edited by E. Giusti. VII, 285 pages. 1985.

Vol. 1162: S.J.L. van Eijndhoven, J. de Graaf, Trajectory Spaces, Generalized Functions and Unbounded Operators. IV, 272 pages. 1985.

Vol. 1163: Iteration Theory and its Functional Equations. Proceedings, 1984. Edited by R. Liedl, L. Reich and Gy. Targonski. VIII, 231 pages. 1985.

Vol. 1164: M. Meschiari, J.H. Rawnsley, S. Salamon, Geometry Seminar "Luigi Bianchi" II – 1984. Edited by E. Vesentini. VI, 224 pages. 1985.

Vol. 1165: Seminar on Deformations. Proceedings, 1982/84. Edited by J. Ławrynowicz. IX, 331 pages. 1985.

Vol. 1166: Banach Spaces. Proceedings, 1984. Edited by N. Kalton and E. Saab. VI, 199 pages. 1985.

Vol. 1167: Geometry and Topology. Proceedings, 1985. Edited by J. Alexander and J. Harer. VI, 292 pages. 1985.

Vol. 1168: S.S. Agaian, Hadamard Matrices and their Applications. III, 227 pages. 1985.

Vol. 1169: W.A. Light, E.W. Cheney, Approximation Theory in Tensor Product Spaces. VII, 157 pages. 1985.

Vol. 1170: B.S. Thomson, Real Functions. VII, 229 pages. 1985.

Vol. 1171: Polynômes Orthogonaux et Applications. Proceedings, 1984. Edité par C. Brezinski, A. Draux, A.P. Magnus, P. Maroni et A. Ronveaux. XXXVII, 584 pages. 1985.

Vol. 1172: Algebraic Topology, Göttingen 1984. Proceedings. Edited by L. Smith. VI, 209 pages. 1985.